Environmental Sociology

Theory and Practice

Edited by

MICHAEL D. MEHTA
York University

and

ÉRIC OUELLET
York University

Captus Press

Environmental Sociology: Theory and Practice

 Correspondence can be sent to: Captus Press Inc., York University Campus, 4700 Keele Street, North York, Ontario M3J 1P3

Canadian Cataloguing in Publication Data
Main entry under title:
Environmental sociology

ISBN 1–895712–80–7

1. Human ecology. I. Mehta, Michael D., 1965–
II. Ouellet, Éric, 1968–
HM206.E58 1995 304.2 C95–930740–0

9 8 7 6 5 4 3 2 1
Printed in Canada

TABLE OF CONTENTS

PREFACE

Environmental sociology is not just a sociology of the environment, but rather a sociology informed by an understanding of ecological limits. The thinning ozone layer, pesticide residues in food, nuclear waste and water pollution, should make it abundantly clear that human activity has a direct effect on the natural world. Conversely, the natural environment influences the social realm as well. Environmental sociology explores these effects by examining the interdependencies between population, organization, technology and environment. As such, this approach departs from traditional sociology in that it emphasizes the natural and built environment as factors influencing human behaviour and social organization.

This book brings together a variety of issues and case-studies that demonstrate the latent diversity of research interests, falling under the umbrella of environmental sociology. Since this field is a newly revitalized one, with material spread across several journals, edited collections and books, this text brings together a wide-ranging group of writers with interests spanning the field of environmental sociology. This series of previously unpublished articles, with a few exceptions, is one of the first texts specifically devoted to this rapidly expanding field of inquiry. As such, we are delighted to offer a collection of articles which help define environmental sociology, trace its historical background, examine its epistemological and ontological status, explore some of its major issues, and illustrate how it can be applied.

We have divided this book into two sections: theoretical and case-study. The theoretical section covers topics such as the history and paradigmatic status of the field; the history of environmental concern and the broader environmental crisis; the relationship between nature and society; social movements and Green social theory; risk and social policy; sustainable development; and the role of the built environment in environmental sociology. The case-study section, although shorter, examines the case of Uniroyal Chemical and its polluting practices in the town of Elmira, Ontario; the organization of Greenpeace Canada; transportation policy in the Greater Metropolitan Toronto area and public participation; a study of the press coverage given to the James Bay hydroelectric project; and a land-use study of watershed management in the Lake Simcoe region of Ontario. Discussion questions are provided following each chapter, to facilitate further probing of the issues and problems arising from the topics addressed. This format should allow educators the opportunity to use this edited collection as a textbook for teaching undergraduate and graduate students.

This book is intended to be used as a foundation for exploring some of the central concepts with which environmental sociologists are now grappling. The theories examined by the various contributors represent current perspectives and approaches to topics and concepts, which have been extensively documented in a variety of venues. As a result, this book should help expand and define the theories that environmental sociologists claim as their own. In addition, the use of case-studies illustrates how such theories can be applied by environmental consultants, policy makers and professionals in a variety of related fields and disciplines. Lastly, this book may be useful for stimulating academics — in departments of sociology, environmental studies, geography and policy studies — to mount courses that deal with some of the key dimensions of environmental sociology. The growing concern of academics and students, in general, on issues related to the environment, make such a textbook a valuable contribution to the literature.

We would like to thank all of the contributors who helped make this book a success. The donation of their work, constant encouragement and quick response made this book not only possible, but a pleasure to bring together. Our thanks also go to Captus Press of North York, Ontario, for their professional, quick and courteous service. Their commitment to this type of publishing will ensure that work in the field of environmental sociology will continue to grow and gain prominence. A debt of gratitude is also owed to the support staff at York University's Department of Sociology. Without the assistance of all these people, this would not have been the first Canadian textbook on environmental sociology.

Michael D. Mehta
Éric Ouellet
Toronto
April, 1995

THE THEORETICAL APPROACH

SOCIOLOGY OF THE ENVIRONMENT: FROM HUMAN ECOLOGY TO ECOSOCIOLOGY

Jean-Guy Vaillancourt

Université de Montréal

INTRODUCTION

Since the end of World War I, the study of the relationship between nature and human societies has gone through three main stages in sociology: human and urban ecology was developed by Chicago School sociologists during the 1920s and '30s; environmental sociology emerged, mostly in the United States, during the 1970s and early '80s; and since the late 1980s, the third phase, "ecosociology," has arisen. It developed after the publication of the Brundtland Report and because of a new interest in the human dimensions of global environmental changes (especially their human causes and their social impacts). These three stages of environmental sociological thought and research more or less correspond to three parallel stages in the development of the natural and social sciences of the environment, and in the rise of an environmental consciousness in the general population, and among groups dedicated to environmental protection.

Sociology is one of the basic disciplines of the social and human sciences but until lately, it has not appeared as one of the core environmental sciences, as have geography or biology. Nevertheless, as the environmental social sciences as a whole have acquired a respectable place alongside the natural sciences of the environment, we have realized that sociology — along with other social sciences, such as economics or history — has an important contribution to make.

Many of the founders of sociology have shown an interest in the relationship between human societies and their biophysical environment. Durkheim, Weber and even Marx and Engels — and before them social thinkers like Ibn Kaldun, Montesquieu and Malthus — wrote many memorable pages on the relationship between human societies and the natural environment.

In his first essay on the principal of population, in 1798, Malthus wrote:

> the power of population is indefinitely greater than the power of the earth to produce subsistence for man. Population, when unchecked, increases in a geometrical ratio. Subsistence increases only with an arithmetical ratio. A slight acquaintance with numbers will show the immensity of the first power in comparison with the second (Malthus [1798] 1909: 7).

In the course of the 20th century, it seems to me that there have been three main stages in the development of what is now called the sociology of the environment. These coincide approximately with the crucial periods in the evolution of ecology and of environmental sciences in general, and with the important phases in the rise of consciousness concerning environmental problems among conservationists, environmentalists, political ecologists and the public. These three types of Greens have been, and are still today, the three major kinds of social actors most prominent in the Green movement. I have no intention of spending much time on this typology here, nor do I intend to insist on the isomorphism between the development of natural and social sciences of the environment and the sociology of the environment, on the one hand, and the rise in environmental consciousness among the public in general and among the Green groups that form the new environmental social movement, on the other. The points that I want to make are both more modest and more focused, since I would rather centre my attention here on the three successive stages of the sociology of the environment, in the North American context especially, in order to show what general models and what paradigms are prevalent during each of these three stages.

From the early 1920s, after the blossoming of plant and animal ecology on the one hand, and of autecology and synecology on the other, a new branch of sociology sprang forth at the University of Chicago, with Park (1916), Burgess (Park and Burgess 1921), McKenzie (1968) and their graduate students. This new subdiscipline was called human ecology, and sometimes also urban ecology. After this golden age of human-urban Chicago School ecology, which flourished during the '20s and '30s, this branch of sociology suffered a certain eclipse, as empirical quantitative sociology and more theoretical approaches like structural-functionalism — and up to a certain point also, conflict sociology and symbolic interactionism — came to positions of prominence in sociological theory, particulary in the United States. After the crisis and the fragmentation of Western sociology in the 1960s, we witnessed the emergence of a panoply of new theoretical approaches and of new empirical subfields in sociology. Environmental sociology was one of them. In my view, the flowering of environmental sociology was due, in good part, to the emergence of new environmental concerns originating from natural scientists like Rachel Carson (1962) and Barry Commoner (1971) among others, and from the founding of numerous new environmentalist groups like Friends of the Earth in many countries around the world.

Some sociologists working in the areas of rural sociology and the sociology of natural resources, developed at the beginning of the '70s, a new subfield of their discipline which they labelled environmental sociology. Catton (1980) and Dunlap (1980) (Catton and Dunlap 1978, 1980; Dunlap and Catton 1979a, 1979b, 1983), on the one hand, with their new ecological paradigm, Schnaiberg (1975, 1980) and Buttel (1978, 1986, 1987), on the other hand, with their more dialectical approach, are among the best known of these contemporary U.S. environmental sociologists who have developed and disseminated this new type of sociological knowledge, and who have discussed the issue of a new ecological paradigm for environmental sociology.

Since the late '80s, the sociology of the environment has received a new shot in the arm, thanks to the raised consciousness concerning the global environmental problems, that has sprung up at the end of the Cold War. Three events mark the end of the second stage and the beginning of a third stage of reflection and of research for sociologists interested in environmental questions and in the future of the biosphere: the publication of the *Report of the United Nations Commission on the Environment and Development* (Brundtland 1987), which put forward the central concept of sustainable development; the launching of the Human Dimensions of Global Environmental Change Programme by the International Social Science Council at the Tokyo Symposium of 1988 on the human causes of global change (Jacobson and Price 1990); and the creation of a working group on sociology of the environment inside the International Sociological Association during its 1990 meeting in Madrid (this group merged later with Research Committee 24 on Social Ecology to give birth to a new Research Committee on Environment and Society).

After having first developed human ecology and social ecology with the Chicago School sociologists and their intellectual opponents, the sociologists with an interest in the relationship between human beings and their material and spatial environment have subsequently widened their horizons in order to create, starting in the early '70s, a truly original environmental sociology.

As far as the third stage of the sociology of the environment is concerned, it is characterized by a great diversity of models and paradigms, and by approaches that are both more global and more interdisciplinary. I think that we should call it "ecosociology," given the important developments of these past few years. In the following pages, I shall refer briefly at first to the human ecology of the '20s and '30s, then I shall focus a little on the environmental sociology of the '70s and early '80s, and finally I shall describe somewhat more extensively, but not at all exhaustively, because of a lack of space, the new ecosociology of the present period which started in the late '80s. Our attention will be centred mostly on the themes, paradigms and models that characterize more specifically each of these three successive periods of the history of the sociology of the environment.

Human Ecology

During the 1920s and 1930s, a group of sociologists from the University of Chicago, including Park, Burgess, McKenzie and others, created a new subfield of sociology which they called human ecology. This human or urban ecology stressed the idea that urban social problems are influenced, and even determined, by urban spatial arrangements. For them, because human beings are the dominant animal species, it was normal that sociologists should apply the latest principles of plant and animal ecology to the study of human communities living in urban settings. The Chicago School of human ecology dedicated itself more precisely to "a study of spatial and temporal relations of human beings, as affected by the selective, distributive and accommodative forces of the environment" (McKenzie 1968: 4).

Social relations between humans, according to this group of sociologists, have a subsocial and material base, and more specifically, a spatial infrastructure, which they characterized as biotic (Willhelm 1964). The accessibility of resources and the transformations of the territorial and biotic base of society, and especially of the transport system and spatial infrastructure, are thus able to have an impact on the structure of the community, the social behaviour of its members, and even the culture of a society. Demographic fluctuations are explained by the search for an equilibrium between available resources and the size of a given population. As for the internal structure of a community, it is determined by the competition and struggle for space in which the processes of selection and adaptation are equally integrated. The delimitations of concentric zones, which stretch from the central business district to the suburban periphery, in fact, constituted one of the best known and most fruitful hypotheses of the human ecologists of the Chicago School.

Human ecologists justly took into consideration the existence of environmental factors in explaining social phenomena. Nonetheless, we can criticize them for exaggerating the impact of those factors, for having postulated a constant equilibrium between human populations and accessible resources, and consequently, for having neglected the retroaction effect — that is, the parallel impact of human communities on the biophysical environment. They

FIGURE 1 Malthus's Concept Graphically Displayed

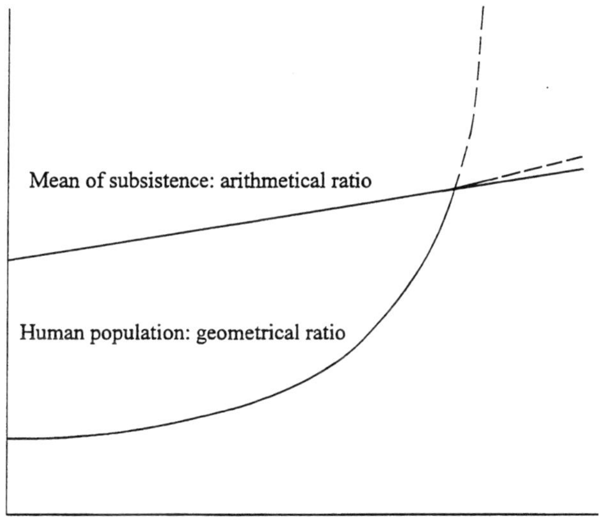

have also failed to notice that competition exists at the sociocultural level, and that consensus and co-operation can be found at the biotic subsocial level.

Other sociologists, many of them from the U.S., later developed, often in reaction to this spatial and biotic determinism of human and urban ecology, a voluntaristic social ecology which tried to show that it was the sociocultural phenomena that had an impact on the spatial environment and the urban milieu (Alihan [1938] 1990; Mukerjee 1945; Wirth 1945; Hollingshead 1947).

The expression "social ecology" had also been utilized later on, but in a quite different sense — on the one hand, by the radical anarchist ecologist Murray Bookchin, and on the other hand, by European sociologists who created the Research Group 24 on Social Ecology inside the International Sociological Association.

We can represent the paradigm put forward by the human and urban ecologists of the Chicago School with a very simple model. This model was substantially reaffirmed, with modifications and specifications, in 1950 by Quinn in his book *Human Ecology*. That same year, Hawley also published a book entitled *Human Ecology*, which presents a global theory of human ecology and of the structure of urban communities, and reaffirms the main ideas of the founders and pioneers of the Chicago School. This is the reason

FIGURE 2 Human Ecology and Social Ecology

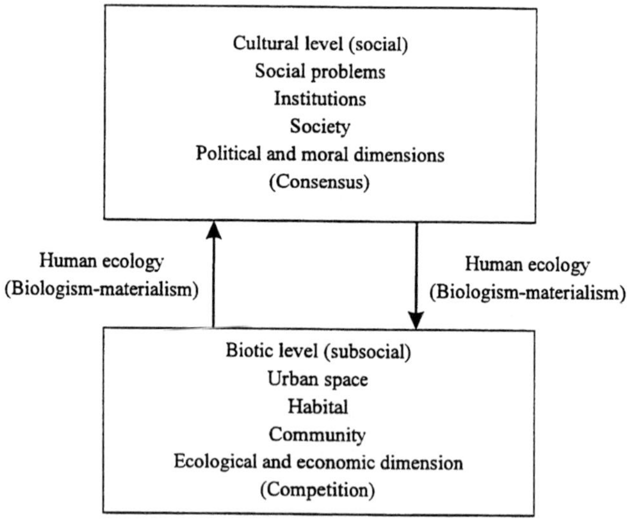

why we usually refer to Quinn and Hawley as neo-orthodox human ecologists (Theodorson 1961), or as neoclassical human ecologists (Willhelm 1964).

Gettys (1940), Theodorson (1961), as well as Duncan (1964), resolved the issue of the direction of causality by going beyond the positions of the Chicago School and the neo-orthodox human ecologists, as well as their sociocultural critics, by suggesting that there should be a greater integration between the biotic and the cultural level, between the human ecology approach and the social ecology approach. In fact, they proposed a merging of the two competing paradigms into a single one (as I have done here in Figure 2 by drawing two arrows instead of only one).

Environmental Sociology

The environmental sociology which sprang forth, especially in the U.S., at the beginning of the '20s, started off by reaffirming, in a somewhat more sophisticated manner, the infrastructural determinism of classical and neo-orthodox human ecologists, not so much in reaction to social ecologists, who gave a lot of attention to sociocultural factors, but rather in opposition to sociology's dominant theoretical approaches in the post-World War II period — namely structural-functionalism and behaviourism, as well as symbolic interactionism and conflict sociology, which had eliminated from sociology any preoccupation with material and infrastructural phenomena. For all these sociological currents, society seemed to function outside its spatial and environmental setting and independently from natural resources and any material substratum. The model that was put forward by the first environmental sociologists in reaction to these types of sociological theorizing, was an extension of the one proposed by the Chicago School, in the sense that the major effort was put into showing the importance and the impact of environmental factors on social organization. This was the case, for example, in Cottrell's 1955 book, *Energy and Society*, which shows the role played by energy in the evolution of society, as well as in the transitional work of Duncan (1961, 1964). Instead of insisting only on the influence of the environment on society, Duncan talks of the reciprocal relationships between **P**opulation, social **O**rganisation, **E**nvironment and **T**echnology. He proposes a new paradigm, which he labels the POET model, in order to represent the four fundamental variables of what he calls the "ecological complex," an analytical tool he devised to study the "ecosystem," in which there are continuous exchanges of matter, energy and information. For Duncan, P (population) and T (technology) are variables that have both human and material aspects.

The POET model was put forward for the first time at the end of the 1950s, when structural-functionalism and quantitative behaviourism dominated sociology. In Duncan's "ecosystemic" approach, the relationship between biological evolution and sociocultural evolution, and more generally between nature and society, were presented as being of a more complex nature than in the simplistic models of the human and the social ecologists. The relationships between the four basic variables of POET are presented as being multidirectional; this means that each can be influenced by the three others, and that it can also have an impact on each of them (Figure 3). In fact, it is mostly the social variables (POT) that are perceived as being determined by the environment (E), which is, in turn, presented as a material and physical variable, and not as a social one. For Duncan and those who adopted the POET model of the ecosystem, the causality generally flowed in the same direction as it did for the human ecologists who preceded them — that is, from the environment seen as an independent variable, to social organisation seen as the dependent variable (Figure 4), even if they tend to affirm that there is a functional interdependence between the four basic variables.

Other types of models have also stimulated a lot of interest in the early '70s in the environmental sciences, for example, the neo-Malthusian models put forward in the first Club of Rome reports by researchers like Meadows et al. (1972). In a recent book that updates their findings and their thinking, Meadows, Meadows and Sanders (1992) use the same kind of models as in 1972, with projections for the five following variables: population growth,

FIGURE 3 The POET Model

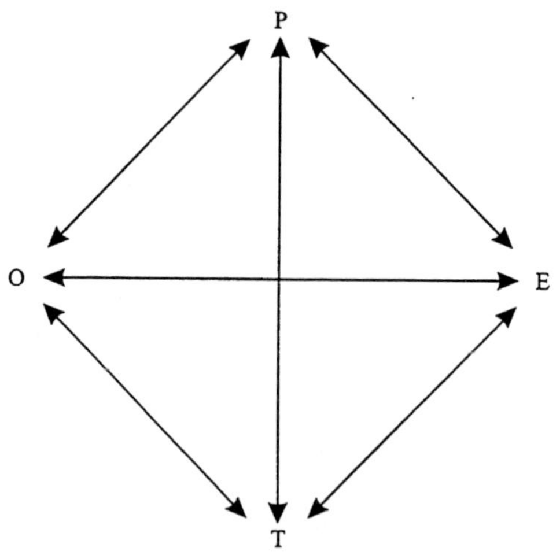

FIGURE 4 The POET Model

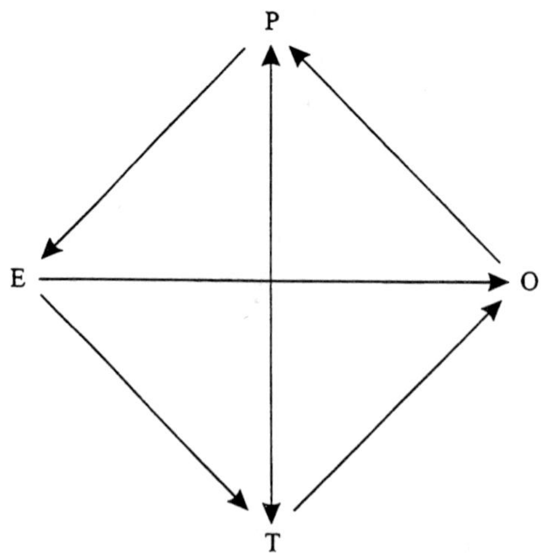

industrial output, the growth of food production, pollution and resource depletion. In their basic scenario, which represents but one possibility among many, namely the one where we do little or nothing to counter present trends, things start to really deteriorate around the years 2015–2020, and we come to some kind of collapse near the end of the 21st century (Figure 5). In scenario 10, which represents a world in which an effort has been made to lower both population and consumption figures starting around 1995, and where mankind has really gotten down to the task of controlling pollution and soil erosion, conserving natural resources, and developing technologies that are more efficient and more respectful of the environment in both industry and agriculture, we arrive at a situation that is nearly in equilibrium around the year 2040 (Figure 6). This book by Meadows et al. tackles some of the global environmental problems such as climate change, and brings forth new perspectives for solutions, like sustainable development strategies.

The expression "environmental sociology" appeared for the first time in North America in 1971, although the idea of a sociology of the environment had been in the air for a long time (Thomas 1925). Five year later, this subdiscipline obtained the status of a specialized subsection inside the American Sociological Association (ASA). In fact, as early as 1964, a research group on forestry was formed inside the Rural Sociological Society. This group then changed its name to the Natural Resources Research Group. The members of these groups created the Environmental Sociology Group in the Society for the Study of Social Problems, in 1972, and in the American Sociological Association, in 1974. In 1976, the ASA finally created an Environmental Sociology Section (Dunlap and Catton 1979a: 245–46).

Catton and Dunlap (1978) argued that standard U.S. sociology is characterized by the human exceptionalism paradigm (later renamed human exemptionalism paradigm), which holds that human beings are perceived independent from nature in their social activities, as social and cultural rather than natural beings. The seriousness of local and regional environmental problems, and more specifically the energy crisis of 1973 and the rise of the Green movement during the 1970s, favoured the emergence of a new environmental paradigm (rebaptized ecological paradigm a little later), which is in opposition to the human exemptionalism paradigm. This latter paradigm, especially in its early formulations, has a certain affinity with the traditional Chicago School human ecology approach, especially with Duncan's POET model.

Catton and Dunlap have thus proposed, in replacement of the human exemptionalism paradigm (HEP), a new ecological paradigm (NEP), where the biophysical environment becomes the crucial causal factor, and consequently has an important impact on other, more social, variables of the ecological complex. In other words, for Dunlap and Catton, human societies

FIGURE 5 Scenario 1: Sustainable Development

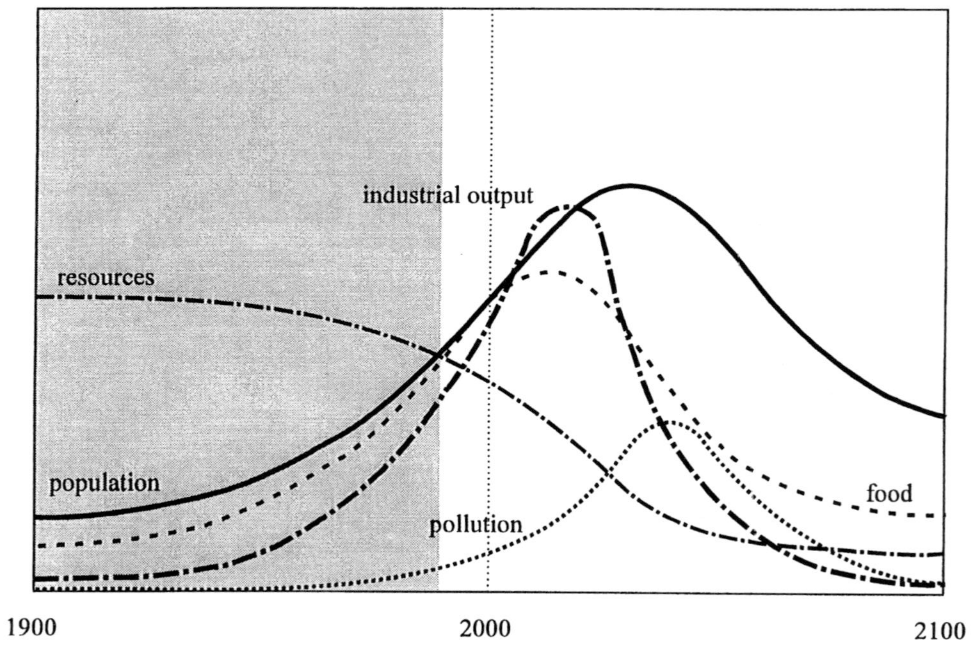

Source: Meadows, Meadows and Sanders 1992.

FIGURE 6 Scenario 10

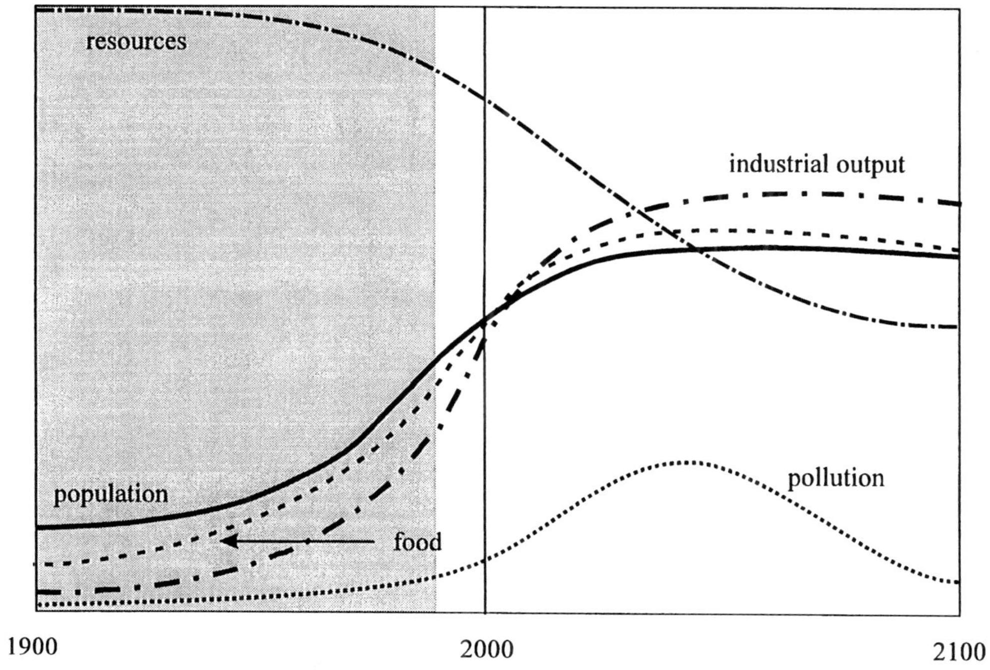

Source: Meadows, Meadows, and Sanders 1992.

depend heavily on their biophysical environment. Like their intellectual ancestors from the University of Chicago, environmental sociologists, who were centred mostly in universities located in the northwest of the United States, refused to take an anthropocentric stance in sociology, preferring to give more importance to environmental factors and to the material and ecological substratum of society, than to cultural and superstructural variables. As Catton and Dunlap (1980: 33) remark: "In sharp contrast to the anthropocentric HEP, the NEP stresses the ecosystem dependence of human societies. The NEP thus clearly points to the importance of considering societies' biophysical context."

In my view, what truly characterizes the new ecological paradigm devised by Catton and Dunlap, is not so much an ecological or an environmental determinism, but rather the taking into account of the existence of the environment and of its importance as an independent variable, and also sometimes as a dependent variable. In other words, Catton and Dunlap are not really environmental determinists but rather, sociologists who want to underline the existence of a reciprocal relationship between environment and society — between nature and culture (Catton and Dunlap 1978, 1980). Environmental sociology is thus interested in topics that are eminently social like the Green movement, public opinion and attitudes concerning the environment, environmental problems, ecological constraints on human activities, the built environment as opposed to the natural environment, environmental impact studies, dwindling natural (especially energy) resources available to humans, and the allocation of these resources between humans (Dunlap and Catton 1979a). The new environmental paradigm seems to give more importance to human dimensions than Duncan's POET model did, since social organization (O) is replaced, in the graphical representation of the new model, by the social, psychological and cultural systems. These three systems, in conjunction with population (P) and technology (T), function as dependent variables affected by environmental change (E), but they also function as independent variables affecting the environment. Thus Duncan's POET model is slightly transformed by Catton and Dunlap, with the environment becoming the key variable (Figure 7). Recently, Dunlap and Catton (1993: 4) admitted that "it was the effects of resource constraints on society that received emphasis rather than the impacts of society on the environment."

In sum, Catton and Dunlap's new ecological paradigm made them present environmental sociology as the study of the interaction between society and its biophysical environment, with a certain insistence given to the idea that every human society is dependent on the ecosystems in which it exists (Dunlap and Catton 1993: 1). This environmental sociology of the '70s, according to these authors, centres its attention on the study of the Green movement, energy issues, risks of catastrophes, public attitudes towards

FIGURE 7 Causes and Consequences of Environmental Change

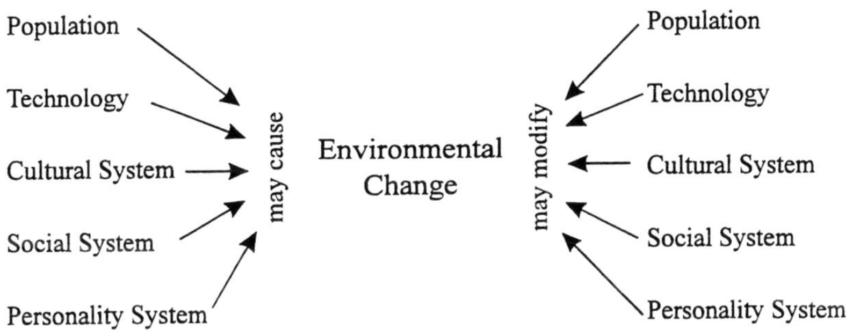

Source: R.E. Dunlap and W.R. Catton Jr. 1979b.

environmental questions, environmental policies, and the quality of the environment as a social problem.

Environmental sociology is also interested, in a more general manner, in human causes and social factors of environmental pollution, as well as in the social impacts of this pollution and of diminishing natural resources — that is, in the reciprocal relations between human societies and their biophysical environment. In a recent article, Laska (1993: 7) also enumerates the main topics of interest of environmental sociologists:

> Environmental sociology has different dimensions to it that have their own research foci: environmental attitudes and the environmental movement, social impact analysis, risk assessment, responses to toxic siting and discovery, natural hazards research and so forth.

Frederick H. Buttel, on the other hand, in his 1987 article on the state of environmental sociology, describes the five major subfields that characterize this field. The first subfield concerns the history, paradigms, models and important variables of environmental sociology, and represents its theoretical core. On this subfield, the effort has been put mostly on criticizing the main currents of traditional sociology, and on trying to invalidate the principle of a perfect equilibrium between human communities and their natural environment. The present paper, for example, would be part of this first subfield.

The second field of research defined by Buttel concerns the attitudes, values and behaviour of people towards the environment. The research in this area, according to Buttel, can be further subdivided in three subcategories. The first includes the studies, mostly empirical and sociocultural, which analyse, for example, the differences of attitudes and values towards the environment between social classes, men and women, and age groups. The second subcategory of this second subfield is centred on the verification derived from social-psychological theories, for example, the hypothesis according to which attitudes and behaviour in environmental protection issues would be the reflection of social and psychological processes implying moral norms. Finally, the third subcategory consists of applied studies that aim at determining which social factors are linked to pro-environmental behaviour, for example, which incentives facilitate the participation in recycling or energy conservation programmes.

The third subfield of research identified by Buttel, is the study of environmental groups — of the Green movements. The early research in this area focused on the analysis of the social origin and composition of these groups. Based on the first results, more theoretical propositions concerning the mobilization inside the Green movement were elaborated. Buttel notes an increasing interest for local or specialized environmental groups, such as anti-nuclear groups or citizens groups formed to cope with an environmental threat.

The evaluation of environmental risks constitutes a fourth subfield of research. According to Buttel, since 1980, sociologists have been involved in social and environmental-impact studies and have tried to influence decision makers responsible for environmental policies. An increased sensitivity and public mobilization against certain "technological risks" (e.g., the Love Canal saga) have greatly stimulated research in this area. A chapter on risk in this collection by Michael Mehta goes into more detail.

Finally, since the early 1970s, environmental sociologists have also manifested much interest, according to Buttel, in the study of interrelations between the economy, politics and the environment. Two main topics dominate this fifth subfield, namely, environmental policies and the influence of the Green movement on these policies.

In the 1980s, a certain number of books and special issues of journals prepared by sociologists active in environmental sociology, tried to assess the state of the situation in this field. Some like Anderson, *The Sociology of Survival: Social Problems of Growth* (1976) and Schnaiberg, *The Environment: From Surplus to Scarcity* (1980) did so in a dialectical and neo-Marxist perspective. Others, like Michelson, *Man and His Urban Environment: A Sociological Approach* (1976), Moos, *The Human Context: Environmental Determinants of Behavior* (1976) and Catton, *Overshoot: the Ecological Basis of Revolutionary Change* (1980) have adopted positions that are closer to those of the Chicago School. The most interesting book, and the most comprehensive in my view, was certainly Humphrey and Buttel, *Environment, Energy and Society* (1982), which presented a general overview of the field, in a perspective that we could call "neo-Marxist-Weberian." My own book *Mouvement écologiste, énergie et environnement: essais d'écosociologie* (Vaillancourt 1982), published two years after my special issue of *Sociologie et Sociétés* on social ecology and ecological movements, also tries to express the state of environmental sociology in North America at the beginning of the 1980s, while the four books which I have co-edited with José Prades and Robert Tessier, and which focus on sustainable development (1991, 1992 and 1994b), and on social dimensions of acid rain (1994b), are representative of the kind of research now being done during the third period of the sociology of the environment, namely that of ecosociology, about which we shall elaborate shortly.

Buttel (1978, 1986) disagrees with Catton and Dunlap (1978) that the major cleavage in sociology is between their new ecological paradigm (NEP) and another paradigm called the human exemptionalism paradigm (HEP), which includes all the other dominant approaches in sociology. Buttel tries to convince Catton and Dunlap to recognize that the NEP–HEP distinction can be found inside, as well as outside, sociologies of order and sociologies of conflict, and that all order or conflict sociologists are not necessary partisans of the human exemptionalism paradigm. Buttel also criticizes Catton and Dunlap's NEP for promoting an exaggerated material determinism. Poverty, according to him, cannot be explained solely by a lack of natural resources, but depends on social, political and economic factors. Society, the state and culture, as well as the economy, have a tremendous impact on the environment and are not completely dependent on it. In his 1986 and 1987 articles, Buttel intimates that environmental sociology is maturing and that it should become more dialectical. He notes that in the most recent research, there exists a greater concern for taking into account the reciprocal relationships between environmental and cultural factors in the explanation of social phenomena. He also notices an increasing interest in interpreting the environmental crisis on the basis of social facts and human factors. The multiplication of subfields of research in environmental sociology is an indication that it is a growing and changing field. But it is also the reflection of a transition from a strictly biophysical or strictly social perspective on the environment, to the more general concept of ecosystem where all the factors, biophysical and human, have some influence on each other. This dialectical approach is also the one put forward by Schnaiberg, especially in his 1975 article on the society-environment dialectics, and in his 1980 textbook, *The Environment: From Surplus to Scarcity*, in which he adopts a neo-Weberian-Marxist political economy approach, similar to that of James O'Connor (1988), to study environmental issues and problems. For him, these problems are caused by our industrial production systems, rather than by demographic or technological factors, or by our consumption patterns only. Buttel seems to agree somewhat with this type of analysis, even if recently he seems to have moved to a constructivist approach, closer to social idealism than to dialectical and historical materialism. Constructivism affirms that social — and in our case environmental — problems are constructed socially, rather than being hard facts existing out there as purely objective realities.

The Third Period: Ecosociology

During the 1980s, more precisely during the eight years of the Reagan presidency in the U.S., there was a certain decline in public interest in environmental issues. Many Green activists in many countries recycled themselves into the disarmament and peace movement, because of the increased danger of a nuclear war between the two superpowers. It was only in 1987, with the gradual ending of the Cold War, the publication of the Brundtland Report (which put forth the concept of sustainable development), and the growing international consciousness concerning global environmental problems (the signing of the Montreal Protocol on ozone depletion in 1987, the Toronto Conference and the Tokyo Symposium on climate change in 1988, the declaration of the Hague in 1989), that the interest in the environment resurfaced again in public opinion and in the social sciences.

In 1988, *Time* magazine chose the "Planet in danger" as planet of the year, in replacement of its usual "Man of the year." Earth Day in 1990 was celebrated with much more emphasis than the original one in 1970. The fact that the 1980s was the hottest decade in recent history, and 1990 was the hottest year in many centuries, drove home, to the public, the fact that global heating really was taking place. The Earth summits of June 1992 in Rio de Janeiro (Vaillancourt 1992) were better organized than the environmental conferences in Stockholm in 1972 and in Nairobi in 1982, which received little public notice. The participants' insistence on biodiversity, climate change, dwindling forests and desertification, and waste of all kinds, prodded the environmental social scientists to follow the example of the environmental natural scientists, in reorienting their research towards the larger global environmental problems of the day, and to better organize themselves at the international and national levels, in order to establish links among themselves and with the natural science specialists and the social movement activists already involved in these areas.

The concepts of biosphere, global environmental change, and of sustainable development thus took a central place, in the minds of experts and involved people, somewhat like the important role the concepts of ecosystem, air, water and soil pollution, environmental protection, and resource conservation had played an important role 20 years earlier, at the end of the '60s, in the birth of the environmentalist movement and environmental sociology. Furthermore, because of the more serious and generalized risks that they represent, the present global environmental problems appear as problems of public health and of common security, somewhat like the ultimate menace to life and collective security of people that the threat of nuclear war had appeared during the Cold War. Most researchers have reoriented their work in these new directions.

Barry Commoner, for example, in his 1990 book, *Making Peace with the Planet*, shows that we now have to make peace not only among ourselves, but also with nature itself. He suggests that we put the technosphere in harmony with the ecosphere, that we have to promote prevention with public participation rather than *ex post facto* pollution control by experts, that we must abolish poverty rather than try to limit population growth directly, and that we must transform the present production system in agriculture, industry, energy and transports. Commoner concludes that the only way to succeed in all these endeavours, is to create a convergence between the various progressive social movements, and by a reconversion of current military budgets towards the solution of today's environmental problems.

Since about 1987, the horizons of environmental sociology have widened considerably, just as interdisciplinary approaches and models have made great strides in the environmental social sciences. It is not possible to attempt here to do a systematic inventory of everything of importance that has been done in the last few years in the environmental social sciences. I would prefer to try to give a few examples of the abundance and quality of the research, the conferences and the publications in this new sector of the sociology of the environment: "ecosociology." I use this term to distinguish it from the environmental sociology of the preceding period (1970 to the late 1980s). I used the term ecosociology for the first time in the title of my book on Green movements, energy and the environment in 1982, and we find the same term in the French version of Buttel's 1986 article in the *Revue Internationale des sciences sociales*, as well as in the title of an article by Klaus Feldmann in the special issue of *Innovation, the European Journal of Social Sciences* on "Environmental Sociology" in 1993. I prefer this term ecosociology to environmental sociology, which appears to me to be too limited historically and geographically. Ecosociology is preferable to the redundant term "ecological sociology" recently adopted by Dunlap and Catton (1993), the expression "study of society and the environment" that G. Szëll seems to favour (1992, 1994) or the even less elegant expression "sociological human ecology" proposed by Micklin and Chaldin (1984).

But what counts, in the end, is not so much the name of this new stage in the sociology of the environment, but rather the underlying reality that started to emerge at the end of the Brezhnev-Reagan era. It is as if the end of the Cold War, as well as concerns about nuclear war and nuclear winter, had been replaced by concerns about global warming and global environmental changes generally, including problems linked to disarmament and to sustainable development at the international, national and regional levels. In the present period, interest is centred on global issues, and on the relationships between economy, ecology and society; hence the pertinence of the term "ecosociology" to characterize the work that sociologists are now doing in these areas.

Ecosociologists and other environmental social scientists, who are interested in the study of global environmental change and especially of human activities that have an impact on it — such as transport, polluted terrestrial and aquatic ecosystems, overconsumption of resources (especially fossil energy resources), the proliferation of various types of waste, deforestation etc. — participate more and more in national and international interdisciplinary programmes of research and action. UNESCO's MAB (Man and Biosphere) programme, for example, has for quite some time covered socio-economic and cultural aspects, along with physical, chemical and biological dimensions of environmental problems. The Human Dimensions of Global Change Programme, which was created in 1990 by the International Social Science Council as a follow up to the 1988 Tokyo Symposium on climate change, has also paid much attention to the anthropogenic causes of global environmental changes. This programme, which consists of an international network of researchers from various disciplines, is presently developing appropriate research methodologies and data banks, and is examining the economic, social, political, juridical and cultural factors that constitute the human dimensions of these changes.

Roberta B. Miller and Harold K. Jacobson, in a recent report, "Research on the human components of global change: next steps," suggest that we broaden the spatial, temporal and disciplinary aspects of our research in this sector. They cite the demographic transition theory as one of the five theories that can serve as a conceptual basis for this type of international, historical and interdisciplinary research (Figure 8). It goes without saying that ecosociologists and others will have to go beyond the stage of quantitative data accumulation in information banks, if they are to reach the level of true sociological analyses of pertinent structures and transformations.

Already, the seven sectors, identified by the Human Dimensions of Global Environmental Change Programme as being crucial for further research, constitute a solid starting point which permits us to cover a lot of ground. These seven sectors are: (1) land use; (2) social dimensions of resource

FIGURE 8 Complete Demographic Transition

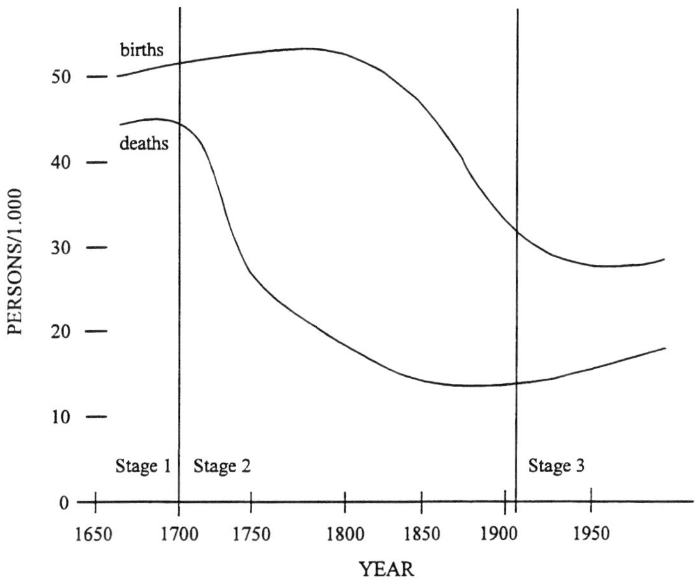

utilization; (3) the perception and evaluation of environmental conditions and changes; (4) the impact of institutions and social, economic and political structures at the local, national and international levels; (5) the production and consumption of energy; (6) industrial growth; and (7) environmental security and sustainable development (Jacobson and Price 1990).

The four major projects of the programme at the present time are devoted to: (1) perceptions and evaluations of global change, and of behaviour related to it; (2) changes in land use and occupation; (3) socio-economic and political changes in relationship to global environmental changes; and (4) social science data and global change (Pivot 1993).

Unfortunately, these projects and this programme in general, with their unilateral focus on quantification and on data banks, leave little room for many sociologists, particularly for those oriented towards more qualitative types of research.

In Canada, the Canadian Global Change Programme of the Royal Society of Canada functions as the Canadian section of the International Human Dimensions of Global Environmental Change Programme. Its research committee, of which I was a member before it was dissolved, was made up of about 20 natural and social environmental scientists. In 1992, this committee chose 22 important themes as research priorities for global environmental change research in Canada. These themes can be grouped in three larger categories: (1) natural and anthropogenic causes of global environmental changes; (2) the biophysical consequences of these factors on the natural environmental, i.e., the global environmental changes themselves; and (3) the impact or effects of these global environmental changes on socio-economic, political and cultural systems.

If we examine these three series of variables, we can show, for example, with our framework, how the existing economic production system brings about deforestation, and how this, in turn, affects the use of air, water and land, or again, how the production of CFCs and CO_2 affects the hole in the ozone layer, which in turn has a negative impact on human health.

It is possible to represent these series of variables with the help of a model or a general framework of analysis which points us to possible cause and effect links between the different variables. This model, or framework, also indicates what type of strategies — mitigation, adaptation, evaluation, behaviour modification or sustainable development practices, for example — can help us do some correction or some prevention (Figure 9). We have even used this framework to survey various groups concerning the degree of importance they give to each of the 22 variables, that is, to find out how these changes and their causes and effects are perceived by various social groups.

The Canadian Global Change Programme publishes *Delta*, a trimestrial bulletin, as well as some very interesting research reports. The report on environmental security, for example, examines the link between environmental security and population movements, in the context of global environmental security, while the famous COGGER report suggests that it would be economically advantageous, and technically possible to see a very substantial decrease of CO_2 emissions from now to 2010, by implementing serious programmes of energy conservation, and by replacing polluting sources of energy with renewable ones that are less nefarious for the environment.

In February 1994, the Canadian Social Science Federation, which regroups the Canadian associations of scholars from the various social science disciplines, organized, in Ottawa, an important interdisciplinary conference on the environment, as an area of study and intervention for Canadian researchers and teachers working in the social sciences. In Canada, research in the environmental social sciences is financed mostly by the Social and Human Sciences Research Council (SHSRC), which also administers the Tri-Council Eco-Research Programme with funds from the Green Plan. The federal and provincial ministries of the environment, many other ministries and government agencies, and some paragovernmental and private enterprises, are

FIGURE 9 A Framework for Global Change

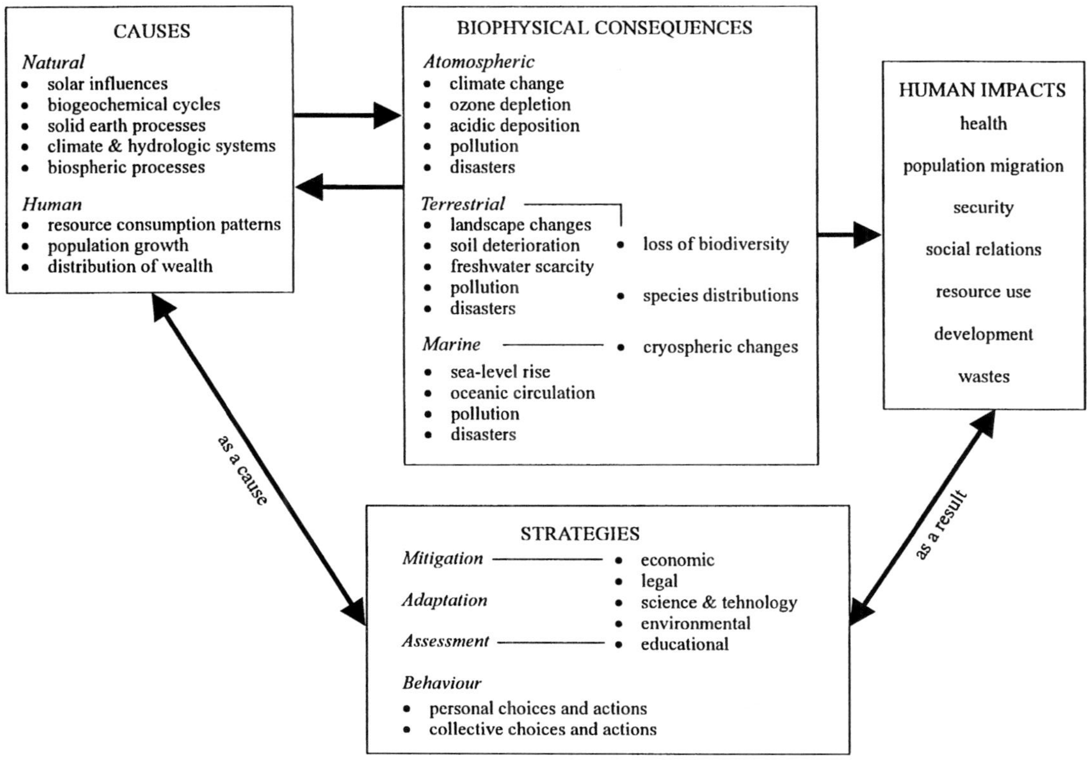

presently quite active in promoting research in the environmental social sciences, including sociology. A new association of researchers, the Environmental Studies Association of Canada (ESAC), has been founded recently by activist university researchers, most of whom are graduate students of York University, to encourage and promote that kind of research.

The 1992 report, *For Earth's Sake*, was published by the Commission on Developing Countries and Global Change, a group of Third World researchers and activists. The report acknowledges the importance of social science research, and focuses on the social dimensions of global environmental problems, and on the need to implement sustainable and equitable development, especially in the Third World. The last section of the report presents a plan of action for research on the social causes and social impacts of major environmental problems. Figure 10 summarizes the main topics of research proposed in this report, topics that can be addressed at various geopolitical levels — local, regional, national, international and global.

In the U.S., the National Academy of Sciences has created a National Research Council to address the issue of global environmental change (GEC). In their book, *Global Environmental Change: Understanding the Human Dimension*, Stern, Young and Druckman (1992) suggest that the National Science Foundation should spend more money on GEC (already the NSF gives more money to GEC than to all of sociology), and should create a system of scholarships to encourage interdisciplinary education in environmental social sciences. They also suggest that more attention be given to the social organization of energy, to land-use planning, evaluation procedure, technological change, environmental decision making, environmental conflicts and international co-operation. They see the five key causes of global environmental change as: population growth, technology, economic growth, political and economic institutions, and attitudes and beliefs of individuals. However, they also note the influence of culture, communities, the family, social networks, business and the state.

Unfortunately, sociologists play a minor role at the international level, and in the U.S. especially, in organizations concerned with the human dimensions of global environmental change. The Executive Committee of the Human Dimensions of Global Change Programme of the International Social Science Council comprises six geographers, five economists, two demographers, two political scientists, and only one sociologist. In the U.S., the Joint National Academy of Science-National Research Council Committee that produced the Stern, Young and Druckman book (1992) was made up of 18 members — economists, psychologists and political scientists (four each), but only one historian, jurist, geographer, anthropologist, ecologist and sociologist. This seems to indicate that either U.S. sociologists are not very active in

FIGURE 10 Framework for the Research Agenda

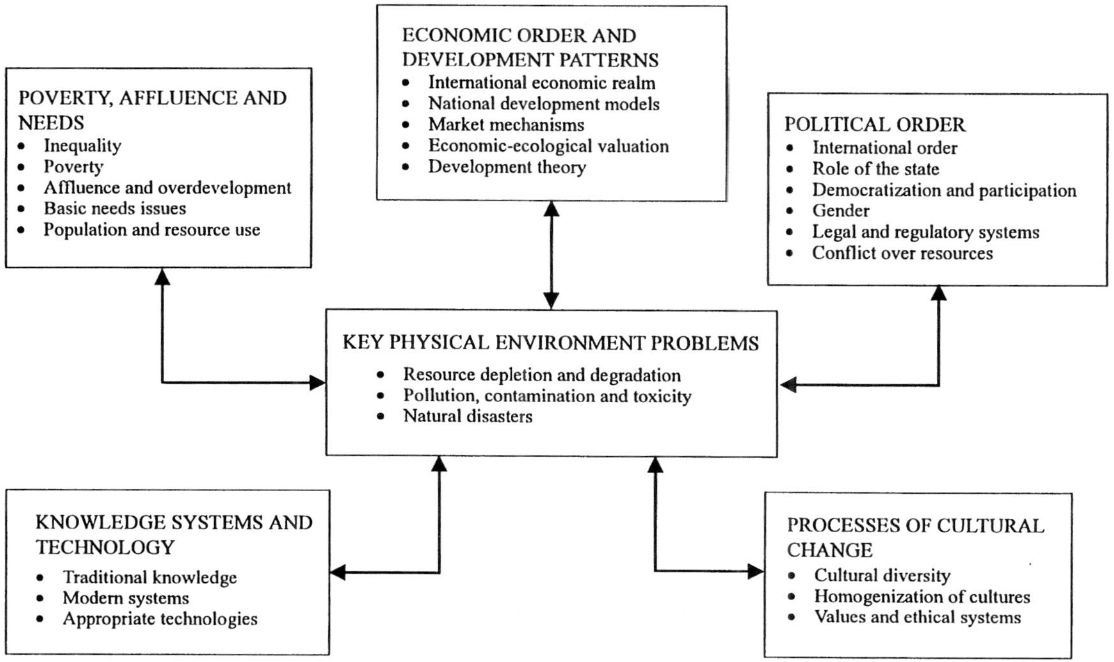

Source: For Earth's Sake, 1992.

this area, or more probably, that their discipline is perceived as being peripheral, by those who create these committees and nominate their members.

This is not the case, however, in Great Britain, which has an important programme on global environmental change in which sociologists are particularly active. The co-ordinator for many years was the famous environmental sociologist, Michael Redclift (1984, 1987, 1992; Redclift and Benton 1994). In this programme, primary importance is given to sustainable development, common security, environmental health and education, economic and political questions concerning the environment, the precautionary principle, and the relationship between nature and culture.

Many other British sociologists, such as De la Court (1990), also work on sustainable development, and many have done research on the Green movement, on its links with sustainable development and on many other environmental social science topics. For example, Leslie Sklair (1992), of the London School of Economics, is trying to develop a global sociology of the environment which would study the impact of the global capitalist system on the major contemporary environmental problems, in comparison with the impact of transnational environmental movements. He is also concentrating on problems of demand and overconsumption. The British journal, *Global Environmental Change: Human and Policy Dimensions*, and *GEC-O*, the bulletin of the Global Environmental Change Programme of the Economic and Social Research Council (ESRC) of Great Britain, along with the U.S. journal *Society and Natural Resources* and *Alternatives* in Canada, are invaluable sources of information on what sociologists and other environmental social scientists, working in these areas, are doing.

In France, the recently published book *L'Etat de l'environnement dans le monde* (Beaud, Beaud, Bouguerra 1993), (like the interesting annual *State of the World* reports of the Worldwatch Institute founded by Lester Brown Jr. in the U.S.) is a good encyclopedia on the present environmental situation of Planet Earth from a social-scientific point of view. The voluminous *Encyclopedia of Conservation and Environmentalism*, edited by Robert Paehlke, will soon be published, and should prove to be a useful tool for ecosociologists. Furthermore, two handbooks on the present state of environmental sociology, one by Dunlap et al. in the U.S. which will be entitled *Handbook of Environmental Sociology*, another by Beato Fulvio in Rome, Italy will be very useful for us in this new period that is opening up for our subdiscipline.

In France again, two periodicals, *Ecologie politique* and *Nature, sciences, sociétés*, publish numerous articles on the topic of the relationship between society and the environment. These two journals express the new tendencies and the points of interest in environmental social sciences, including ecosociology. In France, most of the research on the environment and on human dimensions of environmental global change, is done by researchers linked to the National Centre for Scientific Research (CNRS). In 1993, the CNRS's strategic plan on the environment insisted, among other things, on the importance of the human impact on ecosystems, the need for sustainable development, the perception and use of biodiversity by human societies, the degradation and rehabilitation of resources, the urban environment, energy efficiency and renewable energy, and on "the definition of alternative strategies for sustainable development, drawing from the analysis of their economic, social and political components" (Comité d'orientation de la recherche environnement du CNRS 1993: 6). The journal *Sciences sociales et environnement* is also a very useful source of bibliographical information on the environmental social sciences.

We could thus review what is being done in many other countries in the area of the new ecosociology of the late 1980s and early 1990s. In Italy, Germany, Scandinavia, Spain, in many other European locales, as well as in certain Third World countries, sociologists are building a new sociology of the environment, building on the world of the Chicago School and environmentalists of the '70s and '80s. Foremost among these are those who form the Research Group 24 on "Environment and Society" in the International Sociological Association. They participated in the meetings of the ISA in Madrid in 1990, the Symposium in Woodschooten in Holland in 1992, the two conferences in Chantilly in France in 1993 (one on Geopolitics of the Environment, the other on Nature and Culture), and in the ISA Congress of sociology in Bielefeld in Germany in June of 1994.

This new sociology of the environment — ecosociology — will be eminently international, centred mostly, but not exclusively, on the large global issues presently affecting the future of us all. It will be multidimensional, focusing on the impacts of humans on the environment, and on the impact of the environment on humans. It studies the facts and factors, the effects and impacts, but also the actors, groups, movements, organizations, decision makers, governments and public opinion.

Ecosociology is interested in problems but also in solutions, in theories but also in methods and in action, in empirical data but also in critical perspectives. Its practitioners, models, paradigms and its topics of interest are multiplying, as we were able to see in Bielefeld. So, instead of being something reserved for sociologists, it has rapidly become pluridisciplinary, interdisciplinary and even transdisciplinary. It is open to the contributions of other disciplines which are interested in the same questions, but approach them from different, and sometimes complementary, angles. Sociologists do not imagine that they are the first social scientists to take this road. Thousands of researchers from other disciplines, both social and natural, have made similar journeys in recent years, enriching the common heritage without losing their own disciplinary identity and autonomy.

Some sociologists have sometimes tended, especially in recent years, to insist on the idea that social reality, and even some "natural" realities, are socially constructed. There is some truth in this (for example, the idea that global environmental problems, like social problems, are in part "socially constructed"). But the fact remains, as Steen (1993) and Catton and Dunlap (1993) have justly indicated, that these problems remain phenomena that exist "out there" in the real world, and are not only in mental and social constructs. Sociologists should be interested in the way these problems are perceived and used in social struggles by various social actors — even partially constructed by them — but they must also study the social factors that affect the environment, the social impact of environmental changes, the ways to prevent these impacts by suggesting strategies to act on the anthropogenic causes of these changes, and to mitigate these impacts through adaptation measures. The critical realism of the ecosociological approach permits us to avoid the opposed, twin dangers of naïve materialism and idealistic constructivism that are still so common in sociology generally, and even in the new subdiscipline of ecosociology.

DISCUSSION QUESTIONS

1. What are the three stages in the development of the sociology of the environment?
2. What is the difference between human ecology and social ecology?
3. In Duncan's POET model, what are the four key variables and how do they relate to each other?
4. What are the five major areas of environmental sociology according to Buttel's 1987 article?
5. What are some of the major themes encountered in ecosociology since the late 1980s?

Bibliography

Alihan, Milla A. [1938] 1990. *Social Ecology: A Critical Analysis*. Reprint, New York: Columbia University Press.

Anderson, Charles H. 1976. *The Sociology of Survival: Social Problems of Growth*, Homewood, Ill.: The Dorsey Press.

Bharadwaj, L.K. 1992. Human Ecology and the Environment. In *Encyclopedia of Sociology*, ed. E.F. Borgotta and M.L. Borgotta, 848–67. New York: MacMillan.

Bailey, K.D. and P. Mulcahy. 1972. Sociocultural Versus Neoclassical Ecology: A Contribution to the Problem of Scope in Sociology. *Sociological Quarterly* 13(Winter): 37–48.

Beaud, Michel, Calliope Beaud and M.L. Bouguerra. 1993. *L'Etat de l'environnement dans le monde*. Paris: La Découverte.

Braybrooke, D. and G. Paquet. 1987. Human Dimensions of Global Change: The Challenge to the Humanities and Social Sciences. *Transactions of the Royal Society of Canada* V-II: 271–91.

Brown, Lester R. et al. *State of the World*, New York: W.W. Norton, published annually.
Brundtland, Gro Harlem et al. 1987. *Notre avenir à tous*. Montréal: Editions du Fleuve.
Buttel, F.H. 1978. Environmental Sociology: A New Paradigm. *The American Sociologist* 13: 252–56.
———. 1986. Sociologie et environnement : la lente maturation de l'écologie humaine. *Revue internationale des sciences sociales* 109: 359–79.
———. 1987. New directions in Environmental Sociology. *Annual Review of Sociology* 13: 465–88.
Buttel, Frederick H., Ann P. Hawkins and Alison G. Power. 1990. From Limits to Growth to Global Change. *Global Environmental Change* Vol. I, no. 1 (Dec.): 57–66.
Buttel, Frederick H. and Peter J. Taylor. 1992. Environmental Sociology and Global Environmental Change: A Critical Assessment. *Society and Natural Resources* 5: 211–30.
Caldwell, L.K. 1993. Globalizing Environmentalism: Threshold of a New Phase in International Relations. In *American Environmentalism, The U.S. Environmental Movement, 1970-1990*, ed. R.E. Dunlap and A.G. Mertig, 66–76. Philadelphia: Taylor and Francis.
Carson, Rachel. 1962. *Silent Spring*. Boston: Houghton Mifflin.
Catton, W.R. Jr. 1980. *Overshoot: The Ecological Basis of Revolutionary Change*. Urbana: University of Illinois Press.
———. 1994. Foundations of Human Ecology. *Sociological Perspectives* 37(1): 75–95.
Catton, W.R. Jr. and R.E. Dunlap. 1978. Paradigms, Theories, and the Primacy of the HEP-NEP Distinction. *The American Sociologist* 13: 256–59.
———. 1980. A New Ecological Paradigm for Post-Exuberant Sociology. *American Behavioral Scientist* 24: 15–47.
Comité d'orientation de la recherche environnement du CNRS. 1993. Schéma stratégique "Environnement" du CNRS. *Lettre du Programme Environnement*, 11(octobre): 4–6.
Commoner, Barry. 1971. *The Closing Circle: Nature, Man and Technology*. New York: Knopf.
———. 1990. *Making Peace with the Planet*. New York: Pantheon Books.
Cottrell, Fred. 1955. *Energy and Society*. New York: McGraw Hill.
Daly, H.E. and J.B. Cobb. 1989. *For the Common Good: Redirecting the Economy Toward Community, the Environment and a Sustainable Future*. Boston: Beacon Press.
De la Court, Thijs. 1990. *Beyond Brundtland: Green Developments in the 1990's*. New York: New Horizons Press.
Deléage, Jean-Pierre. 1992. *Histoire de l'écologie*. Paris: La Découverte.
Dickens, Peter. 1992. *Society and Nature: Towards a Green Social Theory*. Philadelphia: Temple University Press.
Duncan, O.D. 1961. From Social System to Ecosystem. *Sociological Inquiry* 31(Spring): 140–49.
———. 1964. Social Organization and the Ecosystem. In *Handbook of Sociology*, ed. Robert E.L. Faris, 36–82. Chicago: Rand McNally and Co.
Duncan, O.D. and L. Schnore. 1959. Cultural, Behavioral and Ecological Perspectives in the Study of Social Organization. *American Journal of Sociology* 65(Sept.): 132–45.
Dunlap, R.E. 1980. Paradigmatic Change in Social Science: From Human Exemptionalism to an Ecological Paradigm. *American Behavioral Scientist* 24: 5–14.
———. 1992–93. Toward and Ecological Sociology: the Development, Current Status and Probable Future of Environmental Sociology. *Annals of the International Institute of Sociology* 3: 263–84.
Dunlap, R.E. and W.R. Catton Jr. 1979a. Environmental Sociology. *Annual Review of Sociology* 5: 243–73.
———. 1979b. Environmental Sociology: A Framework for Analysis. In *Progress in Resource Management and Environmental Planning*, ed. T. O'Riordan and R.C. D'Arge. Vol. 1. Winchester, Eng.: John Wiley and Sons.
———. 1983. What Environmental Sociologists Have in Common (Whether Concerned with "Built" or "Natural" Environments). *Sociological Inquiry* 53-2-3(Spring): 113–35.

———. 1993. Struggling With Human Exemptionalism: The Rise, Decline and Revitalization of Environmental Sociology. Revision of a paper presented at the Annual Meeting of the A.S.A., Miami Beach, August, 1–35.

Dunlap, R.E. G.E. Gallup Jr. and A.M. Gallup. 1993. Of Global Concern: Results of the Health of the Planet Survey. *Environment* 35(Nov.): 6–15, 33–39.

Dunlap, R.E. and A.G. Mertig, eds. 1992. *American Environmentalism: the U.S. Environmental Movement, 1970–1990.* Washington, D.C.: Taylor and Francis.

Dunlap, R.E. and W. Michelson, eds. Forthcoming. *Handbook of Environmental Sociology*. Westport, Conn.: Greenwood Press.

Dunlap, R. E., L. A. Lutzenhiser and E. A. Rosa. Forthcoming. Understanding Environmental Problems: An Environmental Sociology Perspective. In *Socio-Economic Approaches to the Environment*, ed. B. Burgenmeier. Armonk, N.Y.: M.E. Sharpe Inc.

Ehrlich, P.R. and A.E. Ehrlich. 1990. *The Population Explosion.* New York: Simon and Schuster.

Feldmann, Klaus. 1993. Ecosociology: Inside and Outside the Cocoon. *Innovation. The European Journal of Social Sciences* 6: 4.

Firey, W. 1945. Sentiment and Symbolism as Ecological Variables. *American Sociological Review* 10(April): 140–48.

Freudenberg, W.R. and R. Gromling. 1989. The Emergence of Environmental Sociology: Contributions of Riley E. Dunlap and William R. Catton Jr. *Sociological Inquiry* 59: 439–52.

Gettys, W.E. 1940. Human Ecology and Social Theory. *Social Forces* 18(May): 469–76.

Gibbs, Jack P. and Walter T. Martin. 1959. Toward a Theoretical System of Human Ecology. *Pacific Sociological Review* 2(Spring): 29–36.

Gleick, P.H. 1989. Climate Change and International Politics: Problems Facing Developing Countries. *Ambio* 18: 333–39.

Goodland R. et al., eds. 1991. *Environmentally Sustainable Economic Development: Building on Brundtland.* Paris: UNESCO.

Goodman, D. and M. R. Redclift. 1991. *Refashioning Nature: Food, Ecology and Economy*. New York: Routledge.

Gore, Al Jr. 1993. *Earth in the Balance. Ecology and the Human Spirit*. New York: Plume.

Hawley, Amos H. 1950. *Human Ecology. A Theory of Community Structure*. New York: Ronald Press.

———. 1984. Sociological Human Ecology: Past, Present and Future. In *Sociological Human Ecology: Contemporary Issues and Applications*, ed. M. Micklin and H.M. Choldin. Boulder, Colo.: Westview Press.

———. 1986. *Human Ecology: A Theoretical Essay*. Chicago: University of Chicago Press.

Hollingshead, A.B. 1947. A Re-Examination of Ecological Theory. *Sociology and Social Research* 31(Jan–Feb.): 194–204.

Humphrey, C.R. and F.H. Buttel. 1982. *Environment, Energy and Society*. Belmont, Calif.: Wodsworth.

Jacobson, Harold K. and Martin F. Price. 1990. *A Framework for Research an the Human Dimensions of Global Environmental Change*. Paris: ISSC/UNESCO.

Laska, S.B. 1993. Environmental Sociology and the State of the Discipline. *Social Forces* 72: 1–17.

Llewellyn, E. and A. Hawthorne. 1945. Human Ecology. In *20th Century Sociology*, ed. G. Gurvitch and W.E. Moore. New York: Philosophical Library.

Lowe, Philip D. and Wolfgang Rudig. 1986. Political Ecology and the Social Sciences. The State of the Art. *British Journal of Political Science* 16: 513–50.

MacNeill, Jim et al. 1991. *Beyond Interdependence: The Meshing of the World's Economy and the Earth's Ecology*. New York: Oxford M. Press.

Malthus, T.R. [1798] 1909. *An Essay on the Principle of Population*. Reprint, New York: MacMillan.

McKenzie, R.D. 1968. *On Human Ecology*. Chicago: University of Chicago Press.

Meadows D.H. et al. 1972. *The Limits to Growth*. New York: Universe.

Meadows, D.H., D.L. Meadows and J. Sanders. 1992. *Beyond the Limits. Confronting Global Collapse. Envisioning a Sustainable Future*. Toronto: McLelland and Stewart.

Michelson, W. 1976. *Man and His Urban Environment: A Sociological Approach*, Reading. Mass.: Addison-Wesley.

Micklin, M. and H.M. Chaldin, eds. 1984. *Sociological Human Ecology: Contemporary Issues and Applications*. Boulder, Colo.: Westview.

Milbrath, Lester W. 1989. *Envisioning a Sustainable Society: Learning Our Way Out*. Albany: State University of New York Press.

Miller, R.B. 1991. Social Science and the Challenge of Global Environmental Change. *International Social Science Journal* 130: 609–17.

Miller, R.B. and H. Jacobson. n.d. Research on the Human Components of Global Change: Next Steps. CISS/ISSC, Discussion Paper I.

Moos, R.H. 1976. *The Human Context: Environmental Determinants of Behavior*. New York: Wiley Interscience.

Mukerjee, R. 1945. *Social Ecology*. London: Longmans Green.

Mungall C. and D. Mc Laren. 1990. *La Terre en péril*. Ottawa: Société royale du Canada.

O'Connor, James. 1988. Capitalism, Nature, Socialism: A Theoretical Introduction. *Capitalism, Nature, Socialism* 1: 11–38.

Odum, Eugene P. 1975. *Ecology: the Link Between the Natural and the Social Sciences*. 2d ed. New York: Holt, Rinehart and Winston.

Park, R.E. 1916. The City: Suggestions for the Investigation of Human Behavior in the Urban Environment. *American Journal of Sociology* 20: 577–612.

———. 1936. Human Ecology. *American Journal of Sociology* 42(July): 1–15.

Park, R.E. and E.W. Burgess, eds. 1921. *Introduction to the Science of Society*. Chicago: University of Chicago Press.

Pearce, D.W., et al. 1990. *Sustainable Development: Economics and Environment in the Third World*. Aldershot: Hants.

Pivot, Agnès. 1993. Programme sur les dimensions humaines des changement globaux. *Nature, Sciences, Sociétés* 1(2): 164–67.

Prades, J.-A., J.-G. Vaillancourt and R. Tessier, eds. 1991. *Environnement et développement. Questions éthiques et problèmes socio-politiques*. Montréal: Fides.

Prades, J.-A., R. Tessier and J.-G. Vaillancourt, eds. 1992. *Gestion de l'environnement. Éthique et société*. Montréal: Fides.

———. 1994. *Instituer le développement durable. Éthique de l'écodécision et sociologie de l'environnement*. Montréal: Fides.

———. 1994. *Aspects sociaux des précipitations acides au Québec*. Montréal: Fides.

Quinn, J.-A. 1950. *Human Ecology*. New York: Prentice Hall.

Redclift, Michael. 1984. *Development and the Environmental Crisis*. London: Methuen.

———. 1987. *Sustainable Development: Exploring the Contradictions*. New York: Methuen.

———. 1992. Sustainable Development and Global Environmental Change. Implications for a Changing Agenda. *Global Environmental Change* March: 32–42.

Redclift, Michael and Ted Benton. 1994. *Social Theory and the Global Environment. London: Routledge*.

Rockwell, Richard C. and Richard H. Moss. 1992. The View from 1996: A Future History of Research on the Human Dimensions of Global Environmental Change. *Environment* 34(Jan/Feb.): 12–17, 33–38.

Roszak, Theodore. 1992. *The Voice of the Earth — An Exploration of Ecopsychology*. New York: Simon and Schuster.

Schnaiberg, Allan. 1975. Social Syntheses of the Societal-Environmental Dialectic: The Role of Distributional Impacts. *Social Science Quarterly* 56: 5–20.

———. 1980. *The Environment: From Surplus to Scarcity*. New York: Oxford University Press.

Schnaiberg, A. and K.E. Gould. 1994. *Environment and Society: The Enduring Conflict.* New York: St-Martin's.

Schnore, Leo. 1958. Social Morphology and Human Ecology. *American Journal of Sociology* 63(May): 620–34.

Silver, C.S. and R.S Defries. 1990. *One Earth/One Future: Our Changing Global Environment.* Washington, D.C.: National Academy Press.

Sklair, Leslie. 1992. Global Sociology and Global Environmental Change. Paper presented at ESRC Research Workshop on Sociology and Global Environmental Change, University of Kent, 27 March.

Stern, Paul C, Oran R. Young and Daniel Druckman, eds. 1992. *Global Environmental Change: Understanding the Human Dimensions.* Washington, D.C.: National Academy Press.

Stockdale, Jerry. 1989. Pro-growth, Limits to Growth, and a Sustainable Development Synthesis. *Society and Natural Resources* 2: 163–76.

Szëll, György. 1992. "Environment and Society" or "Environmental Sociology?" On the Search for a Paradigm. Presidential address to the symposium, Current Developments in Environmental Sociology of the I.S.A. Thematic Group Environment and Society, Woudschoten, the Netherlands, 17–21 June, 1–15.

———. 1994. Technology, Production, Consumption and the Environment. *International Social Science Journal* 140: 213–25.

Taylor Peter J. and Frederick H. Buttel. 1992. How Do We Know We Have Global Environmental Problems. *Geoforum* 23: 405–16.

Theodorson, G.A. 1961. Human Ecology and Human Geography. In *Readings in Contemporary American Sociology*, ed. J.S Roucek. Patterson, N.J.: Littlefield Adams.

Thomas, Franklin. 1925. *The Environmental Basis of Society. A Study in the History of Sociological Theory.* New York: The Century Co.

Turner II, B.L. et al., eds. 1990. *The Earth as Transformed by Human Action: Global and Regional Changes in the Biosphere Over the Past 300 Years.* New York: Cambridge University Press.

United Nations Conference on Environment and Development. 1992. *Agenda 21.*

Vaillancourt, Jean-Guy. 1980. Ecologie sociale et mouvement écologique. *Sociologie et Sociétés* 13 (Avril).

———. 1982. *Mouvement écologiste, énergie et environnement. Essais d'écosociologie.* Montréal: Editions Saint-Martin.

———. 1992. UNCED and '92 Global Forum: The Rio Summits. *Delta* 3-2(Fall): 1, 3–5.

Willhelm, S.M. 1964. The Concept of the Ecological Complex: A Critique. *The American Journal of Economics and Sociology* 23(July): 241–48.

Wirth, Louis. 1945. Human Ecology. *American Journal of Sociology* 50-6(May): 483–88.

Wittman, J.S. 1973. *Selected Articles in Social Ecology.* New York: MSS Information Corporation.

Yearly, Steven. 1991. *The Green Case: A Sociology of Environmental Issues, Arguments and Politics.* London: Harper-Collins.

Young, Gerald, ed. 1983. *Origins of Human Ecology.* Stroudsburg, Pa.: Hutchinson Press.

Xiaoyan Tang and Harold K. Jacobson. 1988. *Human Dimensions of Global Environmental Change Proposals for Research.* Ann Arbor, Mich.: Institute for Social Research.

WHAT IS ENVIRONMENTAL SOCIOLOGY? AN INQUIRY INTO THE PARADIGMATIC STATUS OF ENVIRONMENTAL SOCIOLOGY

David B. Tindall

University of British Columbia

In this essay, I will discuss theoretical aspects of environmental sociology in order to consider its justification as a separate subfield of sociological inquiry. To do this, I first consider the concept of a paradigm, and then review several paradigms outlined by environmental sociologists. In the discussion section, I will attempt to explicate the core themes of environmental sociology that give it coherence and distinguish it from other areas of sociology. Finally, I develop a metatheoretical scheme for understanding "subdomains" within the subfield.

The Concept of a Paradigm

In his examination of the paradigmatic status of sociology, Ritzer (1975, 1983, 1992) draws upon the work of Kuhn (1962), and suggests that sociology is a multiple paradigm science.[1] According to Ritzer, a paradigm is a fundamental image of the subject matter within a science; this is one of many possible definitions (see Masterman 1970). It serves to define what should

The author would like to thank the following people for their assistance: Loraine Lavallee for providing insightful comments on an earlier draft of this paper; Chantelle Marlor for assistance in preparing the chapter and bibliography; and Aaron Doyle for help with literature searches.

be studied, what questions should be asked, how they should be asked, and what rules should be followed in interpreting the answers obtained. The paradigm is the broadest unit of consensus within a scientific discipline, and serves to differentiate one scientific community (or subcommunity) from another. It subsumes, defines and interrelates the exemplars, theories, methods and instruments that exist within it. In Ritzer's view, a paradigm has four basic components: (1) an exemplar, or piece of work that stands as a model for those who work within the paradigm; (2) an image of the subject matter;[2] (3) theories; and (4) methods and instruments. Based on these criteria, Ritzer (1983) identifies three basic sociological paradigms: social facts, social definitions and social behaviour.[3]

The Development of an Environmental Sociological Paradigm

A sociological focus upon the environment can be traced to the rise of serious environmental problems and general social discontent in American society in the 1960s (Catton and Dunlap 1978a: 42). In particular, Catton and Dunlap (1978a) note that this period marked a break with the 1950s in which social progress, social stability and upward mobility seemed to be a permanent condition. Rapid economic expansion in the post-World-War-II era gave rise to concern about "limits to growth."

In their (1979) review, Dunlap and Catton document the organizational development of environmental sociology from small committees within the Rural Sociology Association, to the eventual formation of an ASA "Section of Environmental Sociology" in 1975 (this section has more recently been renamed "Environment, Technology and Society").

Hannigan (1995a, chap. 16: 3) notes:

> In 1983, *Sociological Inquiry* became the first sociological journal to publish a complete special issue (53-2/3) on "environmental sociology." Although natural-resource issues have long been central to sociology in Canada, it has only been in the last five years that environmental theory and research have come to the fore in this country. The *Canadian Review of Sociology and Anthropology* recently recognized this interest with a special issue entitled "Social Studies of the Environment" (31-3[August 1994]).

Lenski and Lenski's (1970) volume, *Human Societies: An Introduction to Macrosociology*, was one of the earliest general/introductory sociology textbooks to give serious consideration to society-environment linkages. Hannigan's 1995a chapter entitled "Sociology and the Environment" is the latest in a series of chapters on "society and environment/ecology" that have emerged in recent years in introductory sociology textbooks. See also: Spencer (1990), Zeitlin with Brym (1991), Harding (1993) and Warriner (1994).

Arguably, the two most influential early environmental sociology texts were Schnaiberg's *The Environment: From Surplus to Scarcity*, and Humphrey and Buttel's *Environment, Energy and Society*. Recently, there has been a proliferation of books on "society and environment/ecology," ranging from general overviews (e.g., Yearly 1991; Martell 1994; Schnaiberg and Gould 1994; Murphy 1994; Hannigan 1995b; Dunlap and Michelson, forthcoming), to Green social theory (e.g., Dickens 1992; Reclift and Benton 1994; arguably, Beck 1992 could also be included on this list). This is by no means an exhaustive list.

For Catton and Dunlap, the rise of unanticipated new social problems, particularly environmental problems, revealed that dominant sociological perspectives were seriously flawed. Although they acknowledge the existence of so-called competing perspectives in contemporary sociology — such as functionalism, symbolic interactionism, conflict theory and Marxism — Catton and Dunlap argue that these different theoretical perspectives tend to exaggerate their differences: "... their apparent diversity is not as important as the fundamental *anthropocentrism* underlying *all* of them" (1978a: 42).

This common anthropocentrism, Catton and Dunlap argue, constitutes a basic sociological world-view. They refer to it as the "Human Exceptionalism Paradigm" or HEP, later renamed the "Human Exemptionalism Paradigm." (See column 2 of Table 1.) They maintain that the assumptions underlying HEP make it difficult for most sociologists to explain adequately and take into account the social implications of ecological problems and constraints.

According to Catton and Dunlap (1978a: 42), the Human Exceptionalism Paradigm is based on the following assumptions:

1. Humans are unique among the earth's creatures, for they have culture.
2. Culture can vary almost infinitely and can change much more rapidly than biological traits.
3. Thus, many human differences are socially induced rather than inborn, they can be socially altered, and inconvenient differences can be eliminated.
4. Thus, also, cultural accumulation means that progress can continue without limit, making all social problems ultimately soluble.

Catton and Dunlap note that when the public started to become aware and concerned with environmental problems, the scientists, who became important interpreters of these problems, and opinion leaders were not sociologists, but rather, natural scientists — notably biologists.[4] Catton and Dunlap state that the discipline's allegiance to HEP has led sociologists to focus exclusively on the social environment, meanwhile leading to a neglect of the physical environment. (Michelson 1970, 1976 makes a similar argument about human

TABLE 1

A Comparison of Major Assumptions in the Dominant Western World-View, Sociology's Exemptionalism Paradigm and the Proposed New Ecological Paradigm

Assumptions of Different World-Views	*Dominant Western World-View (DWW)*	*Human Exemptionalism Paradigm (HEP)*	*New Ecological Paradigm (NEP)*
Assumptions about the nature of human beings	DWW1: People are fundamentally different from all other creatures on Earth, over which they have dominion.	HEP1: Humans have cultural heritage in addition to (and distinct from) their genetic inheritance, and thus are quite unlike all other animal species.	NEP1: While humans have exceptional characteristics (culture, technology etc.), they remain one among many species that are interdependently involved in the global ecosystem.
Assumptions about social causation	DWW2: People are masters of their destiny; they can choose their goals and learn to do whatever is necessary.	HEP2: Social and cultural factors (including technology) are the major determinants of human affairs.	NEP2: Human affairs are influenced not only by social and cultural factors, but also by intricate linkages of cause, effect and feedback in the web of nature; thus purposive human actions have many unintended consequences.
Assumptions about the context of human society	DWW3: The world is vast, and thus provides unlimited opportunities for humans.	HEP3: Social and cultural environments are the crucial context for human affairs, and the biophysical environment is largely irrelevant.	NEP3: Humans live in and are dependent upon a finite biophysical environment, which imposes potent physical and biological restraints on human affairs.
Assumptions about constraints on human society	DWW4: The history of humanity is one of progress; for every problem, there is a solution, and thus progress need never cease.	HEP4: Culture is cumulative; thus technological and social progress can continue indefinitely, making all social problems ultimately soluble.	NEP4: Although the inventiveness of humans and the powers derived from there may seem for a while to extend carrying capacity limits, ecological laws cannot be repealed.

Source: Catton and Dunlap (1980: 34)

ecology — an approach, ironically, that initially set out to incorporate the environment into its framework for social analysis.) In time, sociologists began to pay attention to the work of biologists concerned with environmental degradation, and it became clear that the Durkheimian principle, "that social facts can only be explained by linking them to other social facts," was insufficient for comprehending contemporary Western societies. One of the first sociological topics to develop with an environmental focus was the study of the environmental movement (see Albrecht and Mauss 1975).

Catton and Dunlap note that a research network began to emerge among sociologists with interests in such environmental topics as the built environment, natural hazards, resource management and outdoor recreation, and social impact assessment.[5] Following the rise of the environmental movement, the energy crisis of the early 1970s spawned a series of investigations on the relationship of energy and other resource constraints to aspects of society, such as stratification, legislation etc.

These diverse interests were connected by a core focus upon *the study of interaction between the environment and society* (Catton and Dunlap 1978a: 44).[6] Catton and Dunlap argue that recognition of the importance of this interaction has resulted in the emergence of a new world-view. This new view is entitled the "New Environmental Paradigm" or NEP, later renamed the "New Ecological Paradigm" (Dunlap and Catton 1979). There are a number of assumptions that underlie this new paradigm which distinguish it from HEP (see column 3 of Table 1). These include:

1. Human beings are but one species among the many that are interdependently involved in the biotic communities that shape our social life.
2. Intricate linkages of cause and effect and feedback in the web of nature produce many unintended consequences from purposive human action.
3. The world is finite, so there are potent physical and biological limits constraining economic growth, social progress and other societal phenomena (Catton and Dunlap 1978a: 44–45).

When considering the problem of scarce resources and the consequences of this for social stratification, Catton and Dunlap focus on three issues in the stratification literature: (1) the recent decline in the standard of living; (2) class conflict; and (3) the distribution of environmental costs. Within this literature, they cite the work of Schnaiberg (1975) as providing a good example of work falling within the domain of NEP (see also Schnaiberg 1980, forthcoming; Schnaiberg and Gould 1994)[7] because his "ecological synthesis" would result in a "steady-state" economy, and he argues that society would need to be far more egalitarian (than the contemporary U.S.) for this outcome to be possible. Class conflicts may emerge from policies to create steady-state economies or, relatedly, to enact economic redistribution. In fact, some of these conflicts may already be manifested in the emergence of a new class and new social movements (see Catton and Dunlap 1978a: 47; Schnaiberg 1980, forthcoming; Kriesi 1989; Offe 1985). Catton and Dunlap note that sociologists working within the NEP have not only questioned the assumed universal benefits of growth, but have also raised the issue that "costs" of growth have been neglected. Costs (or externalities), such as health hazards and exposure to pollution, tend to be regressive, in that they are more likely to be experienced by those with lower SES.

With regard to the question of a steady-state economy, Buttel concurs that (1978: 254) "... *redistribution* in some fashion seems to be a prerequisite for breaking the chains of repeated return to the ecologically disastrous economic synthesis...." However, Buttel criticizes Catton and Dunlap's HEP-NEP distinction. He argues that "neither the NEP nor the HEP embodies a coherent set of domain assumptions, and as a result, the assumptions of each are neither internally consistent nor mutually exclusive" (1978: 252). Buttel remarks that while the HEP/NEP is a valid distinction, "... the HEP/NEP cleavage is part of an ongoing, lively debate within major extant sociological paradigms, and that the resolution of this debate depends upon the paradigm in question" (1978: 252–53). For example, it is not simply a matter of there being a cleavage between, say, Marxist sociology and the new ecological paradigm, but rather there are important divisions and debates between traditional Marxist political economy and neo-Marxist ecological syntheses (e.g., Schnaiberg 1980; Redclift 1984; Bunker 1984, 1985). Catton and Dunlap (1978b) respond to Buttel's criticisms by questioning his definition of a paradigm.[8]

Buttel, in his (1987) review of the field, examines five important areas of environmental-sociological scholarship. These are: (1) the "new human ecology"; (2) environmental attitudes, values and behaviours; (3) the environmental movement; (4) technological risk assessment; and (5) the political economy of the environment and environmental politics. He also acknowledges social impact assessment as a burgeoning subfield (see Finsterbusch, Llewellyn and Wolf 1983; Finsterbusch and Freudenburg forthcoming). Curiously, Buttel neglects to review work in the sociology of the built environment (except for his discussion of the built environment in the context of human ecology, technology and risk, or political economy); but see Michelson's classic 1976 work for an overview of this area (see also Michelson and van Vliet forthcoming; Ahrentzen forthcoming; Popenoe forthcoming; and Kilmartin forthcoming). Buttel (1987) provides a thorough overview of the main areas of research and theory, but does not attempt a substantial metatheoretical analysis of the field.

Human Ecology as a Potential Paradigm for Environmental Sociology

Buttel and Humphrey (forthcoming) see environmental sociology as driven by a "double determination:"

> ... environmental sociological theory continues to revolve around the double specification of traditional social theory and of postures on the nature of humans as a biological species in an ecosystem, in comparison with humans as creators of distinctly social environments.

In order to gain a sense of some of the central features of environmental sociology, it is useful to briefly consider the human ecology perspective. Buttel (1978) refers to the "new human ecology" as constituting the core of environmental sociology. He maintains this perspective has both stimulated empirical research and engaged in theoretical work, that is self-consciously critical of "mainstream sociology" (Buttel 1987: 468).[9] In discussing the contribution of human ecology to the development of environmental sociology, Buttel and Humphrey (forthcoming) state:

> Perhaps the chief contribution of the field from our vantage point is the implicit recognition that human population, as other species, share functional interdependencies in a natural web of life. Neither Park and Burgess nor their students, however, fully examined the implication of this interdependence between human population and the environment, especially the problems associated with biophysical constraints on growth, or with the influence of natural resources on the structure of sustenance organization.

In time, the theoretical limitations of a human ecology which focused almost entirely on "competitive co-operation in the spatial organization of metropolitan populations," was recognized, and a number of sociologists attempted to reinvigorate this conceptual framework (Buttel and Humphrey 1987, forthcoming). Michelson (1976), in his critique of human ecology, refers the reader to Robert Park's definition of human ecology as being symptomatic of why this approach never became a true *environmental* sociology. Park (1952: 165) states:

> Human ecology, as the sociologists would like to use the term, is ... not man, but the community; not man's relation to the earth which he inhabits, but his relations to other men, that concerns us most.

Michelson identifies four reasons why human ecologists never made the physical environment a core feature of their analyses. First, he argues, human ecology conceptualized space as a *medium* of social life, rather than as a *variable* that could be used to explain behaviour. Second, human ecology tended to focus on human populations as *aggregates*. However, as Michelson points out, it is important to study how subpopulations are differentially affected by characteristics of the physical environment. He states (Michelson 1976: 19) "... what is functional for one kind of group at one level may be indifferent or even dysfunctional to other groups." Third, Michelson criticizes the formalizing of human ecology as a unique discipline. He argues human ecology erected barriers around the field, which have proven to be unfruitful by restricting potential avenues of inquiry. Finally, he notes that recognition of the possibilities implied by "rational intervention" in the designing of urban and other built environments had not been incorporated into human ecology thinking. Michelson goes on to elaborate ideas for a revised human ecology, integrating theoretical categories from Parsons, and detailing his own ideas about the concept of *congruence*.[10]

One reformulation of human ecology, referred to as the "POET" model, focuses upon the interrelationships of four key aspects of human ecosystems: population, organization, the environment and technology (see Duncan 1961, 1964; Sly 1972). (Hawley 1986 provides a comprehensive overview and synthesis of the Human Ecology perspective.) Buttel and Humphrey (1987, forthcoming) argue that recent work in human ecology suggests that this perspective can be useful in understanding the relationship between population and the environment, but a number of issues remain unresolved:

> ... the question of whether human populations and their organizations tend towards stability or growth remains unclear. Notions about the carrying capacity of particular environments, the functional as well as dysfunctional forms of adaptation (including technological innovations), and the nature of equilibrating processes remain to be adequately explored.

Metatheory and Environmental Sociology

Buttel and Humphrey (1987, forthcoming) have developed a metatheoretical categorization framework (consisting of two dimensions) for understanding the organization of environmental sociology. The first dimension places theorists into one of three categories corresponding to the founding thinkers of classical sociology: Karl Marx, Max Weber and Emile Durkheim.[11] The second dimension consists of the HEP-NEP dichotomy (discussed above) developed by Catton and Dunlap (1978a).

While the HEP-NEP schema has been widely cited, Buttel and Humphrey argue that it has generated very little work in terms of translating its assertions into more precise theoretical statements and testable hypotheses. Buttel and Humphrey (forthcoming) state:

> Some environmental sociologists question the centrality or primacy of the HEP/NEP distinction (Buttel 1978, 1986; Humphrey and Buttel 1982). The distinction is a valid one at the metatheoretical level, but there will be

intense debate within the NEP and HEP reflecting the classical theoretical perspectives in Table 1. Catton and Dunlap argue that these differences lie only within the HEP. The assumptions of the NEP also are at too high a level of abstraction to stimulate meaningful research. Stated so broadly, each of the classical theorists could be cited as exemplars of both the HEP and NEP. It is not that the HEP/NEP distinction lacks utility. We consider it of such importance that it constitutes an axis in the table. However, the NEP cannot, in and of itself, be employed deductively to construct theory.[12]

Discussion

In this discussion section, I undertake a metatheoretical analysis of work within environmental sociology, with the goal of helping to outline the key features of this field of inquiry. In discussing the role of metatheory in sociology, Ritzer (1992: 651) argues that there are three basic varieties of metatheorizing. These can be defined by their end products as follows:

1. Metatheorizing as a means of attaining a deeper understanding of theory (M_U).[13]
2. Metatheorizing as a prelude to theory development (M_P).[14]
3. Metatheorizing as a source of perspectives that overarch sociological theory (M_O).[15]

My purpose in this chapter, and in this discussion section in particular, is to engage in metatheorizing that combines the first and third definitions (M_U and M_O) described above. To this end, I have produced a diagram (Figure 1) and several tables (Tables 2 and 3) that attempt to highlight the important features of the environmental sociology landscape, and to suggest some tentative subdomains within this area of sociology. I discuss these in greater detail in the latter part of this section.

There are two main issues to be considered: (1) What are the core features of environmental sociological approaches that distinguish this area of inquiry from other specialities within sociology?; and (2) How might we map out the terrain of environmental sociology to make sense of its varying perspectives? Scholars such as Michelson (1976) and Catton and Dunlap (Catton and Dunlap 1978a; Dunlap and Catton 1979) have argued that what distinguishes environmental sociology from other areas is its overriding concern with the interrelation between the *physical* environment and social life. This linkage can take one of two main forms. Michelson and van Vliet (forthcoming) note that "one very basic dichotomy in built environment research involves a focus which may vary between, and on the one hand, the impact of environment on people, on the other, people's impact on environment." This distinction applies to environmental sociology more broadly.[16]

Catton and Dunlap (1978a, 1978b) argue that what distinguishes environmental sociology from other paradigms of sociology is its dedication to the assumption underlying the "new ecological paradigm." NEP, I believe, is an ideal type (see Weber 1949) that researchers/designers might aspire to — but useful contributions to environmental sociology have and can be made without incorporating *all* of the tenets of NEP.

Environmental-sociological approaches can be characterized as giving environment-behaviour/cognition relations a central focus in analysis. Take, for instance, the study of stratification patterns. Neighbourhood status, home ownership, dwelling type and car ownership may be used as variables in a regression equation predicting some form of inequality (or combined to form a measure of inequality). However, such analyses generally make environmental variables peripheral to the main subject of analysis. Environmental-sociological approaches make the environment *figure instead of ground*; that is, they make environmental variables the central focus of analysis — e.g., Michelson's (1977) focus in his study of residential choice, behaviour and satisfaction, as opposed to the Chicago School's version of Human Ecology. *Ideally*, an environmental-sociological approach will not just focus upon environmental variables, but will conceptualize the environment as a system (e.g., Schnaiberg 1980; Schnaiberg and Gould 1994; Bunker 1985; Logan and Molotch 1987) that exists and potentially changes over time. Further, such an approach develops a framework that focuses at least one eye on issues relating to *quality of life*. This latter characteristic does not make environmental-sociological approaches entirely unique, but there is a greater tendency for environmental-sociological approaches to be problem based, and to be concerned with policy and other applied considerations (e.g., advocacy, improved design, mitigation of negative impacts etc.).

One may well ask: What is the difference between environmental and urban sociology? The answer to this question, of course, depends on one's emphasis. Gillis (1990: 503) states that "[a] concern with the physical environment has traditionally characterized urban sociology...." While this may be definitionally true, Michelson (1976) and others have demonstrated that in practice, much urban sociology (rightly or wrongly) has focused upon the *social environment*. Another difference is found in the scope of these fields. Urban sociology is primarily concerned with issues relating to urbanization and the nature of urban social life. While the interests of environmental sociology intersect with those noted above (to the extent that they involve the physical environment), they stretch in directions far outside the normal domain of urban sociology (e.g., the political economy of energy production and consumption, social psychological responses to toxic exposure, nationwide environmental movements, the design of safer playgrounds for children etc.).

Let us return now to the second question: How might we map out the terrain of environmental sociology to make sense of its varying perspectives? Buttel and Humphrey (1987, forthcoming) have developed a two-dimensional metatheoretical schema, in an attempt to provide coherence to the organization of environmental sociology. As noted above, one dimension consists of Catton and Dunlap's HEP-NEP distinction, and the second is based on the major schools of thought rooted in the three main classical figures of sociology (interestingly omitting Simmel).[17] I have talked above about the ideal-typical nature of the HEP-NEP distinction. Buttel and Humphrey argue that while it is of some value, it does not stand on its own. Rather, they assert, there are serious debates within existing paradigms (e.g., between traditional Marxist approaches to political economy versus new political-economy-ecological syntheses). While I tend to agree with Buttel and Humphrey, I think their additional dimension is flawed, and I wonder if further dimensions may be necessary to make sense adequately of work within this field (ignoring, for the moment, the relative utility of such metatheoretical distinctions in the first place; on this question, see Ritzer 1992).

A problem with using the classical theorists to construct categories in a typology, is that their work tended to cross paradigms (as defined and discussed by Ritzer 1983). While there is certainly nothing wrong with this on intellectual grounds, it makes theoretical classification more difficult. For example, if one works within the "Marxist Tradition," does this mean drawing upon "the young existential Marx" who was concerned with social-psychological problems, such as alienation (e.g., Marx 1961), or the "elder Marx" who worked within the tradition of political economy in his endeavours to develop a critique of capitalism (e.g., Marx 1967). Do Weberians draw primarily upon the Weber who defined sociology as "a science which attempts the interpretive understanding of social action" (e.g., Weber 1949), or the Weber who was concerned with social structures, such as bureaucracy (e.g., Weber 1968). Do scholars working in the "Durkheimian tradition" use as an exemplar the Durkheim who defined sociology as the study of "social facts" (e.g., Durkheim 1895), or the Durkheim who was interested in the interaction rituals of religious practitioners (e.g., Durkheim 1912). My point here echoes one advocated by Randall Collins (1988); that is, sociological theory should be organized around problems (and levels of social reality), rather than around schools of thought or classical thinkers (although such distinctions are perhaps easier advocated than made). For instance, instead of organizing work following in the tradition of Marx, it should be categorized as falling within the domain of political economy (e.g., Schnaiberg 1980; Bunker 1985; Redclift 1984; Logan and Molotch 1987). Again, rather than categorizing certain works following in the tradition of Weber, they might be identified as falling within the realm of organizational theory (e.g., see Perrow 1986).

It now becomes obvious that Buttel and Humphrey's (1987, forthcoming) categorization scheme is inadequate (as the authors admit) because it omits microlevel phenomena — a rather significant omission. A number of important works in environmental sociology have become concerned with phenomenological/social construction issues. Most notable are works that have examined victims and potential victims of toxic waste and technological accidents (e.g., Bogard 1989; Edelstein 1988; Brown and Mikkelsen 1990; Levine 1982). Further, constructionist approaches have developed within the social movement literature (Melucci 1980; Cohen 1985; Touraine 1981, Touraine et al. 1983).[18]

At this point, it would seem useful to abandon Buttel and Humphrey's additional dimension, and replace it with the two dimensions proposed by Ritzer. Ritzer, in developing a metatheoretical scheme for analysing sociological theory, distinguishes several *levels of social reality*. He argues that there are two important dimensions of social reality. The first of these is the *macroscopic–microscopic* continuum. Starting at the micro level and ranging through to the macro, the phenomena in this continuum include: individual thought and action, interaction, groups, organizations, societies (and finally) world systems. The second dimension identified by Ritzer is the *objective–subjective* continuum. Subjective refers to phenomena that occur in the realm of ideas, while objective relates to real, material events. According to this scheme, a society is made up of both objective structures, such as governments, bureaucracies and laws, and subjective phenomena, such as norms and values. (Ritzer also recognizes "mixed types" of social phenomena which involve some combination of objective and subjective elements — for instance the family or religion.)

Ritzer then combines these two dimensions to create a typology of the major levels of social reality. The categories include: (1) macro-objective; (2) macrosubjective; (3) micro-objective; (4) microsubjective.[19] Ritzer states that his schema is intended to be used as a metatheoretical tool for the comparative analysis of sociological theories. He asserts that any good sociological theory will incorporate at least two major levels of social reality, and an ideal theory will consider all major levels of the social world.

If we jettison Buttel and Humphrey's focus on the classical theorists, and add Ritzer's two dimensions of social reality to the HEP-NEP criterion, we are left with the following dimensions (for now):

1. HEP-NEP
2. macro-micro
3. objective-subjective

FIGURE 1
A Metatheoretical Scheme for Understanding Environmental Sociology

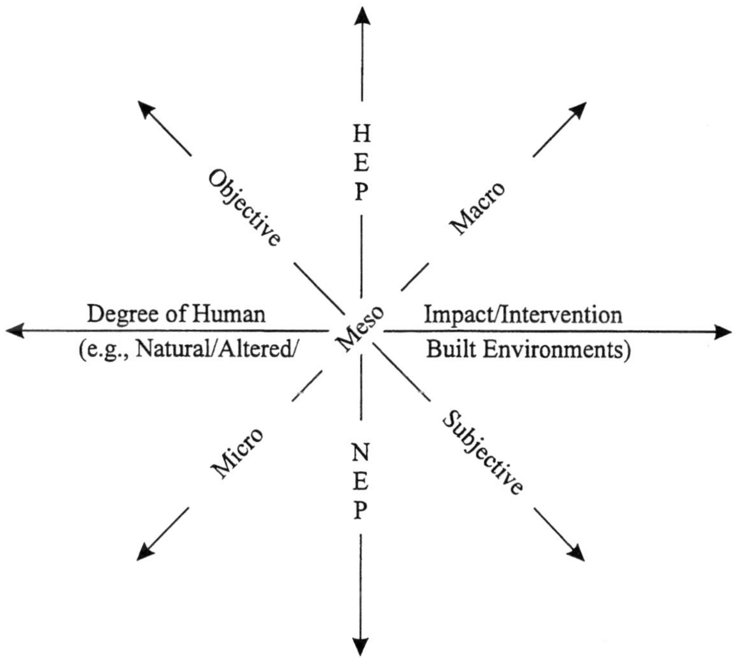

What other basic dimensions might we find? Dunlap and Catton (1979) suggest it is useful to categorize research within environmental sociology into one of three categories: (1) natural environments (e.g., wilderness, air, water etc.); (2) built environments (houses, cities etc.); (3) modified environments (polluted land, water, air and planned landscapes). These distinctions would seem to be of immediate value for creating a fourth metatheoretical distinction (curiously not included in earlier works). In Figure 1, I have attempted to display graphically these dimensions — which are key, I believe, to understanding work within environmental sociology.

There are, however, additional issues, themes and distinctions within the field. These include:[20]

- Basic versus applied approaches
- Prescriptive versus descriptive approaches (e.g., the nature of theory/science, and the role of the sociologist)

- The question of social networks (e.g., the nature of urban versus suburban community life; the mobilization of collective action)
- The differential impacts of the environment experienced by subpopulations (e.g., increased exposure of poor people and other minorities to toxic waste and other types of pollution; the importance of building design for women, children, the handicapped, the elderly etc.)
- Private (self) versus public (collective) interests regarding use and regulation of the environment and the role of government intervention
- The conception and role of social structure (e.g., the relative importance of economic, organizational, network, individual reward, cultural/ideological/phenomenological structures etc.)
- The importance of space
- The importance of time
- The nature of change (e.g., what drives change? What is the relative importance of new technologies, social conflicts etc.)
- The nature of causality (e.g., Michelson and van Vliet's forthcoming discussion of deterministic, probabilistic and possibilistic theories)
- Inductive versus deductive theoretical approaches
- The scope of theory (theories of the middle range versus general theory)

Unfortunately (or perhaps fortunately for the reader), it is beyond the scope of this essay to discuss all of these items, but as can be discerned, there are many issues which could be considered in the classification and development of environmental-sociological theory. From my reading of the literature, I have constructed eight subdomains of the discipline.[21] These include: (1) Social Impact Assessment; (2) Environmental Design Research; (3) Political Economy Approaches; (4) Organizational/Decision-Making Approaches; (5) New Human Ecology/Congruence Perspectives; (6) Social Psychology of Environmental Problems; (7) Social Construction of Environmental Problems; and (8) Theories of Collective Action and Social Movements. In Table 2, I have borrowed from Ritzer (1975, 1983, 1992), and listed some suggested exemplars, images of subject matter, primary methods (and unique methods) and theories (and perspectives) for each of the subdomains I have identified.

The classification strategy I used in constructing these eight subdomains was based initially upon distinguishing between the respective subject matters. Unlike Ritzer's (1992) practice, for each subdomain, I have listed not one, but multiple exemplars. My intention is not to assert that these are the definitive works in each subdomain; rather, I have introduced these exemplars because they provide good illustrations of work in each of these subareas. I will leave arguments about the definitive work in each subparadigm to later debate. I have also listed the primary and unique methods associated with each area, and listed some associated theories and perspectives.

TABLE 2

Subdomains within Environmental Sociology

Subdomain	*Exemplars*	*Image of Subject Matter*	*Methods*	*Theories, Perspectives, Concepts*
1. Social Impact Assessment	*Scientific SIA Approach*: Finsterbusch, Llewellyn and Wolf 1983; Freudenburg 1981, 1985, 1988 *Participatory SIA Approach*: Brown and Mikkelson 1990 *Environmental Justice*: Bullard 1990, 1993 *Social Psychological Approaches to Toxic Exposure*: Levine 1982; Edelstein 1988	Impacts of Development and Technology on Communities, Groups and Individuals	*Primary Methods*: secondary analysis of existing data Other traditional methods such as survey-interview techniques etc. are also employed. Case-study designs (see Wolf's 1983 ten steps of SIA). *Unique Methods*: computer modelling techniques (e.g., the Texas Modelling System).	Adopted from other areas: main distinction is between "scientific" and "participatory" approaches; environmental justice; "Edelstein's stages of toxic disaster"
2. Environmental Design Research	Zeisel 1981; Moore 1984; Marcus and Sarkissian 1986; Barker 1968	The study of behaviour-environment relations with the objective of improving design	*Primary Methods*: Observation, survey research, analysis of archival sources; *Unique Methods*: postoccupancy evaluation	Adopted from other areas: e.g., "defensible space," "cognitive mapping," "ecological pyschology"

TABLE 2

Subdomains within Environmental Sociology (*cont'd*)

Subdomain	*Exemplars*	*Image of Subject Matter*	*Methods*	*Theories, Perspectives, Concepts*
3. Political Economy Approaches	Schnaiberg 1980; Bunker 1985; Redclift 1984; Logan and Molotch 1987; Marchak 1983; O'Connor 1988	Emphasis on economic class structures and their social and environmental consequences	*Primary Methods*: Secondary analysis, historical comparative methods	"The environmental-societal dialectic and the treadmill of production"; "staples/ dependency theory"; Bunker's distinction between "productive" and "extractive" economies
4. Organizational/ Decision-Making Approaches	Perrow 1984; Nelkin 1979	The interface between organizational structure, technology and the environment	*Primary Methods*: Secondary analysis, case-study designs	The theory of normal accidents
5. New Human Ecology/ Congruence Perspectives	Michelson 1976, 1977; Popenoe 1985	Study of the fit between built environments and social behaviour; a particular focus is placed on studying the differential effects of built/altered environments on subpopulations/groups	*Primary Methods*: Secondary analysis, survey designs *Unique Methods*: time budgets, postoccupancy evaluation	The "congruence perspective," "the privatization of urban life"
6. Social Psychology of Environmental Problems	Dunlap and Van Liere 1978; Cotgrove and Duff 1981; Heberlein and Black 1981; Sammuelson and Messick 1986; Sammuelson et al. 1984; Mathews 1993	Focus on the relationship of values/attitudes and behaviour; factors that influence individual tendencies to consume or conserve resources	*Primary Methods*: Survey research, experimentation *Unique Methods*: laboratory game simulations (e.g., research on the "commons dilemma")	Research on the "human exemptionalist (dominant) vs. new ecological (alternative) paradigm," commons dilemma research

TABLE 2
Subdomains within Environmental Sociology (*cont'd*)

Subdomain	*Exemplars*	*Image of Subject Matter*	*Methods*	*Theories, Perspectives, Concepts*
7. Social Construction of Environmental Problems	Mauss 1975; Bogard 1989; Yearly 1991; Benford 1993a, 1993b, Gamson and Modigliani 1989	Focus on how and why certain environmental issues become identified as "problems" including contestation of such claims as problematic	*Primary Methods*: Analysis of exiting documents, e.g., analysis of media accounts, case histories, as well as other methods. Emphasis on the "discourse" of environmental conflicts	"Social constructionism," "social problems," "social movement framing literature"
8. Theories of Collective Action and Social Movements	*Resource Mobilization*: McCarthy and Zald 1977; Gamson 1975; Tilly 1978 *New Social Movement Theory*: Melucci 1980, 1988; Cohen 1985; Touraine 1981; Boggs 1986	*Resource Mobilization*: Focus on social structure and the mobilization of resources for collective action *New Social Movement Theory*: Focus on the symbolic nature of collective action Focus on the macrosocial, structural and historical precipitants of social movements Focus on quality of life Focus on grass-roots participation	*Primary Methods*: Varied — include secondary analysis, survey analysis, participant observation *Unique methods*: sociological intervention, e.g., Touraine et al. 1983; "popular epidemiology"; feminist methodologies	Resource Mobilization; New Social Movement Theory; Not In My Backyard Movement; Appropriate Technology Movement; Ecofeminism

TABLE 2

Subdomains within Environmental Sociology (*cont'd*)

Subdomain	*Exemplars*	*Image of Subject Matter*	*Methods*	*Theories, Perspectives, Concepts*
8. Theories of Collective Action and Social Movements (*continued*)	*Not In My Backyard Movement*: Freudenberg 1984 *Appropriate Technology Movement*: Schumacher 1973 *Ecofeminism*: Plant 1989; Mies and Shiva 1993	*Not In My Backyard Movement*: Focus on the development of technology consonant with sustainable development *Appropriate Technology Movement*: Mobilization and response to actual or potential hazards *Ecofeminism*: Emphasis on the connections between the oppression of women and the oppression of nature. The role of women in creating social and environmental change		*Other contenders include*: deep ecology (e.g., Devall and Sessions 1985); social ecology (e.g., Bookchin 1980, 1986); the concept of "stewardship groups" (e.g., Lerner 1986)

TABLE 3

Subdomains of Environmental Sociology Classified by Dimension of Social-Environmental Reality

Subdomain	*Degree of Human Impact/Intervention*	*HEP/NEP Criterion*	*Micro/Meso/Macro Dimension*	*Subjective/Objective Dimension*
1. Social Impact Assessment	Altered/Built	Varies by Theorist/Analyst	Primary focus on meso and micro	Objective and subjective
2. Environmental Design Research	Built	Varies by Theorist/Analyst	Primary focus on meso and micro	Objective and subjective
3. Political Economy Approaches	Primary focus on Natural (e.g., Resources), but also Altered and Built	Varies by Theorist/Analyst	Primary focus on macro (also meso)	Primarily objective
4. Organizational/ Decision-Making Approaches	Primary focus on Built, but also Altered	Varies by Theorist/Analyst. Primary focus on meso	Objective and subjective	
5. New Human Ecology	Congruence Perspectives	Primary focus on Built (but also altered)	Varies by Theorist/Analyst	Focus varies: all three levels
6. Social Psychology of Environmental Problems	Focus varies: all three levels	Varies by Theorist/Analyst	Primary focus on micro and meso (also macro — e.g., large-scale value shifts)	Primarily subjective (also objective)
7. Social Construction of Environmental Problems	Focus varies: all three levels	Varies by Theorist/Analyst	Focus varies: all three levels	Primarily subjective
8. Theories of Collective Action and Social Movements	Focus varies: all three levels	Varies by Theorist/Analyst	Focus varies: all three levels	Objective and subjective

As can be seen in Table 2, there is some overlap in content between the subdomains. For example, several of the works suggested as exemplars for Social Impact Assessment (e.g., Levine's 1982 work on community responses to toxic exposure; Bullard's 1990 and 1993 works on environmental justice) could also be slotted under Theories of Collective Action and Social Movements. Further, some work listed as falling within the Social Construction of Environmental Problems subparadigm could be classified elsewhere (e.g., Bogard's 1989 work could be classified under Social Impact Assessment; the works of Benford 1993a and 1993b; and Gamson and Modigliani 1989, could be slotted under Theories of Collective Action and Social Movements).

In Table 3, I have attempted to classify each subparadigm in terms of the four dimensions of environmental sociology which I introduced in Figure 1. As quickly becomes evident, these dimension *alone* do not distinguish the eight subdomains — it is only when the "image of the subject matter is included as well, that we can make a qualified claim that these areas constitute unique fields of research within environmental sociology. While the categories in the schema have led to the creation of subdomains that fail the test of "exhaustiveness" and "mutual exclusiveness" — criterion usually judged to be important in the creation of classification systems — the main purpose of this discussion is to illuminate the characteristics of work in the field.[22] With regard to this latter purpose, I believe the classification scheme is of some value — particularly for understanding and evaluating the dimensions of particular theories or pieces of research.[23]

Interestingly, in constructing Table 3, I was unable to distinguish work at the subdomain level by using Catton and Dunlap's HEP-NEP schema. This result is revealing on two fronts. Some critics of the HEP-NEP schema, such as Buttel, have suggested that the called-for "paradigm shift" from the HEP to the NEP has failed to occur at a general level within sociology. Instead, critics have suggested, environmental sociology is well on its way to becoming just another subarea in the discipline. While this may be the case with regard to the discipline, I think the HEP-NEP distinction (along with the other three dimensions I have discussed above) can help one to evaluate work within the field — that is, to discriminate between "sociology of environmental issues" and "environmental sociology" — thus making "societal-ecosystem" linkages central to social analysis.

DISCUSSION QUESTIONS

1. Compare and contrast the HEP and NEP concepts.
2. Tindall lists four dimensions of social-environmental reality. Identify and explain what these are.
3. Define what is meant by the term "paradigm."

4. What are the distinctive features of environmental sociology?
5. Does environmental sociology constitute a paradigm shift? Explain.

Endnotes

1. Thomas Kuhn (1962) has developed a theory of how major changes in science occur. According to Kuhn, science at any one point in time is dominated by a specific paradigm (that is, as a fundamental image of the sciences' subject matter). *Normal science* is a period of accumulation of knowledge in which scientists work to expand the reigning paradigm. Such scientific work inevitably spawns *anomalies*, or findings that cannot be explained by the reigning paradigm. A *crisis* stage occurs if these anomalies mount, and this crisis may ultimately end in a scientific revolution. In such an event, the *reigning* paradigm is overthrown, and a new one takes its place at the centre of the science. A *new dominant paradigm* emerges, and things are set for the cycle to repeat itself. According to Ritzer (1975), sociology is a multiple paradigm science. Interestingly, Kuhn (1970) suggests that the social sciences are in a preparadigmatic stage.
2. For example, according to Ritzer, "social definitionists focus upon the way in which actors define their social situations, and the effect of these definitions on ensuing action and interaction.
3. Ritzer categorized sociological theory as follows:

   The Social Facts Paradigm: (1) Exemplar: Emile Durkheim; (2) Image of the subject matter: social facts, or large-scale social structures and institutions. Those who adhere to this paradigm focus not only on these phenomena, but also on their effect on individual thought and action; (3) Methods: interview questionnaire and historical comparative; (4) Theories: structural functionalism, conflict theory, systems theory.

   The Social Definition Paradigm: (1) Exemplar: Max Weber; (2) Image of the subject matter: the way in which actors define their social situations, and the effect of these definitions on ensuing action and interaction; (3) Methods: interview-questionnaire method, observation. The distinctive method is observation; (4) Theories: action theory, symbolic interactionism, phenomenology, ethnomethodology, existentialism.

   The Social Behaviour Paradigm: (1) Exemplar: the work of B.F. Skinner; (2) Image of the subject matter: the unthinking behaviour of individuals. The rewards that elicit desirable behaviours, and the punishments that inhibit undesirable behaviours; (3) The distinctive method is the experiment; (4) Theories: behavioural sociology and exchange theory.
4. This, incidentally, had its own consequences for "environmental studies." The pioneers of this field, such as biologist Garret Hardin (1968), while having a good understanding of ecological processes, had little knowledge of social structure and social processes. This led to some rather conservatively biased proposed solutions to problems such as destruction of resource commons.
5. The field of social impact assessment was legitimated by the U.S. National Environmental Policy Act of 1969.
6. Catton and Dunlap (1978) state: "This involves studying the effects of the environment on society (e.g., resource abundance or scarcity on stratification), and the effects of society on the environment (e.g., the contributions of differing economic systems to environmental degradation)."
7. Interestingly, while Schnaiberg's work has been identified by a number of authors as falling within the "new ecological paradigm," Schnaiberg himself describes it as being rooted in "anthropocentrism." Schnaiberg and Gould (1994: 17) state: "In this book, we take no moral stance on the values of preserving the natural environment, apart from a

concern for preserving and enhancing the quality of social life. We are unabashedly and unashamedly anthropocentric; that is, this book is oriented to the needs of human society. We have taken this position for several reasons, and most notably because there is no social scientific justification for any particular ecological value, apart from the somewhat utilitarian position we have stated."

8. For instance, Catton and Dunlap (1978: 256) state: "... Buttel mistakenly equates 'theoretical diversity' with 'paradigmatic diversity.'"
9. Buttel (1987: 468) adds: "Moreover, the new human ecology departs significantly from classical human ecology (e.g., Hawley 1986) by elaborating a distinctive argument: Instead of tending toward equilibrium with the natural environments from which they derive sustenance, modern societies tend to exhibit quite the opposite pattern — social dynamics that exacerbate environmental degradation and resource depletion."
10. Michelson (1970, 1976) has constructed a theoretical framework around the concept of *congruence*. This framework tries to provide an understanding of how behaviours can be contingent on their environments. Michelson's analysis demonstrates ways in which environmental design can serve to facilitate or constrain access of people to each other, and to use of spaces. Michelson also notes that individuals and groups can be examined in terms of the demands they make on the environment. "The concept of congruence is central, signifying whether the people at hand can find it possible to realize their preferred or mandated behaviours in the specifically conceptualized setting, or not (incongruence)." (Michelson and van Vliet forthcoming: 17–18) Michelson (1977) has further elaborated this framework in his investigation of people's choices, behaviours and evaluation regarding their urban and suburban residences. He argues that *values* and *choice of physical structure* of individual environments are important in creating opportunities for different *lifestyles*.
11. They do note, however, that "this framework is most applicable to macrosociology, and is not adequate for representing microsociological theories such as phenomenology, symbolic interactionism, behaviourist social psychology" etc.
12. In the broader context of sociological theory, this last criticism should be taken lightly. For as Freese 1980 has noted, practically no theory in sociology is truly based on deductive logic. Only formal theories of closed systems fit the requirements for deductive logic to be properly implemented.
13. According to Ritzer (1992: 651), this "involves the study of theory in order to produce a better, more profound understanding of extant theory.... M_U is concerned, more specifically, with the study of theories, theorists and communities of theorists, as well as the larger intellectual and social contexts of theories and theorists."
14. According to Ritzer (1992: 651), this "entails the study of extant theory in order to produce new sociological theory." He adds that this type of metatheorizing was common among classic sociological theorists, such as Marx and Parsons.
15. Ritzer (1992: 651) states that this variety of metatheorizing is "oriented towards the goal of producing a perspective, one could say a metatheory, that overarches some part or all of sociological theory."
16. Hannigan (1995) notes that in addition to serving as an independent (cause), and dependent (effect) variable, the environment can serve as a contextual (or background) variable.
17. Like other scholars (notably Collins 1988), I feel that the organization of theory around "schools of thought" is of limited utility. A more useful approach is to organize theory around "theoretical problems" and "levels of social reality."
18. In particular, this work has examined the ways in which identities are actively constructed through interaction, negotiation, resistance etc.
19. Examples of *macro-objective* phenomena include: society, law, bureaucracy, architecture, technology and language. Examples of *macrosubjective* phenomena include: culture, norms

and values. Examples of *micro-objective* phenomena include: patterns of behaviour, action and interaction. Examples of *microsubjective* behaviour include: the social construction of reality.

20. These are not all exclusive to environmental sociology.
21. Loraine Lavallee (personal communication) has suggested that both Ritzer's and Catton and Dunlap's use of the term "paradigm" differs substantially from Kuhn's. While I agree with this observation, I feel that both Ritzer's and Catton and Dunlap's conceptual schemes are of metatheoretical utility. On this note, Lavallee has persuaded me to use the term "subdomain," rather than "subparadigm" — a term which I used in a previous draft of this chapter. In Kuhn's sense, a paradigm is an overarching perspective about how the world works. While it is a debatable question as to whether "environmental sociology" constitutes a paradigm, the subdomains I have identified are too narrow to be true "paradigms."
22. In other words, I have used metatheory strategies M_U and M_O to help develop an understanding of current research and theory, and to organize subareas of environmental sociology.
23. The subparadigms I have identified are *not* exclusive to environmental sociology, but merely help to organize perspectives within the field (e.g., works within each subparadigm have a similar "subject matter" and, to varying degrees, constitute groups of related research and theories).

Bibliography

Ahrentzen, Sherry. Forthcoming. Socio-Behavioral Impacts of the Built Environment. In *Handbook of Environmental Sociology*, ed. Riley Dunlap and William Michelson. Greenwood Press.

Albrecht, Stan L. and Armand L. Mauss. 1975. The Environment as a Social Problem. In *Social Problems as Social Movements*, ed. A.L. Mauss, 556–605. Philadelphia: Lippincott.

Barker, Roger G. 1968. *Ecological Psychology*. Stanford, Calif.: Stanford University Press.

Beck, Ulrich. 1992. *Risk Society: Towards a New Modernity*. London: Sage Publications.

Benford, Robert D. 1993a. Frame Disputes within the Nuclear Disarmament Movement. *Social Forces* 71: 677–701.

———. 1993b. You Could Be the Hundredth Monkey: Collective Action Frames and Vocabularies of Motive Within the Nuclear Disarmament Movement. *The Sociological Quarterly* 34: 195–216.

Bogard, William. 1989. *The Bhopal Tragedy: Language, Logic, and Politics in the Production of a Hazard*. Boulder, Colo.: Westview Press.

Boggs, Carl. 1986. *Social Movements and Political Power*. Philadelphia: Temple University Press.

Bookchin, Murray. 1980. *Toward an Ecological Society*. Montreal: Black Rose Books.

———. 1986. *The Modern Crisis*. Philadelphia: New Society Publishers.

Brown, Phil and Ed Mikkelson. 1990. *No Safe Place: Toxic Waste, Leukemia and Community Action*. Berkeley, Calif.: University of California Press.

Bullard, Robert D. 1990. *Dumping in Dixie: Race, Class and Environmental Quality*. Boulder, Colo.: Westview Press.

———. ed. 1993. *Confronting Environmental Racism: Voices from the Crossroads*. Boston, Mass.: South End Press.

Bunker, Stephen G. 1984. Modes of Extraction, Unequal Exchange, and the Progressive Underdevelopment of an Extreme Periphery: The Brazilian Amazon, 1600–1980. *American Journal of Sociology* 89(5): 1017–64.

———. 1985. *Underdeveloping the Amazon*. Urbana: University of Illinois Press.

Buttel, F. 1978. Environmental Sociology: A New Paradigm? *The American Sociologist*, 13: 252–56.

———. 1986. Toward a Rural Sociology of Global Resources: Social Structure, Ecology, and Latin American Agricultural Development. In *Resources and People: Disciplinary Attempts to Bridge the Natural and Social Sciences*, ed. K.A. Dahlberg and J.W. Bennett, 129–64. Boulder, Colo.: Westview Press.

———. 1987. New Directions in Environmental Sociology. *Annual Review of Sociology* 13: 465–88.

Buttel, Frederick H. and Craig Humphrey. 1987. Sociological Theory and the Natural Environment. Unpublished paper prepared for the presentation at the annual meetings of the American Sociological Association, Chicago, August.

———. Forthcoming. Sociological Theory and the Natural Environment. In *The Handbook of Environmental Sociology*, ed. Riley Dunlap and William Michelson. Westport, Conn.: Greenwood Press.

Catton, W. and R. Dunlap. 1978a. Environmental Sociology: A New Paradigm. *The American Sociologist* 13: 41–49.

———. 1978b. Paradigms, Theories, and the Primacy of the HEP-NEP Distinction. *The American Sociologist* 13: 256–59.

Cohen, J. 1985. Strategy or Identity: New Theoretical Paradigms and Contemporary Social Movements. *Social Research* 25: 4.

Collins, Randall. 1988. *Theoretical Sociology*. New York: HBJ Publishers.

Cotgrove, S. and A. Duff. 1981. Environmentalism, Values and Social Change. *British Journal of Sociology* 32: 92–110.

Devall, B. and G. Sessions. 1985. *Deep Ecology: Living As If Nature Mattered*. Salt Lake City, Utah: Perigrine Smith Books.

Dickens, Peter. 1992. *Society and Nature: Towards a Green Social Theory*. New York: Harvester Wheatsheaf.

Duncan, O.D. 1961. From Social System to Ecosystem. *Sociological Inquiry* 31: 140–49.

———. 1964. Social Organization and the Ecosystem. In *Handbook of Modern Sociology*, ed. R.E.L. Faris, 36–82. Chicago: Rand McNally.

Dunlap, R. and W. Catton. 1979. Environmental Sociology. *Annual Review of Sociology* 5: 243–73.

Dunlap, Riley. and K.D. Van Liere. 1978. The New Environmental Paradigm: A Proposed Measuring Instrument and Preliminary Results. *Journal of Environmental Education* 9: 10–19.

Dunlap, Riley and William Michelson, eds. Forthcoming. *Handbook of Environmental Sociology*. Westport, Conn.: Greenwood Press.

Dunlap, Riley E. and Robert Emmett Jones. Forthcoming. Public Opinion and Attitudes Toward Environmental Issues. In *The Handbook of Environmental Sociology*, ed. Riley Dunlap and William Michelson. Westport, Conn.: Greenwood Press.

Durkheim, Emile. 1933. *The Division of Labor in Society*. New York: Free Press.

———. [1912] 1954. *The Elementary Forms of the Religious Life*. New York: Free Press.

———. [1895] 1982. *The Rules of Sociological Method*. New York: Free Press.

Edelstein, Michael R. 1988. *Contaminated Communities: the Social and Psychological Impacts of Toxic Exposure*. Boulder, Colo.: Westview Press.

Edney, Julian. J. 1980. The Commons Problem: Alternative Perspectives. *American Psychologist* 35: 131–50.

Elder, Klaus. 1982. A New Social Movement? *Telos* 50: 5–20.

Eyerman, Ron and Andrew Jamison. 1989. Environmental Knowledge as an Organizational Weapon: the Case of Greenpeace. *Social Science Information* 28: 99–119.

Finsterbusch, Kurt. 1980. *Understanding Social Impacts: Assessing the Effects of Public Projects*. Beverly Hills, Calif.: Sage.

Finsterbusch, K., L.G. Llewellyn, C.P. Wolf. 1983. *Social Impact Assessment Methods*. Beverly Hills, Calif.: Sage.

Finsterbusch, Kurt. and William R. Freudenburg. Forthcoming. Social and Technological Impact Assessment. In *The Handbook of Environmental Sociology*, ed. Riley Dunlap and William Michelson. Westport, Conn.: Greenwood Press.
Freese, Lee. 1980. Formal Theorizing. *Annual Review of Sociology* 6: 187–212.
Freudenberg, Nicholas. 1984. *Not in our Backyard!: Community Action for Health and the Environment*. New York: Monthly Review Press.
Freudenburg, W.R. 1981. Women and Men in an Energy Boomtown: Adjustment, Alienation and Adaptation. *Rural Sociology* 46: 220–44.
Freudenburg, W.R. 1984. Boomtown's Youth: The Differential Impacts of Rapid Community Growth on Adolescents and Adults. *American Sociological Review* 49: 697–705.
———. 1985. Nuclear Reactions: Public Attitudes and Public Policies Toward Nuclear Power Plants. *Policy Studies Review* 5(August): 96–110.
———. 1986a. Social Impact Assessment. *American Review of Sociology* 12: 451–78.
———. 1986b. The Density of Acquaintanceship: An Overlooked Variable in Community Research? *The American Journal of Sociology* 92(July): 27–63.
———. 1988. Perceived Risk, Real Risk: Social Science and the Art of Probabilistic Risk Assessment. *Science* 242(October): 44–49.
Gamson, William A. 1975. *The Strategy of Social Protest*. Chicago: The Dorsey Press.
Gamson, William A. and Andre Modigliani. 1989. Nuclear Discourse and Public Opinion on Nuclear Power: A Constructionist Approach. *American Journal of Sociology* 95: 1–37.
Gillis, A.R. 1990. Urbanization and Urbanism. In *Sociology*, 4th ed., ed. Robert Hagedorn, 465–505. Toronto: Holt, Rinehart and Winston of Canada Limited.
Gould, Allan and Kenneth A. Gould 1994. *Environment and Society: The Enduring Conflict*. New York: St. Martin's Press.
Hannigan, John. 1995a. Sociology and the Environment. In *New Society: Sociology for the 21st Century*, ed. Robert Brym. Toronto: Harcourt, Brace and Co. Canada.
———. 1995b. *Environmental Sociology: A Social Constructionist Perspective*. London and New York: Routledge.
Hardin, Garrett. 1968. The Tragedy of the Commons. *Science* 162: 1243–48.
Harding, Jim. 1993. Ecology and Social Change. In *Contemporary Sociology: Critical Perspectives*, ed. Peter S. Li and B. Singh Bolaria, 439–66. Toronto: Copp Clark and Pitman Limited.
Hawley, Amos H. 1986. *Human Ecology*. Chicago: University of Chicago Press.
Heberlein, T.A. and J.S. Black. 1981. Cognitive Consistency And Environmental Action. *Environmental Behavior* 13: 717–34.
Humphrey, C. and F. Buttel. 1982. *Environment, Energy, and Society*. Belmont, Calif.: Wadsworth.
Kilmartin, Leslie. Forthcoming. Designing the Built Environment. In *Handbook of Environmental Sociology*, ed. Riley Dunlap and William Michelson. Westport, Conn.: Greenwood Press.
Kriesi, H. 1989. New Social Movements and the New Class in the Netherlands. *American Journal of Sociology* 94: 5.
Kuhn, Thomas. 1962. *The Structure of Scientific Revolution*. Chicago: University of Chicago Press.
———. 1970. *The Structure of Scientific Revolutions*. 2d ed. Chicago: University of Chicago Press.
Lenski, Gehard E. and Jean Lenski. 1970. *Human Societies: An Introduction to Macrosociology*. New York: McGraw-Hill.
Lerner, Sally C. 1986. Environmental Constituency — Building: Local Initiatives and Volunteer Stewardship. *Alternatives* 13(3): 55–60.
Levine, Adeline Gordon. 1982. *Love Canal: Science, Politics, and People*. Lexington, Mass.: Heath.
Logan, J.R. and H.L. Molotch. 1987. *Urban Fortunes*. Berkeley: University of California Press.
Marcus, Clare Cooper and Wendy Sarkissian. 1986. *Housing As If People Mattered: Site Design Guidelines for Medium Density*. Berkeley: University of California Press.

Marchak. M. Patricia. 1983. *Green Gold: The Forest Industry in British Columbia*. Vancouver: UBC Press.

Martell, Luke. 1994. *Ecology and Society: An Introduction*. Cambridge, Mass.: Polity Press.

Marx, Karl. 1961. *Economic and Philosophic Manuscripts of 1844*. Moscow: Foreign Language Publishing House.

———. 1967. *Capital*. 3 vols. New York: International Publishers.

Masterman, Margaret. 1970. The Nature of the Paradigm. In *Criticism and the Growth of Knowledge*, ed. I. Lakatos and Alan Musgrove, 59–89. Cambridge: Cambridge University Press.

Mathews, David Ralph. 1993. *Controlling Common Property: Regulating Canada's East Coast Fishery*. Toronto: University of Toronto Press.

Mauss, A.L. 1975. The Environment As a Social Problem. In *Social Problems As Social Movements*, ed. A.L. Mauss, 556–605. Philadelphia: Lippincott.

McCarthy, John D. and Mayer Zald. 1977. Resource Mobilization and Social Movements: A Partial Theory. *American Journal of Sociology* 82: 1212–41.

Melucci, Alberto. 1980. The New Social Movements: A Theoretical Approach. *Social Science Information* 19: 199–226.

———. 1988. Getting Involved: Identity and Mobilization in Social Movements. In *International Social Movement Research*, ed. B. Klandermans, H. Kriesi, and Sidney Tarrow, Greenwich, Conn.: London: JAI Press.

Michelson, William. 1970. *Man and his Urban Environment: A Sociological Approach*. New York: Random House.

———. 1976. *Man and his Urban Environment: A Sociological Approach*. 2d ed. Reading, Mass.: Addison-Wesley.

———. 1977. *Environmental Choice, Human Behavior, and Residential Satisfaction*. New York: Oxford University Press.

Michelson, William and Willem van Vliet. Forthcoming. Sociological Theory and the Built Environment. In *The Handbook of Environmental Sociology*, ed. Riley Dunlap and William Michelson. Westport, Conn.: Greenwood Press.

Mies, Maria and Vandana Shiva. 1993. *Ecofeminism*. Halifax, N.S.: Fernwood Publications.

Moore, Gary T. 1984. *Environmental Design Research Directions: Process and Prospects*. New York: Praeger.

Murphy, Raymond. 1994. *Rationality and Nature: A Sociological Inquiry into a Changing Relationship*. Boulder, Colo.: Westview Press.

Nelkin, Dorothy, ed. 1979. *Controversy: Politics of Technical Decisions*. Beverly Hills, Calif.: Sage Publications.

O'Connor, James. 1988. Capitalism, Nature, Socialism: A Theoretical Introduction. *Capitalism, Nature, Socialism* 1: 11–38.

Offe, Claus. 1985. New Social Movements: Challenging the Boundaries of Institutional Politics. *Social Research* 52: 817–68.

Park, Robert. 1952. The Urban Community as a Spatial Pattern and a Moral Order. In *Human Communities*. New York: The Free Press of Glencoe.

Perrow, Charles. 1984. *Normal Accidents: Living With High-Risk Technologies*. New York: Basic Books, Inc.

———. 1986. *Complex Organizations: A Critical Essay*. Glenview, Ill.: Scott, Foresman.

Plant, Judith, ed. 1989. *Healing the Wounds: The Promise of Ecofeminism*. Santa Cruz, Calif.: New Society Publishers.

Popenoe, D. 1985. *Private Pleasure, Public Plight: American Metropolitan Community Life in Comparative Perspective*. New Brunswick, N.J.: Transaction Books.

———. Forthcoming. Urban and Suburban Environments. In *The Handbook of Environmental Sociology*, ed. Riley Dunlap and William Michelson. Westport, Conn.: Greenwood Press.

Redclift, Michael. 1984. *Development and the Environmental Crisis — Red or Green Alternatives.* London: Methuen.

Redclift, Michael and Ted Benton, eds. 1994. *Social Theory and the Global Environment.* London: Routledge.

Ritzer, George. 1975. *Sociology: A Multiple Paradigm Science.* Boston: Allyn and Bacon.

———. 1983. *Contemporary Sociological Theory.* New York: Alfred A. Knopf, Inc.

———. 1992. *Sociological Theory.* 3d ed. New York: McGraw-Hill, Inc.

Sammuelson, C.D., D.M. Messick, C.G. Rutte and H. Wilke. 1984. Individual and Structural Solutions to Resource Dilemmas in Two Cultures. *Journal of Personality and Social Psychology* 46: 94–104.

Sammuelson, C.D. and D.M. Messick. 1986. Inequalities in Access to and Use of Shared Resources in Social Dilemmas. *Journal of Personality and Social Psychology* 51: 960–67.

Schnaiberg, Allan. 1975. Social Syntheses of the Societal-Environmental Dialectic: The Role of Distributional Impacts. *Social Science Quarterly* 56(June): 5–20.

———. 1980. *The Environment: From Surplus to Scarcity.* New York: Oxford University Press.

———. Forthcoming. The Political Economy of Environmental Problems and Policies. In *The Handbook of Environmental Sociology*, ed. Riley Dunlap and William Michelson. Westport, Conn.: Greenwood Press.

Schnaiberg, Allan and Kenneth A. Gould. 1994. *Environment and Society: The Enduring Conflict.* New York: St. Martin's Press.

Schumacher, E.F. 1973. *Small is Beautiful.* New York: Harper.

Spencer, Metta. 1990. Population and Ecology. In *Foundations of Modern Sociology*, 5th ed., ed. Metta Spencer, 531–64. Scarborough, Ont.: Prentice-Hall Canada.

Sly, D.F. 1972. Migration and the Ecological Complex. *American Journal of Sociology* 37: 615–28.

Tilly, Charles. 1978. *From Mobilization to Revolution.* Reading, Mass.: Addison-Wesley.

Touraine, Alain. 1981. *The Voice and the Eye: An Analysis of Social Movements.* New York: Cambridge.

Touraine, Alain et al. 1983. *Anti-Nuclear Protest: The Opposition to Nuclear Energy in France.* New York: Cambridge University Press.

Warriner, G. Keith. 1994. Environment and Social Change. In *The Social World*, ed. L. Tepperman, J. Curtis, and J. Richardson, 780–817. Toronto: McGraw-Hill.

Weber, Max. 1949. *The Methodology of the Social Sciences.* New York: Free Press.

———. 1968. *Economy and Society*, ed. by Guenther Roth and Claus Wittich. New York: Bedminster Press.

Wolf, C.P. 1983. Social Impact Assessment: A Methodological Overview. In *Social Impact Assessment Methods*, ed. K. Finsterbusch, L.G. Llewellyn and C.P. Wolf, Chapter 1. Beverly Hills, Calif.: Sage.

Yearly, Seven. 1991. *The Green Case: A Sociology of Environmental Issues, Arguments, and Politics.* London: Harper Collins Academic.

Zeisel, John. 1981. *Inquiry by Design.* Monterey, Calif.: Brooks/Cole Publishing Co.

Zeitlin, Irving M. with Robert J. Brym. 1991. Saving Our Planet: The Environmental Crisis and What We Can Do About It. In *The Social Condition of Humanity*, Canadian Edition, ed. Irving M. Zeitlin with Robert J. Brym, 362–86. Toronto: Oxford University Press.

ENVIRONMENTAL SOCIOLOGY: A DISTINCT FIELD OF INQUIRY?

Donna E. Hughes

York University

The author speculates that, despite Catton and Dunlap's claim that environmental sociology is a distinct field of inquiry, there is insufficient evidence that they have developed defensible claims regarding their characterization of mainstream sociology, the role of sociological specialism and the significance of paradigmatic interpretation. As a result of these deficiencies and over a decade of activity within the field, there is mounting evidence that environmental sociology is increasingly orthodox in its internal diversity and role as a specialism, and unable to generate sufficient force to attract researchers and to modify the foundational assumptions of sociology.

Introduction

Purportedly arising from the inability of mainstream sociology to explain bewildering societal change, environmental sociology, as set out in foundational papers by Catton and Dunlap (1978a, 1978b, 1980) and Dunlap and Catton (1979a, 1979b, 1983), is the study of the complicated and diverse symbolic and non-symbolic interactions and reciprocal influences between society and an environment, which includes not only social and cultural aspects, but also physical and biological ones. Moreover, Catton and Dunlap — regarded as "... most influential in the emergence, development, and institutionalization of ... environmental sociology ..." (Freudenburg and Gramling 1989: 440) — characterize environmental sociology as a distinct field of inquiry because, in including environmental variables, environmental sociologists depart from mainstream sociology and adhere to a new shared and developing paradigm regardless of their diverse research interests (Catton and Dunlap 1978a: 42; Dunlap and Catton 1983: 114). But what do Catton and Dunlap mean by a distinct field of inquiry, and what might have caused them to assert this? Is their assertion justifiable, and has time shown it realizable? By examining mainstream sociology, sociological specialisms and paradigmatic interpretations, this paper is an attempt to answer concurrently these intriguing questions, which contribute to knowledge not only about environmental sociology, but also about sociology in general.

In setting out "... to delineate the essential characteristics that qualify this new specialization as a distinct area of inquiry" (1979a: 243), Dunlap and Catton's choice of terms and lack of explanation prove problematic to the investigator. Dunlap and Catton (1979a, 1983) used the terms "area" and "field" interchangeably with reference to environmental sociology,[1] and compounded this terminological imprecision by using the term "area" and later "subarea" to describe diverse research pursuits within environmental sociology (1979a: 255). Further, they also used the term "field" to describe sociology (Catton and Dunlap 1980: 23), which causes one to wonder if they also used the terms "field," "area" and "discipline" interchangeably and, because a field is "... a topic, subject or area of academic interest or specialization" (Morris 1980), the terms "field" and "specialization" may also have been used interchangeably.

However we understand this unfortunate terminological muddle, it is clear from their thought-provoking writings, that Catton and Dunlap regarded their nascent environmental sociology as distinctly different from other sociological specializations — in fact, not just "... another specialty within the discipline" (1978a: 42) — and from other sociological fields (or areas or subareas) of inquiry. We know that by their very nature, fields of inquiry are separate and distinguishable from one another, encapsulated by purposeful self-delineated parameters with varying degrees of permeability and overlap. Yet Dunlap and Catton chose to emphasize the singularity of the field of environmental sociology, in particular, to show how its relative distinctiveness both characterizes and isolates it from what they term "mainstream sociology," and to hold environmental sociology forth as genuinely influential — the wake up call to sociology's slumber. In sum, environmental sociology was to be neither conventional nor impotent.

Mainstream Sociology

In stating that "... all elements in environmental sociology represent a departure from mainstream sociology ..." (1983: 115), Dunlap and Catton continued casting the singular environmental sociology by detailing the ways in which it is different. But what exactly is the so-called "mainstream sociology" from which environmental sociologists are withdrawing? Literally the study of companionship, at the most general level of analysis, sociologists analyse the structure of social relationships, which are constructed through social interaction (Abercrombie, Hill, and Turner 1988: 232). For Catton and Dunlap, this dominant sociology was mainly Durkheimian (1980: 18–19; Dunlap and Catton 1983: 116) and characterized by: the reactive belief — to threats of disciplinary indistinction — that only social variables present in social settings influence the structure of social relationships and adherence to the Human Exemptionism Paradigm (formerly the Human Exceptionalism Paradigm), an all-embracing anthropocentric mega-paradigm (Catton and Dunlap 1980: 22), which is the source of precluded ecological considerations (Freudenburg and Gramling 1989: 442). For Catton and Dunlap, environmental sociology considers, in addition to social variables, non-social variables, which also influence the structure of social relationships (1978a: 45, 1980: 19, 22; Dunlap and Catton 1979a: 255, 1983: 116) and follows the New Ecological Paradigm, an emerging mega-paradigm, which replaces, amends (Catton and Dunlap 1978a: 42) and rejects (Dunlap and Catton 1983: 8) the obsolete (Catton and Dunlap 1978b: 258) Human Exemptionalism Paradigm.

Assessment of the validity of Dunlap and Catton's claim requires closer scrutiny: Can the empirical question at hand — that is, the distinctiveness of environmental sociology — be settled through this key term definition of mainstream sociology? Is there counter evidence? Catton and Dunlap did recognize that in responding to the disciplinary pressures of his day, Durkheim stressed the *sui generis* nature of social facts — that is, their irreducibility to non-social facts — in order to de-emphasize the significance of psychological variables, which reduced social facts to individual consciousness, and to lull threats of psychological reductionism, methodological individualism and psychologisms (Durkheim 1895; Catton and Dunlap 1980: 19; Dunlap and Catton 1983: 116; Ritzer 1975: 25–26; Thompson 1982: 105; Abercrombie, Hill, and Turner 1988: 154, 198). However, Dunlap and Catton did not acknowledge that in developing a voluntaristic framework, which imbued the social actor with motive and intention, Durkheim himself de-emphasized the pre-eminence of social facts (*sui generis*) as his work matured (Abercrombie, Hill, and Turner 1988: 79, 265), and did appreciate environmental variables to some degree (Tschannen and Hainard 1993).

While most of Catton and Dunlap's analysis is devoted to Durkheim, the Weberian tradition, expanded upon by Mead, Cooley, Thomas and others — in which sociologists use a subjective social-definition perspective — is also included in discussion of mainstream sociology. Weber, too, was under pressure to distinguish sociology, asserting, for example, that "... sociology does not bear any closer logical relationship to psychology than to any other science" (Weber 1962: 50–51). While Durkheim (1858–1917) — who may be considered the pre-eminent developer of sociology (Thompson 1982: 23), and Weber (1864–1920), whose work broke ground in sociology, history and politics (Secher 1962: 22), are certainly considered among the founders of modern-day sociology, it is inaccurate to suggest that they alone inspired what Catton and Dunlap term the disciplinary traditions in sociology (1980: 18) from which so-called "mainstream" sociology has developed. Certainly Saint-Simon (1760–1825), Comte (1798–1857), Marx (1818–1883), Spencer (1820–1895), Engels (1820–1895) and Simmel (1858–1918), whose role was eventually acknowledged by Dunlap and Catton (1983: 115), also made crucial contributions to the sociological tradition.

Despite their espoused remoteness from Durkheimian sociology, Dunlap and Catton did come to recognize Durkheim as being at the "intellectual roots" of environmental-sociological investigations focusing on the natural environment, and Weber and Simmel for built environments (1983: 115); and others agree (Tschannen and Hainard 1993). Buttel noticed that in general, environmental sociologists have had difficulties in recognizing the contributions of classical theorists to the development of environmental sociology, speculating that they have had limited exposure to a variety of their primary works, and greater exposure to secondary works by sociologists who were not interested in elaborating an enlarged concept of environment (1986; 1987: 469 in note 3). To be solid in arguments regarding sociological treatment of the environment, primary works may require a thorough visit; however, as time passes, this may be largely a reconstructive exercise of conjecture. We can never ask of Durkheim, Weber and others: What exactly did you mean by "environment," and how can I use it to make sense of my world?

From these early foundational contributions to the theoretical development of sociology, a number of schools — groups of sociologists sharing method, opinion and teaching — have developed, including: conflict theorists (Coser 1956; Dahrendorf 1959; Lockwood 1964), ethnomethodologists (Garfinkel 1967), exchange theorists (Homans 1961; Blau 1964; Ekeh 1974), functionalists (Parsons 1966), evolutionary theorists (Saint-Simon in Taylor 1975; Comte 1838; Spencer 1876–96), Marxists (Marx 1867, 1885, 1894; Marx and Engels 1845), phenomenologists (Schutz 1972; Schutz and Luckmann 1974; Berger and Luckmann 1967), positivists (Comte 1838; Durkheim 1897), symbolic interactionists (Goffman 1959, 1963, 1971; Rock 1979) and systems theorists (Parsons 1951). As well, various schools of sociology represent varying viewpoints, including the Chicago School (1920s, 1930s) — urban ecology, ethnography, symbolic interactionism and the Frankfurt School (1923–1950) — neo-Marxism, epistemology, critical theory, realism. So varied are these schools that one would be hard-pressed to define definitively what is meant by mainstream sociology, or to distinguish environmental sociology from the crowd.

Further, whether or not we accept their characterization — and question-begging definition (Damer 1980) — of mainstream sociology, Dunlap and Catton's writings must be construed as a challenge to the discipline as a whole; what Buttel termed a "self-consciously fashioned ... critique of 'mainstream' sociology" (1987: 468) sparked by perceptions of its deficiencies. Can internal critique of the discipline be considered another distinguishing feature of environmental sociology? As a discipline, sociology has been open to both external and internal critique. Externally (and even internally), for example, critics have challenged the validity of its scientific approach, the impact of its Marxist bias, and the utility of jargon-laden prose (Abercrombie, Hill, and Turner 1988: 96–97). Internally, other sociological specialisms have adopted a similar stance. A most striking example is new feminist scholarship and theory, which has been critical of the gender bias of mostly male sociologists (Abercrombie, Hill, and Turner 1988: 96–97). It has also been influential in challenging existing theory, generating knowledge, and redefining social practices and arrangements as social problems, through acknowledgement of the inherent gender stratification of society (Komarovsky 1988). A critical stance is not, therefore, a distinguishing feature of environmental sociology.

Sociological Specialisms

Theoretical diversity ultimately challenges the distinctiveness, indeed the authenticity, of the discipline as a whole. There is substantial evidence that sociology is a well-established, independent, organized, elaborated, multiperspective, multinational and vibrant discipline dominating investigation of the particular predicament of industrial and secular civilization within the larger discipline of the social sciences (Abercrombie, Hill, and Turner 1988). Nonetheless, one could argue that sociology has never been a distinct discipline, but is rather an integrative approach, selecting from among the findings of economics, politics and psychology; because, despite Durkheim's methodological claim, social facts cannot be treated independently, since they are formed at the convergence of economics, politics, geography, history and psychology. Further, sociologists share the transdisciplinary perspective of examining individuals in social contexts along with historians, geographers, economists and journalists. Recently Becker and Rau (1992) argued that sociology lacks a shared intellectual nucleus because it was constructed from the remnants of other disciplines including history, anthropology, psychology, economics and political science — a literal patchwork quilt. Finally, a dearth of methodology, definite object and scientific analysis suggests — at least to some Marxists — that sociology is an ideology particular to a specific stage of capitalist development, and distorted by class interests (Abercrombie, Hill, and Turner 1988: 118, 233).

The persistent problem of establishing and conserving disciplinary distinction, and of balancing the role of specialisms within the discipline, is compounded by the continuing tendency of groups of sociologists to identify their approaches strongly with a number of other disciplines. Economic sociology, for example, was particularly popular during the 19th and early 20th centuries, and again in the 1970s (Abercrombie, Hill, and Turner 1988: 82), and Marxism explores the unrelenting relationship between economy and society. Clinical sociologists (Wirth 1931) joined the psychiatric team in the treatment of patients, by explaining the social realities which contribute to illness (Abercrombie, Hill, and Turner 1988: 41) and, in accepting the psychological approach to illness, integrated this methodology into sociological practice. Historical sociology has been increasingly prevalent in the last decades as sociologists and historians share expanding empirical interests, especially change, process and development, and methodological approaches "... despite acrimonious debates which proclaim the necessary divisions between the two disciplines" (Abercrombie, Hill, and Turner 1988: 114). Political sociology has long concentrated on the multiple dimensions of power relations in society (Abercrombie, Hill, and Turner 1988: 188). Exchange theory functionalism, evolutionary theory and organic analogy draw heavily on social anthropology. Systems theory parallels the social system of the social sciences with the physiological systems of medicine. Even sociobiology (Wilson 1975), while largely discredited by the community, was an attempt to incorporate principles of biology within a sociological framework, and may yet be revitalized (Benton 1991).

If we accept sociology as a *bona fide* discipline, what does sociology's characteristic disciplinary convention of extradisciplinary excursion and integration tell us about environmental sociology? A scan of disciplinary possibilities suggests a possible relationship with ecology. If so, environmental sociology could hardly be considered distinct in this respect, in light of the sociological-interdisciplinary hybrids, which characterize some sociological approaches. Among other responses, Catton and Dunlap might argue two main points here: the ineffectiveness of such a hybrid as a response to pressing societal problems, and the overriding paradigmatic similarity of seemingly diverse sociological theory irrespective of its influences.

First, if ecology — the study of the relationship between organisms and their environments — is integrated into the discipline of sociology, ecological sociology would inquire into the relationship between society and organisms and their environments. How is this hybrid different from environmental sociology? Dunlap and Catton argue that whereas sociological human ecology focuses on organization independently (1979a, 1983: 121), it ignores the physical environment (1983: 21) and adheres to the dominant mainstream sociological mega-paradigm, Human Exemptionism Paradigm (Dunlap and Catton 1983: 5; Dunlap 1980: 7). Environmental sociology concentrates on the reciprocally influential relationships between the environment — social and physical — and human behaviour (Catton and Dunlap 1978a; Dunlap and Catton 1979b; Schnaiberg 1972; Zeisel 1975) and follows — indeed develops — the emerging ecological paradigm termed the New Ecological Paradigm (Dunlap and Catton 1983: 5). The difference, then, is that as a traditional human ecology (Park 1936; Hawley 1973), new sociological human ecology is constrained in efforts to inquire into pressing social problems, while environmental sociology, both nestled in and developing the New Ecological Paradigm, is the emancipated core of environmental sociology as a whole.

Second, Catton and Dunlap argued that while competing sociological theories appear to be diverse — exaggeratedly diverse (1978a: 42) — in fact they are similar, particularly "... in their shared anthropocentrism," (1978a: 41) drawn together by the overriding paradigmatic commonality of the Human Exemptionalism Paradigm, which is rejected by environmental sociologists. But are sociological theories diverse or not? While patches of sociological theoretical diversity are supposedly stitched together into an unified whole by the golden thread of sociology, Laska, for example, has argued that commonality is mostly conceptual. Practically, she and others have observed, the diversity of substantive sociological areas is very real, resulting in grave intradisciplinary problems including: teaching incoherent subject matter and applying inadequate course standards (Becker and Rau 1992); poor departmental relations across fragmented and autonomous groups (Laska 1993: 1); assignment of secondary educational and research projects due to poor intrauniversity relations (DeFleur 1992); and difficulties demonstrating sociology's relevance in solving critical societal problems (DeFleur 1992; Porter 1992; Laska 1993). We must at least investigate, then, Catton and Dunlap's claim that paradigmatic similarity overrides theoretical diversity.

As a specialism, environmental sociology purports an enlarged definition of the environment as a distinguishing feature. Dunlap and Catton proposed that the environment should be ignored in both Durkheimian and Weberian sociology, and in environmental sociology it might be relevant (1983: 118). Rather than employing a range of possible causal explanations, pressures to achieve disciplinary distinction caused sociologists (and others) to focus exclusively on emerging causal explanations, and to neglect existing, presumably inferior ones. In this way, sociocultural explanations eclipsed physical ones. For sociologists, the term environment came to be understood as something different than the generally accepted usage (Dunlap and Catton 1979a: 244). Despite Young's charge that social scientists, including sociologists (anthropologists, geographers and anthropologists) "... have become so careless in the use of environment without adornment ... that the success or failure of this indispensable word may come increasingly to depend on the adequacy of its adjectival modifiers" (1986: 96; *cf.* Poplin 1972; Michelson 1976; Berry and Kasarda 1977). Dunlap and Catton have elucidated a definitive "... tripartite distinction of built/modified/natural" (1983: 124) physical environments. Young believes that the term "built" is emerging as the dominant one — *cf.* artificial (Roszak 1972); man-made (Dubos 1973); synthetic (Bookchin 1974) — although he noted that we must question why human habitats are not natural to humans (Young 1986).

To examine societal-environmental interactions (1983: 120–22), Dunlap and Catton suggested an ecological framework, which applies Duncan's human ecology complex (1959: 681–684, 1961) to the deficient sociological human ecology hybrid. Duncan's human ecology model narrowed concepts of the ecosystem — the interacting biotic community and the environment (Odum 1975: 4) — specifically to humans. The result was a complex of "web-like interdependence" (Dunlap and Catton 1983: 12) among population (P), technology (T), organization (O) (Park 1936: 15) and environment (E) (Duncan 1959: 684). Because they perceived that sociological human ecologists have used this complex in a way that continues to interpret environment in social (Choldin 1978: 355), or at best spatial, terms (Michelson 1976: 13–23), Dunlap and Catton proposed a framework emphasizing both the environment and the reciprocal nature of the complex. The physical environment — expressed by E, includes natural (n), modified (m) and built (b) environments — and the organizational environment — expressed as O includes cultural (cs), social (ss) and personality (ps) systems (1979a: 252, 1979b) — allowed the investigator to try to answer two types of questions: "How do interdependent variations in population, technology, culture, social systems and personality systems influence the physical environment? ... How do resultant changes (and other variations) in the physical environment modify population, technology, culture, social systems and personality systems, or any of the interrelations among them?" (Dunlap and Catton 1979a: 252). In sum, environmental sociology sets out to examine "... reciprocal relationships between the physical environment and the social complex ..." (Dunlap and Catton 1983: 121).

Does the use of this ecological framework distinguish environmental sociologists from other sociologists? Certainly most sociological approaches have not ascribed to a wider definition of environment as a variable. In formal sociology, through which the form of relationships is examined, for example, certain characteristics of these relationships are considered constant, regardless of the setting; competition is competition regardless of whether it occurs in the marketplace or on the sports field (Simmel 1900; Abercrombie, Hill, and Turner 1988: 98–99). But sociologists have concerned themselves with the physical environment in explaining social interaction. Goffman (1959, 1963, 1971), for example, whose work is associated with symbolic interactionism, acknowledges the influence of spatial attributes of the built environment on the quality (and credibility) of an individual's social performance by locating situations in physical settings. In urban ecology, the city is compared to environments found in nature by looking at competition, equilibrium and natural areas (Abercrombie, Hill, and Turner 1988: 261), using natural forces to explain social interaction, and, although its neglect of cultural influences has presently disabled it, nonetheless, is an example of incorporation of physical environmental variables (Reissmann 1964; Saunders 1981). Neo-evolutionist perspectives were drawn into functionalism to explain social change and adaptation in social and natural environments (Eisenstadt 1968; Peel 1969; Abercrombie, Hill, and Turner 1988: 93). Catton and Dunlap argued, however, that works such as these are examples of a restrictive spatial focus and of the underlying assumption of ecological homeostasis, ignoring what they perceive as an underlying condition of disequilibrium.

Laska observed that "[s]uperficial treatment of the concept [of environment] is no longer permissible for sociology," (1993: 8) suggesting, indeed prescribing, change for the discipline as a whole. If this is so, then an enlarged definition of the environment could not be a distinguishing mark of environmental sociology and a few other specialisms; rather, it would be a shared methodological approach of sociology. Young (1986), however, points out that the extremes of the continuum of environmental definition and their potentially deleterious impact on meaningful research. On the one hand, "[t]hinking about environment in the comprehensive sense — with the implication that *everything* is the environment, extending a long way beyond on every side with each entity connected to each of a multitude of others, can mire a person in helplessness, condemning one to a sort of wallowing in an incomprehensible complexity, unable to move or do anything ..." (Young 1986: 95). On the other hand, Young pointed out, "[a]n operational definition can help us find the mechanisms to deal with problems immediate and local, but can, if we let it, limit us to an unacceptable mechanistic and unfeeling approach to problems in the environment-at-large and long term" (1986: 96). Success, however, lies in the balance, and there is no evidence that environmental sociology has achieved this. In fact, its methodological complexity may explain to some degree the sparse elaboration of its theoretical tenets.

Paradigmatic Interpretations

Catton and Dunlap consistently offer that as environmental sociology emerges, a new paradigm develops (1978a: 42, 1978b, 1980; Dunlap and Catton 1983), which they claim is amending, replacing (Catton and Dunlap

1978a: 42) and rejecting (Dunlap and Catton 1983: 8) the dominant and obsolete (Catton and Dunlap 1978b: 258) paradigm of mainstream sociology. But what do they mean by this? Are distinct fields of inquiry by their very nature paradigmatic? Is each theory a paradigm or are paradigms metatheoretical? Is a discipline in itself paradigmatic or multiparadigmatic? Some answers to these questions are strictly terminological, others practical, while others lie at the crux of bringing to words the nature of the discipline and its place in the scheme of knowledge.

The term "paradigm" came to be used in sociology as a result of Kuhn's seminal work (1970), which explained scientific change by looking at the unresolvable anomalies that cause one paradigm to be abruptly displaced by another, under which routine scientific investigation can proceed. There are varying views as to the extent of paradigmatic diversity within sociology: many sociologists view theories and schools as in themselves paradigmatic (Abercrombie, Hill, and Turner 1988: 176), while others have identified a narrower range of dominant paradigms. Ritzer (1975), for example, deduced three main paradigms in sociology: Durkheim's social-fact paradigm, Weber's social-definition paradigm and Skinner's social-behaviour paradigm.

Describing a paradigm as "... an image shared by members of a scientific community telling them the nature of their science's subject-matter" (Catton and Dunlap 1978b: 256), Dunlap and Catton, however, asserted that within (so-called) mainstream sociology, there is, in fact, only one all-encompassing paradigm; the broadest of Kuhn's three usages (1970) and analogous to the German *Weltanschauung* (or world-view) (Dunlap 1980: 13 in note 5) referring to "... the set of beliefs constituting an outlook on the work characteristic of a particular social group, be it a social class, generation or religious sect" (Abercrombie, Hill, and Turner 1988: 275). A combination of sociology's disciplinary traditions and the dominant Western world-view, the Human Exemptionalism Paradigm (formerly the Human Exceptionalism Paradigm) is an "... anthropocentric, optimistic, and profoundly unecological ..." paradigm, which has been expanded upon by the majority of 20th century sociological theories. (Dunlap 1980) Despite Ritzer's triparadigmatic analysis, Catton and Dunlap asserted that Ritzer alluded to a "common core of agreement," but never specified it (1975: 32, 191, 211). Moreover, environmental sociology is not located within this all-embracing paradigm; in fact, it is characterized by its own paradigmatic distinctiveness, the New Environmental Paradigm (Catton and Dunlap 1978a: 42–43) renamed the New Ecological Paradigm (Dunlap and Catton 1979a: 250).

To focus on this paradigmatic image, which is broader than most sociological considerations, a big "P" paradigm, Ritzer's definition of the "knotty concept of paradigm" (1975: 7) synthesizing Masterman's threefold typology (1970) is useful.

> A paradigm is a fundamental image of the subject matter within a science. It serves to define what should be studied, what questions should be asked, how they should be asked, and what rules should be followed in interpreting the answers obtained. The paradigm is the broadest unit of consensus within a science and serves to differentiate one scientific community (or *subcommunity*) from another. It subsumes, defines, and interrelates the exemplars, theories, and methods and instruments that exist within it (Ritzer 1975: 7).

Briefly, the Human Exemptionalism Paradigm (Klausner 1971; Catton and Dunlap 1978a: 42–43; Dunlap 1980) is an unecological and optimistic model of four implicit assumptions (Catton and Dunlap 1980: 24) placing singular man at the centre of a world, with culturally surmountable environmental constraints and offering unlimited progress, constrained only by social barriers. Comparatively, the New Ecological Paradigm (NEP) (Anderson 1976; Burch 1971, 1976; Buttel 1976; Catton 1976a, 1976b; Morrison 1976; Schnaiberg 1972, 1975) is an ecological model of three explicit assumptions placing man in an interdependent relationship with a multifaceted environment, which sometimes produces unexpected results, in a world laden with biological and physical constraints to progress. Historically, the HEP arises from a 400-year history of abundance, expansion and contrived scarcity, while the NEP develops from the growing recognition of boundaries, limits and uncontrived scarcity.

How does locating environmental sociology within a developing ecological paradigm distinguish environmental sociology? In political science, for example, a new political ecological paradigm (Rodman 1980; Miles 1976) is eclipsing the modern paradigm, which, while recognizing limits on human societies, was nonetheless strongly anthropocentric (the differential imperative). In economics, a steady-state paradigm is nudging the assumptions of mainstream economics, especially growthist ones (Daly 1973, 1977). Even in anthropology, which has, according to Dunlap, the most fully elaborated ecological paradigm (Dunlap 1980: 9), human exemptionist assumptions are being challenged in the face of the prevailing prominence and promise of culture adaption (Hardesty 1977). Placing environmental sociology within an emerging environmental paradigm cannot, therefore, be considered a unique distinguishing feature of environmental sociology; rather, it appears to be part of paradigmatic change in the larger discipline of the social sciences.[2]

Nonetheless, environmental sociology's espoused paradigm is important. In arguing that exposition of the operating paradigm, and its underlying inherent assumptions, is vital to critical thought, Catton and Dunlap (1980) give us a tool to assess the consistency and validity of environmental sociology. Moreover, in inextricably linking field, discipline and paradigm, the soundness of environmental sociology is dependent on the legitimacy of the New Ecological Paradigm. The persistent reference to the NEP tends to prop up the theoretical significance of environmental sociology and spark curiosity, but, in so obviously placing environmental sociology within a paradigm, Catton and Dunlap force us not only to examine the soundness of environmental sociology, but also that of the HEP and NEP. If the paradigms crumble, so too may the status of the distinct field of inquiry that environmental sociology claims to be. Possible weaknesses have not escaped notice. Whereas Buttel (1978) looked inward discerning theory where Catton and Dunlap found paradigm, Satterfield looked outward descrying ideology. Interestingly, neither dispute the validity of the HEP or NEP, but rather the ascribed terminology and its relative influence.

If a paradigm is that which produces "cumulative, coherent literature," Buttel argued, then environmental sociology is not and will not become a paradigm (1978: 255). Its internal theoretical (paradigmatic) diversity is evidence of this. Both the HEP and the NEP are deficient in elucidating "... key social forces leading to change and transformation" (1978: 255) and the cleavage between the two "... is part of an ongoing, lively debate *within* major extant sociological paradigms" (1978: 252). As Buttel, who has demonstrated sustained interest in environmental sociology, pointed out recently, determining whether HEP-NEP or traditional theoretical cleavages are more important is the central issue in the evaluation of Catton and Dunlap's writings (and of new human ecology) (1978, 1987: 469; Catton and Dunlap 1978b; Dunlap and Catton 1983; Humphrey and Buttel 1982). Even Catton and Dunlap (1980) have recognized that both are important (Buttel 1987: 469), although they argue that the HEP-NEP cleavage is more fundamental (Catton and Dunlap 1978b: 258). Buttel's argument is buttressed by the fact that even Dunlap and Catton recognized that "[a]s an emerging field of inquiry, environmental sociology has not yet become a fully coherent area; it comprises diverse research interests with little overlap in personnel and insufficient cross-fertilization of ideas" (1979a: 255); however, their anticipation of substantial development in the field was undoubted.

Almost five years later, Dunlap and Catton expressed this same concern: "... one may wonder whether environmental sociology will mature into a coherent area of inquiry or whether it will evolve into a variety of distinct factions ..." (1983: 115). Further, Buttel argued that being "... more than just another sub-area such as political sociology ... does not in itself confer paradigmatic importance upon environmental sociology" (1978: 255). But Catton and Dunlap argued that, like others, "... Buttel mistakenly equates 'theoretical diversity' with 'paradigmatic diversity' " (1978b: 256–59). If every theory arises from a unique paradigm, they contended, then "... little has been achieved toward understanding the intellectual impact of unstated assumptions on the field of sociology" (1980: 23). Here they reaffirm Ritzer's suggestion that paradigms are "... handy constructs for understanding the nature of sociology or any other field ..." (1975: 200); variety becomes unmanageable and eradicates convenient utility.

If environmental paradigms are "... ideas that refute the ideology of progress," argued Satterfield (1983: 136), then, as an environmental paradigm, the NEP is not a paradigm at all; rather, it is an ideology; that is, a "... system of interconnected ideas and beliefs supported by rules of behavior that maintain those ideas' 'goodness' or 'rightness' " (Bierstedt 1957: 162–63). Because they answer historical, existential, destinational and ethical questions (Satterfield 1983: 148) at the interdisciplinary level and beyond, environmental paradigms are becoming cultural and hence ideological (Satterfield 1983: 138). Satterfield went so far as to liken environmental paradigms to Christian and Marxian ideologies, while proclaiming the conceptual weakness of the idea of progress (1983: 147). This view would again call into question the validity of Catton and Dunlap's claim that environmental sociology both develops and resides in the NEP if we accept that the New Environmental Paradigm is part of a much larger ideological system. Further, this might explain in part the attention and intensity initially generated by environmental sociology.

Discussion

How has environmental sociology developed over the past decade and a half? Has it proven itself to be a distinct field of inquiry? A look at specialism, influence, research interests and disciplines viewed in conjunction with the proceeding, will assist in the answering of this question. While Catton and Dunlap recognized environmental sociology as a specialty, they clearly believed it to be not just "... another specialty within the discipline" (Catton and Dunlap 1978a: 42). Buttel assessed that environmental sociology had become "... a conventional specialty area within sociology ..." (Buttel 1987: 484). Fletcher stressed that while specialisms are "... theoretically distinguishable areas of investigation and enquiry...," they are part of "*one* sociology, not *many* ..." and not " 'special sciences' " (Fletcher 1970: 1). Laska proposed that the "... new intellectual direction of '*de*specialization is likely to be more fruitful than continued accentuation of differences" (1993: 8), because "... the core of the discipline is threatened by specialties that borrow from sociology but distance themselves from it" (p. 6). We see, then, that the extreme position of the specialization of environmental sociology does not necessarily contribute to sociology as a whole.

Was Catton and Dunlap's enormous enthusiasm for environmental sociology — what Buttel has described as "... a vibrant *esprit de corps* that a new sociology was being nurtured" — and its specific methodologies sufficient to set environmental sociology on its desired course; that is, have sociology's foundational assumptions, methodologies and utility in explaining societal problems changed? Buttel proposed that the discipline of sociology has "... handily withstood the challenges to its theoretical assumptions posed by environmental sociologists" (1987: 466) as evidenced by its "... relatively little influence in redirecting social theory as a whole" (Buttel 1987: 484). Among environmental sociologists, Schnaiberg (1980), who employs Marxist, neo-Marxist, and neo-Weberian approaches, has had greater influence within the discipline of sociology as a whole (Buttel 1987). Other sociological specialisms have managed greater influence. New feminist scholarship and theory, for example, has successfully criticized both and influenced mainstream sociology, specifically the long-standing gender-stratification bias (Komarovsky 1988), and has even been influential in other disciplinary domains, including social history, literary criticism and education (Benton 1991).

Dunlap and Catton recognized the need to incorporate findings from other sociological specialisms including social change, deviant behaviour, stratification, demography and social psychology, and to continue to "cross disciplinary boundaries" (1979a: 266). They also acknowledged the "strong interdisciplinary orientations" (Dunlap and Catton 1983: 114) of environmental sociology. Laska pointed out that this is not a process of one-way flow; in fact, interdisciplinary assimilations have a two-way nature as evidenced, for example, by political scientists and historians who use sociological theories (1993: 5). Yet one might question whether the particular research interests and method of (so-called) environmental sociologists might be better pursued by disavowing the sociological method — instead of trying to rewrite it — and embracing multidisciplinary and interdisciplinary studies. Almost 25 years ago, Wohlwill doubted that there are any fields that do not touch on man and environment, but noted that once such relationships are examined, study "... inevitably takes on a strong interdisciplinary flavor ..." (1974: 5).

After all, Catton and Dunlap should recognize that the way we organize the pursuit of knowledge was borne of the same expansionist paradigm, if you will, from which they are repelling. The result has been an "... industrial era intellectual infrastructure, dominated by major research universities and their severe imbalance that favors small and narrow perspectives, while discouraging the increasingly-needed [sic] multi-disciplinary skills of breadth and integration" (Future Survey Annual 1994: vi; see, for example, Boyer 1990). As the new technologies continue their spectacular development and enable us to organize knowledge in new ways — hypermedia, for example — this infrastructure may be eclipsed rendering traditional disciplinary boundaries increasingly restrictive and ineffective. A trend toward researching issues across disciplinary boundaries is already evident. Laska pointed out that "... interdisciplinary research can improve the quality of the involved disciplines rather than merely challenging their integrity ..." (Laska 1993: 2), but also cautions that specialisms within sociology can weaken the discipline if they are placed before sociology.

Where Dunlap and Catton (1979a) identified two major environmental sociology research areas, less than a decade later Buttel (1987) identified five. Dunlap and Catton (1979a) recognized (yet another) *sociology of* the environment — both the precursor to and part of the emerging environmental sociology core — and environmental sociology.[3] Buttel itemized five "... major clusters of inquiry" (1987: 466) — the theoretical core, environmental attitudes and behaviour, environmental movements, risk assessment and political economy — evidence of what he termed increasing specialization, fragmentation and dualism within environmental sociology. It appears that environmental sociology is following the sociological convention of diversity.

Dunlap and Catton conceded that within environmental sociology, "... its most conspicuous cleavage ... [is] the built and natural groups" (Dunlap and Catton 1983: 114); a condition which Buttel (1987) termed "dualism." The results of relationships constrained by "conceptual and methodological difficulties" and varying intellectual roots are discrete groups of environmental sociologists with differing research interests, divergent membership in professional groups, and low intersubarea crossover (Dunlap and Catton 1983: 114). Perhaps recognizing the seriousness of the growing bifurcation and its potential to limit the growth of the field, Dunlap and Catton prescribed strategies aimed at increasing coherence and interchange through sharing of an ecological perspective. This strategy will apparently allow environmental sociologists to increasingly acknowledge the complex reciprocal relationship in societal-environmental interactions, escape from accusations of antideterminism and antireductionism,[4] surmount methodological problems, and benefit from its "integrative potential" (1983: 127–29): a tall order.

But is this troublesome bifurcation simply a problem of coherence and interchange — or lack thereof — or did Catton and Dunlap cast their net too wide in order to increase the critical number of environmental-sociology devotees, which they may have assessed was implicitly critical to its continued development and prominence? Perhaps while those concerned with the natural environment are human ecological sociologists, those with a built environment focus are more like architectural sociologists with a twist of ecology. Their fundamental differences may be irreconcilable. The answer to this question probably lies in the longer-term outcomes of Dunlap and Catton's coherence and interchange strategies — the ecological perspective and others — aimed at bridging this apparently entrenching span.

Although Dunlap and Catton perceived the small number of environmental sociologists as further evidence of sociology's difficulty "... in coming to grips with environmental problems and ecological constraints" (1979a: 244) and hence environmental sociology's relevance, Willmott (1990) remarked that we must seriously question why Catton and Dunlap's work has not generated the disciplinary devotion they predicted. In Canada there are only a handful of environmentally oriented sociologists including Don Willmott, Michael Clow, Pat Marchak, Louis Guay and Melodie Hessing and others whose names appear in this collection, a small, yet tenacious group who in the later 1980s presented papers to each other at the Learned Society meetings with the presenters sometimes outnumbering the audiences (Willmott 1990). Alluding to Mills's (1959) grand theory — a pejorative characterization of abstract concepts — Willmott (1990) wondered if we are asking any of the right questions; that is, in focusing on biospheric limitations, Catton and Dunlap distanced themselves too far from sociology and the useful and pertinent questions it engenders, such as those concerned with conflicting interests and ideologies (Buttel and Flinn 1978). Perhaps this explains Schnaiberg's greater influence within the discipline.

What were Catton's and Dunlap's motivations in asserting environmental sociology's distinctiveness? Did they wish to distinguish themselves or secure research dollars? More likely, they were stimulated by a profound concern over existing and emerging environmental issues, and a desire to sound the wake-up bells of zero hour in a predominantly conservative community, insensitive to their sense of an impending crisis. The call was for the methodology and research agenda of environmental sociology to be the number one priority of the discipline. The very magnitude of environmental problems was to obliterate distinctions among specialisms (Willmott 1990) and overturn a century and a half of tradition.

Catton and Dunlap's work defined a mission of messianic-like proportions (Willmott 1990), yet in over a decade we have heard little from them. While it would be unfair to say that their interest in ecological issues has dwindled (Van Liere and Dunlap 1980, 1981; Dunlap and Martin 1983; Van Liere and Dunlap 1983; Dunlap and Van Liere 1984; Scaminaci and Dunlap 1986; Olsen et al. 1988; Beus and Dunlap 1990; Dunlap 1991; Dunlap and Mertig 1991; Dunlap and Scarce 1991; Mitchell, Mertig, and Dunlap 1991; Beus and Dunlap 1991; Jones and Dunlap 1992; Beus and Dunlap 1992; Catton 1980, 1984, 1986, 1989, 1993), there is no evidence of direct elaboration on the theoretical tenets of the environmental sociology they sought to define. Catton and Dunlap may have recognized that real work aimed at averting environmental catastrophe could not be accomplished at the abstract metatheoretical level.

Conclusions

Is environmental sociology unconventional and influential? Time has shown it to be increasingly orthodox in its internal diversity and role as a specialism, and unable to generate sufficient force to attract researchers and to modify the foundational assumptions of the discipline. Environmental sociology may be a field of inquiry — and a noble place to start — but its distinction appears to exist more in the minds of its well-intentioned, yet waning fathers, than in the work of its sparse and disparate disciples.

DISCUSSION QUESTIONS

1. Can the empirical question at hand — that is, the distinctiveness of environmental sociology — be settled through Catton and Dunlap's key term definition of mainstream sociology?
2. What does sociology's characteristic disciplinary convention of extradisciplinary excursion and integration tell us about environmental sociology?
3. Is each theory a paradigm or are paradigms metatheoretical? Are distinct fields of inquiry by their very nature paradigmatic? Is a discipline in itself paradigmatic or multiparadigmatic?
4. Does locating environmental sociology within a developing ecological paradigm distinguish environmental sociology?
5. In your opinion, has environmental sociology proven to be a distinct field of inquiry; that is, is it unconventional and influential?

Endnotes

1. For example, "distinct area of inquiry" (243); "emerging field of inquiry" (255); "vital field of inquiry" (265); "distinct field of inquiry" (266); "[a]s an emerging field of inquiry, environmental sociology has not yet become a fully coherent area" (255); "area of environmental sociology" (Catton and Dunlap 1980: 42); "viable area of inquiry" (Dunlap and Catton 1983: 113); "coherent area of inquiry" (Dunlap and Catton 1983: 115); "field of environmental sociology" (Dunlap and Catton 1983: 119).
2. For a special volume on paradigmatic change in the social sciences, including sociology, political science, economics and anthropology, see *American Behavioral Scientist*, Vol. 24, no. 1, September/October 1980.
3. Sociology of the environment includes the subareas of: wildland recreation, resource management problems, environmental movement and the future of environmentalism. Environmental sociology includes the subareas of: built environment; organizational, industrial, and governmental response to environmental problems; natural hazards and disasters; social impact assessment; impacts of energy and other resource scarcity; and resource allocation problems. There is no shortage of "The Sociology of ..."; see, for example, development, education, family, gender, health and illness, knowledge, law, mass media, medicine, race, religion, work (Abercrombie, Hill, and Turner 1988: 234–40).
4. When a number of phenomena are explained by a single determining factor, the explanation is said to be reductionist (Baldwin and Baldwin 1980; Abercrombie, Hill, and Turner 1988: 203). When a theory emphasizes the causal primacy of one explanation, it is said to be deterministic. Examples include social structure (over individual free will), economic structure (Marxism), technology and biology (Franck 1980; Abercrombie, Hill, and Turner, 1988: 67).

Bibliography

Abercrombie, N., S. Hill, and B.S. Turner. 1988. *The Penguin Dictionary of Sociology*. 2d ed. Markham, Ont.: Penguin Books.

American Sociological Association. 1975. *Environmental Sociology*, 6(April).

Anderson, C.H. 1976. *The Sociology of Survival*. Homewood, Ill.: Dorsey.

Baldwin, J.D. and J.I. Baldwin. 1980. Reduction and the Future of Sociology. *Sociological Inquiry* 50: 93–101.

Becker, H.S. and W.C. Rau. 1992. Sociology in the 1990s. *Society* 30: 70–74.

Benton, T. 1991. Biology and Social Science: Why the Return of the Repressed Should be Given a (Cautious) Welcome. *Sociology* 25: 1–29.

Berger, P.L. and T. Luckmann. 1967. *The Social Construction of Reality*. London: Allen Lane/The Penguin Press.

Berry, B.J.L. and J.D. Kasarda. 1977. *Contemporary Urban Ecology*. New York: Macmillan.

Beus, C.E. and R.E. Dunlap. 1990. Conventional Versus Alternative Agriculture: The Paradigmatic Roots of the Debate. *Rural Sociology* 55: 590–616.

———. 1991. Measuring Adherence to Alternative vs. Conventional Agricultural Paradigms: A Proposed Scale. *Rural Sociology* 56: 432–60.

———. 1992. The Alternative-Conventional Agriculture Debate: Where Do Agricultural Faculty Stand? *Rural Sociology* 57: 363–80.

Bierstedt, R. 1957. *The Social Order*. New York: McGraw-Hill Book Co.

Blau, P.M. 1964. *Exchange and Power in Social Life*. New York: Wiley.

Bookchin, M. 1974. *Our Synthetic Environment*. 2d ed. New York: Harper and Row.

Boyer, E.L. 1990. *Scholarship Reconsidered: Priorities of the Professoriate*. Princeton, N.J.: The Carnegie Foundation for the Advancement of Teaching.

Burch, W.R., Jr. 1971. *Daydreams and Nightmares: A Sociological Essay on the American Environment*. New York: Harper and Row.

———. 1976. The Peregrine Falcon and the Urban Poor: Some Sociological Interrelations. In *Human Ecology: An Environmental Approach*, ed. P.J. Richerson and J. McEvoy III, North Scituate, Mass.: Duxbury.

Buttel, F.H. 1976. Social Science and the Environment: Competing Theories. *Social Science Quarterly* 57: 307–23.

———. 1978. Environmental Sociology: A New Paradigm? *American Sociologist* 13: 252–56.

———. 1986. Sociology and Environment: the Winding Road Toward Human Ecology. *International Social Science Journal* 109: 337–56.

———. 1987. New Directions in Environmental Sociology. *Annual Review of Sociology* 13: 465–88.

Buttel, F.H. and W.L. Flinn. 1978. The Politics of Environmental Concern: The Impacts of Party Identification and Political Ideology on Environmental Attitudes. *Environment and Behavior* 10: 17–36.

Catton, W.R., Jr. 1976a. Can Irrupting Man Remain Human? *Bioscience* 26: 262–67.

———. 1976b. Toward Prevention of Obsolescence in Sociology. *Social Focus* 9: 89–98.

———. 1980. *Overshoot: The Ecological Basis of Revolutionary Change*. Chicago: University of Illinois Press.

———. 1983. Need for a New Paradigm. *Sociological Perspectives* 26: 3–15.

———. 1984. Probable Collective Responses to Ecological Scarcity: How Violent? *Sociological Perspectives* 27: 3–20.

———. 1985. Emile Who and the Division of What? *Sociological Perspectives* 28: 251–80.

———. 1986. Homo Colossus and the Technological Turn-Around. *Sociological Spectrum* 6: 121–47.

———. 1989. Choosing Which Danger to Risk. *Society* 27: 6–8.

———. 1992. Otto Larsen's Grand View of Social Science. *Sociological Inquiry* 62: 1–10.

———. 1993. Carrying Capacity and the Death of a Culture: A Tale of Two Autopsies. *Sociological Inquiry* 63: 202–23.

Catton, W.R., Jr. and R.E. Dunlap. 1978a. Environmental Sociology: A New Paradigm. *The American Sociologist* 13: 41–49.

———. 1978b. Paradigms, Theories, and the Primacy of the HEP-NEP Distinction. *American Sociologists* 13: 256–59.

———. 1980. A New Ecological Paradigm for Post-Exuberant Sociology. *American Behavioral Scientist*. 24: 15–47.

Choldin, H. 1978. Social life and the Physical Environment. In *Handbook of Contemporary Urban Life*, ed. D. Street. San Francisco: Jossey-Bass.

Comte, A. [1838] 1986. *The Positivist Philosophy of Auguste Comte*. London: Bell.

Coser, L.A. 1956. *The Functions of Social Conflict*. New York: Free Press.

Dahrendorf, R. 1959. *Class and Class Conflict in an Industrial Society*. London: Routledge and Kegan Paul.

Daly, H.E. 1973. *Toward a Steady-State Economy*. San Francisco: Freeman.

———. 1977. *Steady State Economics*. San Francisco: Freeman.

Damer, T.E. 1980. *Attacking Faulty Reasoning*. Belmont, Calif.: Wadsworth Publishing Company.

DeFleur, L.B. 1992. Strengthening the Position of Sociology Within the University. *Footnotes* 20(9): 3.

Dubos, R. 1973. Environment. In *Dictionary of the History of Ideas*, ed. P.P. Wiener. New York: Charles Scribner's Sons.

Duncan, O.D. 1959. Human Ecology and Population Studies. In *The Study of Population*, ed. P.M. Hauser and O.D. Duncan. Chicago: University of Chicago Press.

———. 1961. From Social System to Ecosystem. *Sociological Inquiry* 31: 140–49.

Dunlap, R.E. 1980. Paradigmatic Change in Social Science. From Human Exemptions to an Ecological Paradigm. *American Behavioral Scientist* 24: 5–14.

———. 1991. Trends in Public Opinion toward Environmental Issues: 1965–1990. *Society and Natural Resources* 4: 285–312.

Dunlap, R.E. and W.R. Catton, Jr. 1979a. Environmental Sociology. *Annual Review of Sociology* 5: 243–73.

———. 1979b. Environmental Sociology: A Framework for Analysis. In *Progress in Resource Management and Environmental Planning*, ed. T. O'Riordan and R.C. d'Arge, vol. 1. Chichester, Eng.: John Wiley.

———. 1983. What Environmental Sociologists Have in Common (Whether Concerned With "Built" or "Natural" Environments). *Sociological Enquiry* 53: 113–35.

Dunlap, R.E. and K.E. Martin. 1983. Bringing Environment into the Study of Agriculture: Observations and Suggestions Regarding the Sociology of Agriculture. *Rural Sociology* 48: 201–218.

Dunlap, R.E. and A.G. Mertig. 1991. The Evolution of the U.S. Environmental Movement From 1970 to 1990: An Overview. *Society and Natural Resources* 4: 209–18.

Dunlap, R.E. and R. Scarce. 1991. The Polls-Poll Trends: Environmental Problems and Protection. *Public Opinion Quarterly* 55: 651–72.

Dunlap, R.E. and K.D. Van-Liere. 1984. Commitment to the Dominant Social Paradigm and Concern for Environmental Quality. *Social Science Quarterly* 65: 1013–28.

Durkheim, E. [1897] 1951. *Suicide: A Study in Sociology*. Glencoe: Free Press.

———. [1895] 1958. *The Rules of Sociological Method*. Glencoe, Free Press.

Eisenstadt, S.N. 1968. Evolution: Social Evolution. In *International Encyclopedia of the Social Sciences*, ed. D.L. Stills, 5: 228–34. New York: Macmillan and Free Press.

Ekeh, P. 1974. *Social Exchange Theory: The Two Traditions*. London: Heinemann Educational Books.

Fletcher, R. 1970. Introduction to *Sociology and Industrial Life*, ed. J.E.T. Eldridge. Don Mills, Ont.: Thomas Nelson and Sons (Canada) Ltd.

Franck, K.A. 1980. Exorcising the Ghost of Physical Determinism. Paper presented at the Annual Meeting of the American Sociological Association, New York.

Freudenburg, W.R. and R. Gramling. 1989. The Emergence of Environmental Sociology: Contributions of Riley E. Dunlap and William R. Catton, Jr. *Sociological Inquiry* 59: 439–52.

Garfinkel, H. 1967. *Studies in Ethnomethodology*. Englewood Cliffs, N.J.: Prentice-Hall.

Goffman, E. 1959. *The Presentation of Self in Everyday Life*. Garden City, N.Y.: Doubleday & Co., Inc.

———. 1963. *Behavior in Public Places: Notes on the Social Organization of Gatherings*. New York: The Free Press.

———. 1971. *Relations in Public: Microstudies of the Public Order*. New York: Basic Books, Inc.

Hardesty, D.L. 1977. *Ecological Anthropology*. New York: John Wiley.

Hawley, A.H. 1973. Human Ecology. In *International Encyclopedia of the Social Sciences*, ed. D.L. Sills, 4: 328–37. New York: Macmillan and Free Press.

Homans, G.C. 1961. *Social Behavior: Its Elementary Forms*. New York: Harcourt Brace.

Humphrey, C.R. and F.H. Buttel. 1982. *Environment, Energy, and Society*. Kreiger.

Jones, R.E. and R.E. Dunlap. 1992. The Social Bases of Environmental Concern: Have They Changed Over Time? *Rural Sociology* 57: 28–47.

Klausner, S.Z. 1971. *On Man and His Environment*. San Francisco: Jossey-Bass.

Komarovsky, M. 1988. The New Feminist Scholarship: Some Precursors and Polemics. *Journal of Marriage and the Family* 50: 585–93.

Kuhn, T.S. 1970. *The Structure of Scientific Revolutions*. 2d ed. Chicago: Chicago University Press.

Laska, S.B. 1993. Environmental Sociology and the State of the Discipline. *Social Forces* 72(1): 1–17.

Lockwood, D. 1964. Social Integration and System Integration. In *Explorations in Social Change*, ed. G.K. Zollschan and W. Hirsch, 244–57. London: Routledge & Kegan Paul.

Marchak, M.P. 1980. Sociology, Ecology and a Global Economy. *Cahiers de recherche sociologue* 14: 97–109.

Mauen, M., ed. 1994. *Future Survey Annual*. World Future Society.

Marx, K. [1867, 1855, 1894] 1970. *Capital*. London: Lawrence & Wishart.

Marx, K. and F. Engels. [1845] 1965. *The Germany Ideology*. London: Lawrence & Wishart.

Masterman, M. 1970. The Nature of a Paradigm. In *Criticism and the Growth of Knowledge*, ed. I. Lakatos and A. Musgrave, 59–89. London: Cambridge University Press.

Michelson, W. 1976. *Man and His Urban Environment: A Sociological Approach*. Reading, Mass.: Addison-Wesley.

Miles, R.E., Jr. 1976. *Awakening From the American Dream: The Social and Political Limits to Growth*. New York: Universe Books.

Mills, C.W. 1959. *The Sociological Imagination*. New York: Oxford University Press.

Mitchell, R.C., A.G. Mertig and R.E. Dunlap. 1991. Twenty Years of Environmental Mobilization: Trends Among National Environmental Organizations. *Society and Natural Resources* 4: 219–34.

Morris, W. ed. 1980. *The Houghton Mifflin Canadian Dictionary of the English Language*. Markham: Houghton Mifflin Canada Limited.

Morrison, D.E. 1976. Growth, Environment, Equity and Scarcity. *Social Science Quarterly* 57: 292–306.

Odum, E.P. 1975. *Ecology: The Link Between the Natural and the Social Sciences*. 2d ed. New York: Holt, Rinehart and Winston.

Olsen, M.E., E. Rosa, R.E. Dunlap, R.E. Howell, and D.A. Dillman. 1988. Public Opinion Versus Governmental Policy on National Energy Issues. *Research in Political Sociology* 1: 189–210.

Park, R.E. 1936. Human Ecology. *American Journal of Sociology* 42: 1–15.

Parsons, T. 1951. *The Social System*. New York: Free Press.
———. 1966. *Societies: Evolutionary and Comparative Perspectives*. Englewood Cliffs, N.J.: Prentice-Hall.
Peel, J.D.Y. 1969. Spencer and the Neo-Evolutionists. *Sociology* 3: 173–92.
Poplin, D.E., ed. 1972. Human Ecology. *Communications: A Survey of Theories and Methods of Research*. New York: Macmillan.
Porter, J.N. 1992. The Death of Sociology. *SSSP Newsletter* 23(3): 17–19.
Reissmann, L. 1964. *The Urban Process*. New York: Free Press.
Ritzer, G. 1975. *Sociology: A Multiple Paradigm Science*. Boston: Allyn and Bacon, Inc.
Rock, P. 1979. *The Making of Symbolic Interactionism*. London: Macmillan.
Rodman, J. 1980. Paradigm Change in Political Science: An Ecological Perspective. *American Behaviourial Scientist* 24: 49–78.
Roszak, T. 1972. *Where the Wasteland Ends*. Garden City, N.Y.: Doubleday.
Satterfield, D.R. 1983. A Change in Academic Environmental Thought: From Paradigms to Ideology. *Sociological Enquiry* 53: 113–35.
Saunders, P. 1981. *Social Theory and Urban Questions*. London: Hutchinson.
Scaminaci, J. III and R.E. Dunlap. 1986. No Nukes! A Comparison of Participants in Two National Antinuclear Demonstrations. *Sociological Inquiry* 56: 272–82.
Schnaiberg, A. 1972. *Environmental Sociology and the Division of Labor*. Unpublished manuscript, Department of Sociology, Northwestern University, Evanston, Illinois.
———. 1975. Social Syntheses of the Societal-Environmental Dialectic: The Role of Distributional Impacts. *Social Science Quarterly* 56: 5–10.
———. 1980. *The Environment*. New York: Oxford University Press.
Schutz, A. 1972. *The Phenomenology of the Social World*. London: Heinemann Educational Books.
Schutz, A. and T. Luckmann. 1974. *The Structures of the Life-World*. London: Heinemann Educational Books.
Secher, H.P. 1962. Introduction. In *Max Weber, Basic Concepts in Sociology*. Translated by H.P. Secher. Secaucus, N.J.: The Citadel Press.
Simmel, G. [1900] 1978. *The Philosophy of Money*. London: Routledge & Kegan Paul.
Spencer, H. 1876–96. *Principles of Sociology*. New York: Appelton.
Taylor, K. 1975. *Henri Saint-Simon 1760–1825: Selected Writings on Science, Industry and Social Organisation*. London: Croom, Helm.
Thompson, Kenneth. 1982. *Emile Durkheim*. New York: Tavistock Publications.
Tschannen, O. and F. Hainard. 1993. Sociology and Environment: Tropisms of the Discipline or New Paradigm? *Revue Suisse de sociologie* 19: 421–43.
Van Liere, K.D. and R.E. Dunlap. 1980. The Social Bases of Environmental Concern: A Review of Hypotheses, Explanations and Empirical Evidence. *Public Opinion Quarterly* 44: 181–97.
———. 1981. Environmental Concern: Does It Make a Difference How It's Measured? *Environment and Behavior*, 13: 651–84.
———. 1983. Cognitive Integration of Social and Environmental Beliefs. *Sociological Inquiry* 53: 333–41.
Weber, Max. 1962. *Basic Concepts in Sociology*. Translated by H.P. Secher. Secaucus, N.J.: The Citadel Press.
Willmott, D.E. 1990. Discussions with author. (Dr. Willmott is a senior scholar at Glendon College, York University, Toronto and course director of The Sociology of Environmental Issues.)
Wilson, E.O. 1975. *Sociobiology*. Cambridge, Mass.: Harvard University Press.
Wirth, L. 1931. Clinical Sociology. *American Journal of Sociology* 44: 1–24.
Wohlwill, J.F. 1974. The Emerging Discipline of Environmental Psychology. In *Readings in Environmental Psychology*, ed. D. Stokols. New York: MSS Information Corporation.

Young, Gerald L. 1986. Environment: Term and Concept in the Social Sciences. *Social Science Information* 25: 83–124.
Zeisel, J. 1975. *Sociology and Architectural Design*. New York: Russell Sage Foundation.

MEETING THE CHALLENGE OF THE 21ST CENTURY: SOCIOLOGY AND ENVIRONMENTAL DEGRADATION

Michael Clow

Saint Thomas University

Current knowledge of the relationship between society and the natural environment points unmistakably to the existence of ecological constraints on human endeavour. We must reject the cornerstone assumption of 19th- and 20th-century sociology, that the march of material progress is a lasting inheritance of our historical era. We must alter our basic understandings of contemporary society, its path of development and its future. Sociological theory, both conservative and radical, must be rethought.

Most environmentalists think sociology has little or nothing to contribute to a society facing an ecological crisis which will leave it transformed. Most sociologists have conducted their professional lives as if environmental problems are a minor headache, which will leave society, and the assumptions of sociological theory, unaltered. Both groups are wrong.

Environmentalists have been pursuing piecemeal technical solutions to specific environmental problems with little regard for the social, political and economic causes which are responsible for the problems in the first place. Environmentalists have focused on the *symptoms*[1] of destruction like acid rain, global warming and deforestation, instead of the structural causes of environmental degradation, such as how the modern economy is organized. The study of these causes lies at the heart of sociological inquiry. However, most sociologists have been very complacent about environmental degradation, paying little attention to its implications for sociological theory. Environmental problems throw into question the very *cornerstone assumption* of both Functionalism and Marxism. Both theoretical perspectives have been built on the assumption that the march of material progress is a permanent feature of our society. Environmentalists have been very successful in pointing out the fallacy of this assumption.

This paper argues that contemporary sociological theory must reconsider the relationship between humans and the natural environment. The paper is divided into five sections. The first section examines three predominant assumptions about humans and the natural environment: one, humans have infinite control over nature; two, development can be sustained if only we manage it properly; and three, there are "limits to economic growth." The next section examines the biophysical plausibility of these perspectives and concludes the "limits to growth" perspective is most plausible. In the third section, the implications of ecological constraints for Functionalism and for Marxism are spelled out, while the fourth section reaffirms the relevance of sociology in an era of economic contraction. Finally, the challenges presented to sociological theorizing in the next century are outlined in a later section.

PREDOMINANT ASSUMPTIONS ABOUT SOCIETY AND THE ENVIRONMENT

Nature and the natural world have never been omitted from sociological theory. Theorizing on the relationship between humanity and the rest of the global ecosystem, however, needs to be explored more critically. Three classical and contemporary understandings of the society/ecology relationship are worth examining closely: (1) the traditional "man over nature" perspective, (2) the "sustainable development" perspective, and (3) the "limits to growth" perspective.

The Traditional "Man over Nature" Perspective

Traditionally, nature has been perceived as a body of resources and natural forces which can be channelled and reshaped by science and technology, to provide for, what Functionalists label continuous and unlimited "economic growth," and Marxists, the "development of the forces of production." Neither Functionalists nor Marxists foresaw any widespread negative consequences from economic activity or any limits to economic growth and increasing control over nature.

Within Functionalism, the explanation of social organization and social change is based on the notion of continual "modernization" and "development." Functionalism pictures our society as marked by the institutional capacity to overcome any technological, natural or organizational limitation on economic expansion and increasing affluence. All our societal arrangements facilitate material progress; new ones will evolve to facilitate future progress.

Marx openly identified increasing abundance and power to appropriate nature as the natural goal of humanity, and the key to continuing general human improvement (Marx 1973: 409, 541). Capitalist industrialization was a breakthrough in the social development of humanity because it created unprecedented expansion of the productive capacity of mankind and (allegedly) transformed our relationship with nature (Marx 1973: 325). Under socialism and communism, Marx expected the elimination of class conflict to remove a brake on society's efforts to create more rapid technological innovation and economic expansion, as well as spread their benefits to all of society. Most Marxists believe the future holds, and will be shaped by, the expansion of the forces of production (Grundmann 1991).

Despite their differences, both the Marxist and the Functionalist traditions are rooted in a similar understanding of the relationship between society and the natural environment. They originated when it was almost universally believed nature would provide an endless supply of raw materials and an endless waste "sink" to sustain expanding economic activity.

"Sustainable Development" Perspective

By the late 1960s it became clear that economic activity was causing damage to the earth, damage which was affecting human health and economic activities. For economic growth to continue, we would have to curb the most damaging excesses of our degradation of the environment. The result was the emergence of a new conventional wisdom: environmental management. It assumed *reckless exploitation of renewable resources and dirty technology* were responsible for the disruption of the global environment. It argued that environmental measures were necessary, but economic growth could be sustained indefinitely with proper management of renewable resources and pollution control (WCED 1987: 1). Today this view is best known as the "sustainable development" formula presented by the United Nation's 1987 Brundtland Commission report, *Our Common Future.*

Brundtland predicted better management of the world's environment would produce more renewable resources from the land and sea. It predicted that technological innovation would improve efficiency in the use and reuse of energy and materials, and reduce pollution. Reasoning there would then be no resource shortages or pollution buildup, the report claimed the way was clear for indefinite economic expansion if business, governments and the public co-operate to take the necessary measures to sustain development (WCED 1987: 8).

The "sustainable development" perspective fits comfortably within the Functionalist picture of society. Brundtland lays the blame for environmental problems at the feet of no particular social actors, arguing instead that we are all responsible for "common problems." The report argues that no fundamental flaw exists in modern industrial society, for the "wrinkles" represented by ecological problems can be solved with the right mix of technology and resource management, arising from a common effort of business, governments and the public (WCED 1987: xiv). No change in "free enterprise" society and the market economy, and no radical reorganization of political systems or change in the distribution of political power, is required.

Marxists would accept the analysis that technological efforts and resource management are the means to solve environmental problems while sustaining economic growth (e.g., Grundmann 1991). The political and economic analysis of the Brundtland Commission, however would be seen as naive and misleading. Corporations will be quite uninterested in wiser resource management, which will raise the price of renewable resources, or in cleaner technology, which will require massive new investments, most of which will not increase profitability. Capitalist governments will be most reluctant to "worsen the business climate" in this era of increased international competition, by forcing companies to adopt more expensive resource management practices and cleaner technology. Even if one relaxes these political obstacles to sustainable development, the economic impact in a capitalist economy would be crippling. Sustainable development measures would reduce the general rate of profit and throw the capitalist economy into a serious depression. Sustainable development, they would argue, will have to wait until socialism puts the common interests of society ahead of the profits of capitalists.

While disagreeing on the need for prior social change to facilitate sustainable development, both Functionalism and Marxism are compatible with the technical arguments of the sustainable development thesis. This is not really surprising. Sustainable development is an attempt to specify the technological and resource management measures sufficient to sustain economic growth, not an effort to challenge the assumptions of the traditional "Man over Nature" perspective.

"Limits to Growth"

This perspective argues there are ecological limits to the scale and kinds of economic activities in which humans can engage. We need to maintain the integrity of the earth's habitats and its global air, water and soil cycles to sustain life and economic activity. Mere commitments to *minimize* the environmental impact of particular development projects, or even of economic expansion in general, will not be enough to avoid a spiral of ecological degradation. Society must pull back within the natural limits to the *amount of renewable resources which can be drawn out* of the earth, and to the *kind and volume of wastes* the environment can tolerate. Anything less will allow further degradation which will increasingly impair earth's ability to reproduce "renewable" resources and "naturally recycle" wastes. Ecological constraints will inevitably limit the size of economic activity and the kinds of technologies we can employ.

For more than 20 years in magazines such as *The Ecologist*, proponents of the "limits to growth" perspective have dismissed traditional socio-economic assumptions as irrelevant and perversely idealist understandings of the human condition (e.g., Fry 1976). This rejection of sociological theories by environmentalists is matched by the antipathy of the defenders of capitalist society towards "limits to growth" environmentalists. Traditional Marxists also recognize the opposition between themselves and those who argue there are natural limits on human activity (e.g., Grundmann 1991: 120).

JUDGING THE PLAUSIBILITY OF CONTEMPORARY PERSPECTIVES

Proponents of "sustainable development" and "limits to growth" perspectives actually agree on much. Environmental problems arise from the disruption of the earth's *biosphere*: its habitats (its living plant and animal communities) and its supporting oceanic, atmospheric and land cycles (Raskin and Bernow 1991: 87; WCED 1987: 2–8; Daly and Cobb 1989: 1–2; Benton 1989: 74). They agree that only a healthy biosphere can supply the renewable resources and natural-waste handling capacity necessary for large-scale economic activity. "Renewable" resources are only renewed, and wastes naturally "reprocessed" by the biosphere, if the habitats of the planet thrive; only if the ocean, atmosphere and other physical and chemical cycles operate normally; and only if the wastes of our production do not poison the habitats and are recyclable back into the elements of the biosphere. Both perspectives agree that "business-as-usual" is destroying the ecological underpinnings of economic activity by destroying habitats and polluting the earth's ecosystems.

The contest between these two perspectives can be boiled down to one fundamental disagreement: whether or not technological innovation and resource management can prevent a growing economy's increasing destruction of the ecological basis of production. In order to sustain economic growth *for a period of time* one must engage in a properly synchronized combination of six tasks:

1. Somehow get more materials and energy from the biosphere, the sun, and the Earth's crust without increasing the disruption of the biosphere in the course of increased resource extraction;

2. Somehow get more useful work from a given flow of energy and similarly more product from materials in the production process;
3. Somehow produce more durable products (which need to be replaced less often) and make them easier to reuse, rebuild and recycle;
4. Somehow produce products with less wastes, with no wastes that cannot be recycled or returned safely to the biosphere, and with wastes that are easiest to reuse or recycle;
5. Somehow reuse more waste energy, recycle more industrial waste, and recycle thrown away or junked products, without increasing energy and material use in the process; and
6. Somehow dispose of wastes in the biosphere in forms, locations and quantities which can be returned to the natural flow of such materials and energy, without further disruption of the habitats and air, water and soil cycles of the biosphere.

These technological innovations would have to be ongoing, synchronized and co-ordinated to keep a growing economy within the biosphere's capacity to sustain. Big leaps forward in one area, energy production for instance, would be no help if matching gains were not made in the other areas of technological innovation. To do so *indefinitely*, this set of improvements will have to continue on an *indefinite* basis.

Unfortunately, this is quite impossible. Technological innovation and resource management cannot possibly do what is biophysically required to sustain continuing expansion of the forces of production.

The insoluble problem in tasks 2–6, is that endless rounds of technological innovations (to become ever more and more efficient at using energy and materials and in reducing the volume and toxicity of wastes in a growing economy[2]), require a violation of the laws of thermodynamics (Georgescu-Roegen 1980a, 1980b). According to these laws, the efficiency of our devices and processes are inherently restricted.[3] In practice, our machines, devices and industrial processes seldom approach anything like their theoretical limits of efficiency, because of the exponentially increasing cost of developing and utilizing what are only marginally more efficient processes and machinery. The potential discovery of new natural forces would not change this situation.[4]

The inherent problem in accomplishing task 1 is equally fundamental. Renewable resource extraction is *parasitic* upon the natural processes of the world's habitats and their supporting oceans, air and land ecosystems. As in any parasitic behaviour, too much appropriation from the host, so interferes with its processes that it cannot maintain itself.[5] Since the extraction of materials from the Earth often destroys overlying habitats, and is associated with massive pollution of the biosphere in mining and refining, we cannot expect a growing supply of non-renewable resources without increasing environmental disruption, either.

"Sustainable development" is therefore merely wishful thinking. Nothing less than *magic* is required to conjure up ever more product from the same amount of materials, ever more effort from the same quantity of energy, ever more renewable resources from the Earth, ever less wastes from industrial processes, and wastes ever more integrated into the natural flows of energy and materials in the environment. Pollution controls, improved energy efficiency, extensive recycling of resources, the production of durable goods and aggressive restoration of renewable resources, etc., can reduce the degradation of the biosphere at a given level of production, but only to a certain degree. Or they can increase production for a time with fixed volumes of materials, energy and waste disposal capacity. But they cannot create the basis for continuing economic expansion. Ecological constraints *will* inevitably limit the size of the economy and the constitution of its productive forces,[6] *in spite of* our best possible efforts at technological innovation. The real issue is *how narrow* are ecological constraints, and *when* they will bite into the growth of economic activities?

Unfortunately, material constraints are not a distant, but a near-future prospect. And our ecological constraints are narrowing rapidly. Given that it takes some 20 years for the consequences of an insult to the environment to work its way through the biosphere and show its full effects, current spiralling levels of environmental disruption are compelling evidence *present* levels of demand for renewable resources and waste production are already considerably larger than can be sustained. It is quite unreasonable to suppose the Earth can tolerate substantial increases in our demand on the planetary ecosystem. And it is probably wishful thinking to believe even the best efforts of a determined society could reduce "ecological demand" (Editors of *The Ecologist* 1972: 16–17) per unit of production, enough to sustain current levels of production. As the current overload of ecological demand continues, it reduces the remaining productive capacity of earth and ecological constraints become ever narrower. We must therefore plan on a substantially reduced, not just a limited, scale of economic activity by the middle of the next century.

IMPLICATIONS FOR CONTEMPORARY SOCIOLOGICAL THEORY

Sociological theory, built on an erroneous assumption about what is possible for humans to accomplish, must be revised to acknowledge our diminished prospects. We begin with the implications of ecological constraints for our existing theories of the structure of "modern" society.

Ecological constraints pose a particularly acute problem for Functionalism. The existence of natural constraints undermines the central Functionalist assumption, that our "free enterprise" society "works" without need of any fundamental change and will continue to do so indefinitely. This is now clearly untrue. "Modern industrial society" cannot sustain its characteristic economic expansion. Current conservative triumphalism — proclaiming the end of history, i.e., the end of fundamental social change and the everlasting triumph of capitalism — is hubristic. Capitalism and its era of economic expansion is coming to a close, like it or not. If "modern" society ignores ecological constraints, it will overshoot a sustainable scale of production, drastically reduce the ecological base of future economic activity and suffer an uncontrolled reduction in the scale of those economic activities.

Marxism's progressivist assumptions are also clearly erroneous, and certain cherished verities of the historical materialist tradition must be jettisoned in the light of present knowledge of the ecology/society relationship:

1. *Capitalism will not end as Marx expected*: The concentration and centralization of capital on a world scale will most probably be cut short by ecological constrictions. The end of capitalism will probably be precipitated by the social conflicts arising from ecological degradation.
2. *The progressive phase of capitalism is now part of history*: The propensity of capitalism toward expansion of the forces of production is not the clear "silver lining" to the black cloud of capitalist development pictured by Marx. Material progress is not an inherent feature — both at once the goal and the driving force — of social development. The steady expansion of the forces of production is a feature of capitalism, but only under temporarily facilitating ecological conditions.
3. *It is not the destiny of the proletariat to preside over indefinite expansion of the forces of production*: Future socialist societies may face very limited productive possibilities, particularly if capitalists can stall society's response to ecological degradation.

Does Marxism's notion of class struggle retain any analytic power, given Marx's expectation it would create a socialism which we now know cannot materialize? Does class analysis illuminate or misdirect our attention from the possibilities and agents of social change? Is either conservative or radical sociology relevant to a society about to undergo economic contraction?

SOCIOLOGY'S CONTINUED RELEVANCE?

Many environmentalists believe sociology has little or nothing to tell a society facing ecological collapse. They have turned to biology and philosophy for guidance. Quite correctly, they argue that sociologists did not foresee logical problems or constraints, and that few have acknowledged their importance or impact. However, this is only part of the picture. For while environmentalists have accurately diagnosed the immediate cause of environmental degradation (The Editors of *The Ecologist* 1972; Meadows et al. 1972), they have made wrong-headed efforts to explain modern society's peculiar tendency for economic expansion.

Environmentalists have offered a hodgepodge of "causes" for society's monomania for economic expansion. Misplaced faith in progress and in an illusory ability to dominate nature (Editors of *The Ecologist* 1972), unrestrained human greed, deviant human "instincts" derived from an evolutionary error (Everndin 1985), uncontrolled population growth, and "technology" have all been suggested as the cause of our ecologically self-destructive obsession with economic expansion. All these putative explanations are dead ends. The idea that it is "instinctual" for humans to have a growing economy, or population, is nonsense.[7] Ecological problems on a large and growing scale began only when we adopted economic systems which both allow, and require, us to keep the economy expanding.[8] Population growth becomes exponential only under the conditions created by global economic expansion over the past couple of centuries. The ideology of progress and the pursuit of affluence do not produce, but reflect the kind of socio-economic system which shapes our conceptions of social necessity and personal possibility. Technological development is not an autonomous force but a human endeavour which has served the underlying social project of economic expansion.

Environmentalists' problems began with inaccurately referring to the crisis we face as an "ecological" one. The natural environment has not malfunctioned. It is the *destructive behaviour of humans* which is the problem. Why is our society in the midst of destroying the biosphere? Why is our society not prudently reducing the scale and altering the character of our economic activities? Who is responsible for this mess? Who is resisting the reduction of our economic activities? Who will be at the spearhead and the foot soldiers of change? These problems are properly within the domain of sociologists, not that of biologists and philosophers.

The thinking of most environmentalists concerning social change is no better than their diagnosis of the cause of the problem. Environmentalists have focused on individual conversion and education as the main agency for change. Many environmentalists believe nature itself provides us with the blueprint for the reconstruction of society (Editors of *The Ecologist* 1972), according to a set of "ecological principles" drawn from the study of the biosphere. What does sociology have to say on these matters?

The sociological approach to explaining human behaviour is to examine what social endeavours people are organized to accomplish, and how they are organized to do so. We believe that the outcome of human activities will have both intended and unintended consequences. What insights can sociology throw on the cause and cures to the "ecological" crisis which other disciplines cannot?

Functionalism's explanation for the unprecedented economic expansion of the past 300 years follows in the larger liberal tradition, where economic growth and the pursuit of affluence are understood as driven by inherently unlimited human wants. The Industrial Revolution simply occasioned the right combination of expanding markets, new economic institutions and technology, to finally begin to address those unlimited needs. If this were true, our situation would truly be hopeless, since our ecological problems would be rooted in an invariant "human nature."

In the Marxist tradition, the pursuit of material expansion in our society is driven by the pursuit of the accumulation of wealth by capitalists in one round of investment and profit making after another. It is not driven by some inherent human situation of unlimited need. Environmental problems are not the outcome of runaway drives or individual values, but are socially constructed — the outcome of the features of a particular kind of society. Advertising has to be used constantly on a massive scale to stoke the population into pursuing new consumer wants at the required pitch. But there is no clearer imperative in this society than the creation of "new business opportunities." Everyone and everything is expendable in this society if they stand in the way of an opportunity for capitalists to make more and more money. The pursuit of material expansion *is* most plausibly understood in this manner, since the spiral toward ecological exhaustion is particular to the era of capitalist development.

The social, economic and political guidance most Green thinkers (e.g., Porritt 1984; Bahro 1982, 1984; Fry 1976) believe is found in "ecological principles," has been grossly overestimated. Ecological constraints are just that, a complex envelope of constraints on human action which tell us what we *cannot* do. Nature does not tell us how to socially organize the economic activities which can be biophysically sustained, or how to distribute their benefits. There is more than one way to downsize economic activity, and each has different social consequences. "Ecological principles" are just an admixture of inferences from social practices which have caused ecological problems, and ideas imported from a variety of class, religious, disciplinary and philosophical backgrounds. They are of very limited use in envisioning an ecologically sustainable society or how to get there from this one.

What would sociological theory suggest by way of social change necessary to adapt to ecological constraint?

Functionalism's advice would obviously be limited. It would suggest the preservation of "free market" principles and the development of a value consensus around the course of action decided on by business and state élites. Conservatism is deeply challenged in adapting to as fundamental a failure of modern society as we now face.

A Marxist approach to the question of ecological sustainability would initially draw attention to the influence of class politics on the kinds of "solutions" offered to the environmental crisis. It would suggest class interests will shape all attempts to fit society within the collapsing envelope of ecological constraints. Marxists would suggest the conservative solution to ecological constraints would be designed to preserve the power and privileges of the capitalist class, if not their cycle of accumulation. Martin Ryle believes that the conservative solution would reduce economic activity by forcing the bulk of the population into Haitian levels of poverty and political repression (Ryle 1988: 7). Indeed, he argues we may already be heading in that direction, if unintentionally, as the result of "pre-crisis" class politics (Ryle 1988: 7). For an egalitarian, ecologically sustainable society to emerge, Marxism would suggest, it must be fought for by segments of society with an interest in an egalitarian outcome. Where else but in the working class and various fragments of the *petite bourgeoisie* will those social segments be found? It is unimaginable that society could be changed by a handful of privileged "middle class" environmentalists given the opposition of capitalists to the abandonment of economic expansion. A Marxist would suggest that the support of workers and the bulk of the *petite bourgeoisie* is essential.

Can a *positive and leading* role, rather than a reluctant and defensive one, be found for class politics in the pursuit of a limited but sustainable economy? This is the most interesting and challenging question for Marxist sociologists who have accepted ecological constraints and begun to explore how they recontextualize class politics. The answer probably hinges on whether the causes of increasing misery are understood accurately by subordinate classes, and on the degree to which they respond to be their objective interests. These are not questions which can be answered here, even tentatively.

REQUIREMENTS FOR A 21ST CENTURY SOCIOLOGY

By the early decades of the 21st century, we can expect many areas of economic activity to be visibly affected by the undermining of their ecological basis. To the downward pressure on living standards imposed by NAFTA, GATT, structural adjustment and other redistributions of wealth to the rich, will be added the effects of ecological degradation. Incidents, such as the collapse of the Newfoundland fishery, where whole sectors of regional economies are wiped out, will be commonplace. Symptoms of overshoot and collapse of the global economy will be clearly visible.

New sociological ideas which account for the natural limits of human action, plot a likely and desirable alternate course of social development, and provide an understanding of social process, which will point in that direction, will be needed. Whatever their ideological assumptions, all sociological responses to the coming era of economic contraction, face certain common challenges.

Any 21st-century sociological theory must systematically incorporate an understanding of the dependence of economic activity on the biosphere. We can no longer leave those "details" of economic activities to technical specialists such as chemists and engineers, and focus solely on the social organization of human activities. The ecological consequences of our activities, and their feedback upon human activities, must become an integral part of all sociological theory. We must understand the *interplay* between the ecological and sociological dimensions of economic activity to understand humanity, its opportunities and its limitations.

We have a fairly good idea of what general "technical" changes societies must make to accommodate ecological constraints: (1) reduce the scale of economic activity to what can be supported by the environment's ability to supply renewable resources and reprocess our wastes over the indefinite future; (2) limit the use of many technologies (e.g., the petrochemical and automotive industries) and shed others (e.g., nuclear power) which are grossly damaging to the environment; and (3) restore earth's habitats and physical cycles, as best we can, to strengthen the biosphere's productive capacity.

Where a conservative perspective leads one in the face of the failure of capitalist society to be sustainable, eludes me. Dramatic and far-reaching change in the nature, organization and goals of contemporary capitalist society are unavoidable. Conservative social theory will have to adapt itself to increased conflicts over a shrinking economic pie, the need to abandon business's central goals of investment and expansion, and the need to curb our society's taste for affluence. Conservative sociologists will have to answer many difficult questions. What is the social origin of the ecological crisis? What was the fundamental flaw of industrial capitalist society such that an apparently "functional" society could create such an intractable problem for itself? How can the conservative understanding of the economy, i.e., market economics, be revised to deal with ecological constraints on the scale and technologies of production? How will the impact of a falling GNP be shared amongst various socio-economic groups within society? What kinds of changes will be required in political and economic institutions? What will be the new goals and developmental direction of society? How will these political, social and economic changes occur without massive social instability? The temptation for conservatives may well be to embrace a greater and greater measure of social inequality, repressive political solutions to dissent and poverty, and pseudo-Darwinian programmes of "lifeboat ethics."

Developing a Marxism shorn of its progressivist assumptions about capitalist development and socialism will also not be easy. Many questions remain to be answered. How do Marxists revise their understanding of the production and socio-economic process to account for the central role of the ecological basis of economic activity and ecological constraints?[9] Why was Marx so wrong about the central role of increasing forces of production in the path of human development, and what is to replace that ontological presumption? How does the traditional focus on class struggle relate to the pursuit of ecological sustainability, which requires a reduction in per capita income and an abandonment of the pursuit of progress? What place must be accorded to "non-class" dimensions of human life in the process of social change and the construction of an alternative to capitalism?

Neither conservative nor radical sociology can expect a continuation of intellectual business-as-usual in the next century.

CONCLUSION

The emergence of ecological constraints presents as great a challenge to social theory as the upheaval of the French and Industrial Revolutions which gave birth to the discipline. As of yet, that recognition has not diffused widely within the discipline. Sociologists have largely ignored the implications of the newly appreciated dependence of human society on the biosphere. Until now, it has been far easier to mentally sweep the problem away, with little shared acts of faith in the putative ability of science and technology, to facilitate continued economic expansion, and maintain the integrity of the Earth's living systems. As the increasing effects of ecological disruption begin to "bite" in the early 21st century, however, the need to rethink both the future and the past in the light of ecological constraints, will become unavoidable.

Conservative social theory is left in tatters by the inability of a supposedly functional and enduring society to sustain its social project. Conservative social theorists are faced by the need to confront and adapt to the inevitability of massive social change, and desperate struggles over a declining per capita production. The dangerous temptation will be to move distinctly to the right.

Marxists must confront the fact that the master was in error about the direction of future social change, and that socialism cannot be the expansive society envisioned in the last two centuries. The place of class politics in the transition to a sustainable society, and the relations of the workers' movement to "other" social movements, is no longer clear.

Sociologists face an interesting new century, as we begin to grapple with the coming era of massive social change, forced upon humankind by the unintended consequences of our success at expanding the forces of production beyond their natural limits.

DISCUSSION QUESTIONS

1. Why does Clow argue that conventional environmentalism requires an input from sociology, if environmentalists are to make progress in changing economic practices?
2. Do you think sociology has any special insights into environmental problems?
3. Why does Clow argue that the "sustainable development" perspective proposed by government, business and certain elements of the environmental movement, is only a variation on the discredited traditional conception of society/nature relations?
4. Why does Clow find the notion of a technological means to solve environmental problems and sustainable economic growth implausible? How do you respond to his argument?
5. Why does Clow argue that the recognition of ecological constraints on economic growth is vital for sociology as an academic discipline?

Endnotes

1. I want to thank Susan Machum, part-time lecturer in sociology at St. Thomas and former executive director of the Conservation Council of New Brunswick, for pointing out the environmental movement's focus on *symptoms* rather than the *causes* of environmental problems.
2. To more closely match larger and larger volumes of biodegradable wastes with the natural flows of materials in the biosphere is impossible. And the toxicity of the wastes produced by many modern industries are simply incompatible with the health of living things.
3. There is a limit to how much of a flow of energy can be tapped and turned into useful effort. Similarly, there are limits to the efficiency of our use of materials.
4. There would be limits to the useful effort new technologies could harness "new" natural forces to do. The use of the new technologies would consume resources and produce wastes which would both add to ecological disruption.
5. Neither natural habitats, nor the artificial habitats of agriculture, silviculture and aquaculture can be successfully driven to greater and greater levels of production. One can force-feed into illness and death the goose which lays golden eggs. Plants and animals consumed by the economy, and all the materials and energy in the air, water and soil cycles appropriated by the economy, are being pulled out of their role in the normal operations of the biosphere.
6. As noted in note 2, many technologies create wastes which are simply incompatible with the continued health of the biosphere. Many harvesting technologies are too destructive of habitats, etc.
7. Human behaviour and social organization are too diverse to be genetically determined. If humans had significant instincts we would be trapped by them in a fixed habitat, in a

particular ecological niche, and with a singular social organization, like all other animals. We would be no more trouble for the rest of the biosphere than wolves, lions, deer, horses or even the other great apes. (See Clow 1992: 8–9)

8. Environmental disruption did occur in preindustrial societies; for example, the ancient Greeks deforested their hillsides and then raised sheep, causing devastating soil erosion. The Babylonians made their agricultural land into a desert through misuse of irrigation. These, however, were limited episodes of overexpansion of economic activities, not a broad-spectrum problem.
9. Continuing Ted Benton's work is probably the best place to begin (Benton 1989).

Bibliography

Bahro, R. 1982. *Socialism and Survival.* London: Heretic Books.

———. 1984. *From Red to Green.* London: Verso Books.

Benton, T. 1989. Marxism and Natural Limits: An Ecological Critique and Reconstruction. *New Left Review* 178: 51–86.

Clow, M. 1986. Marxism and the "Environmental Question": An Assessment of Bahro. *Studies in Political Economy* 20(Summer): 171–86.

———. 1990. Sustainable Development Won't Be Enough. *Policy Options/Options Politiques* 11(9): 6–9.

———. 1992. Ecological Exhaustion and the Crisis of Global Capitalism. *Our Generation* 23(1): 1–25.

Daly, H.E. and J.B. Cobb Jr. 1989. *For the Common Good: Redirecting the Economy Towards Community, the Environment and a Sustainable Future.* Boston: Beacon Press.

Editors of *The Ecologist.* 1972. *A Blueprint for Survival.* Harmondsworth: Penguin Books.

Everndin, N. 1985. *The Natural Alien.* Toronto: University of Toronto Press.

Fry, C. 1976. Marxism vs. Ecology. *The Ecologist* 6(9): 328–32.

Georgescu-Roegen, N. 1980a. The Entropy Law and the Economic Problem. In *Economics, Ecology, Ethics: Essays Towards a Steady-State Economy,* ed. H.E. Daly, 49–60. New York: Anchor Books.

———. 1980b. Selections from "Energy and Economic Myths." In *Economics, Ecology, Ethics: Essays Towards a Steady-State Economy,* ed. H.E. Daly, 61–81. New York: Anchor Books.

Grundmann, R. 1991. The Ecological Challenge to Marxism. *New Left Review* 187: 103–120.

Marx, K. 1973. *Grundrisse: Foundation of the Critique of Political Economy.* Translated by Martin Nicholas. New York: Vintage Books.

Meadows, D.H., D.L. Meadows, J., Randers, and W.W. Behrens. 1972. *The Limits to Growth.* New York: New American Library.

Porritt, J. 1984. *Seeing Green: The Politics of Ecology Explained.* Oxford: Blackwell.

Raskin, P.D. and S.S. Bernow. 1991. Ecology and Marxism: Are Red and Green Complementary? *Rethinking Marxism* 4(1)[Spring]: 87–103.

Ryle, M. 1988. *Ecology and Socialism.* London: Radius Books.

World Commission on Environment and Development, WCED, (The Brundtland Commission). 1987. *Our Common Future.* Oxford: Oxford University Press.

NATURE: ADVERSARY OR VICTIM? THE PROBLEM OF CONTROL IN THE RELATIONSHIP BETWEEN NATURE AND SOCIETY

Ralf Starkloff

Brandeis University

Human activity in, and knowledge of, nature and society are informed not only by our practical experience of reality, but by socially constructed cognitive models which circumscribe the horizon of our knowledge and action. In everyday activity, as much as in specialized scientific discourses, these frames of mind, world-views or cosmologies remain largely unreflected, tacit and deeply ingrained presuppositions. The experience of ecological crises motivates us to reflect critically upon fundamental assumptions about our relationship with nature. Our models of knowing are of vital importance in the making of ecological crises, in our ability or failure to understand their making, and for our capacity to exercise ecological prudence.

In this chapter, I will review the cognitive models underlying the production of knowledge about the interface of nature and society in two contemporary intellectual currents: Marxism and ecology. Both of these figure significantly in sociology's undoubtedly belated concern with "the environment." While Marxism constructs nature as humanity's adversary, ecology constructs nature as a victim of human activity. I will argue that both perspectives are fundamentally concerned with conceptual and material control of reality. The limits of control on account of volatility and complexity in social and ecological systems will be problematized. My understanding of nature and society as open interactive networks, and of our knowledge as necessarily fragmented and incomplete, precludes the presentation of yet another cognitive control system or new paradigm. In the spirit of open discourse, alternative concepts are proposed throughout the text as arguments with the cognitive models reviewed.

Nature as an Adversary

The idea of nature as an adversary of society is explicitly stated in classical Marxist theory and reflected in the material practices of the socialist societies of the 20th century. Images of nature as discontinuous with culture, and as an opponent of labouring and knowing human actors, are, however, dominant in the cognitive models of modernity in general. Classical economics and its contemporary successors, most radically in development theory and practice, as much as in Marxism, perceive of human progress as an ongoing struggle to overcome scarcity, defined as a lack of production and consumption. The betterment of the human condition is intimately tied to an increasing capacity to appropriate and transform natural resources and to circulate the products as commodities.

In the *German Ideology*, Marx and Engels propose that human beings are distinguished from nature because they "produce" their livelihoods.

> They begin to distinguish themselves from animals as soon as they begin to produce their means of living, a step which is based on their bodily organisation (Marx and Engels 1969: 21).

In *Capital One*, Marx considers both conceptual knowledge and tool use as the distinguishing properties of the human species (Marx 1973: 192–94). Once human actors devise and use technological means of selectively controlling living and inanimate processes, the "means of living" are no longer a "gift of nature."

Conceptual knowledge and tool use, instead of marking a radical qualitative distinction, may be perceived as a particular state of development in a continuum of evolutionary possibilities. All living beings require a capacity for orientation in a volatile environment on which they depend for their sustenance. While "instinct" may circumscribe the boundaries of a species' capacities, it does not reduce the action of individual organisms to mechanistic reflexes. Marx has characterized tools as "means of labour" which are "means of power," and the human body as a "natural power" which transforms nature (Marx 1973: 192, 194). Hence, the original means of labour is the body, a faculty which human beings share with all natural actors. The transformative capacities of bodies — i.e., their power — link all actors in nature. Tools are extensions of the body, which increase the body's transformative capacities. They can never be entirely substituted for the body, as they require judgement and handling. Phenomena such as water, fire and air are inanimate and thus hardly wilful. They are nevertheless highly volatile — indeed active — and transform their environment. Moreover, given the continuity of transformative capacities among all actors and phenomena in nature, it is scarcely possible to pinpoint the historical moment when human technology turns human participation in nature into "production." Rather, nature may be conceived of as a multitude of interdependent productive relations. Differential cognitive and physical capacities render the various species differentially powerful in determining these relations.

Marx recognized the fundamental material interdependence between society and nature, but constructed human history as a continuous struggle against dependency. For Marx, human relations with nature entail both unity and adversity (Schmidt 1978: 19–23). Unity is grounded in the necessity of (re)producing human life through "metabolic exchanges" with nature, which "ties" human beings to the inherent properties of "natural matter." The activity mediating human beings and nature, accomplishing metabolic exchange, is labour (Schmidt 1978: 76–78).

> As the producer of use values, as useful labour, labour is an existential condition of human beings independent of all social forms, an eternal natural necessity, mediating the metabolism between humans and nature, i.e., human life (Marx 1973: 57).

Yet, while human beings are grounded in nature and reproduce their existence through labour, they do so in opposition to nature: "man and his work on one side, nature and matter on the other" (Marx 1973: 198). Metabolic exchange takes the form of unilateral acquisition. Human beings are provided with a resource pool, "the earth ... as the general object of human labour" (Marx 1973: 193; Schmidt 1978: 8). Furthermore, the struggle in which human labour appropriates nature is conceived of as inevitable and eternal:

> Just as the savage must wrestle with nature to satisfy his wants, to maintain and reproduce life, so must civilised man, and he must do so in all social formations and under all possible modes of production (Marx 1978: 441).

Consequently, the process of human civilization is depicted as the social history of the reversal of power relations between human beings and nature through the development of productive forces — i.e., human labour and its artefacts. Hunting and gathering societies are viewed as "ruled by ... the blind forces of nature," while industrial societies are credited with "rationally regulating their interchange with nature" by "bringing it under their common control" (Marx 1978: 441). In *Capital*, the structuration of the common control of nature by technical-industrial means is analysed as commodity production. As Marx limited his analytical concerns to social production, he failed to pursue the dramatic ecological implications of commoditization. From an ecological perspective, industrial capitalism accomplished the transformation of interdependent metabolic exchanges among complex sets of actors and phenomena (i.e., of material links in ecological systems) into commercial exchanges (i.e., the circulation of fragmented objects as commodities). Apart from fleeting references to the destruction of the "metabolism between man and earth" by capitalist production (Marx 1973: 528–29), Marx celebrated the productivity of industrial labour. He analysed the historical process as a dialectical movement of the development of forces and relations of production. A succession of modes of production in distinct social formations generates increasingly productive means of creating and appropriating surplus labour and products. Freedom, Marx projected, would be attainable once the control of nature by technical means is matched by the control of social relations by associated producers. They would abolish the appropriation of surplus labour by a ruling class and utilize the capacity for surplus production to abolish scarcity.

Nevertheless, the desire for freedom would continue to confront necessity. Marx argued that in social production, which ties human beings to necessity and the struggle with nature, "freedom" is attained through the optimal domination of nature by technical-economic means. "The true realm of freedom," however, lies beyond necessity and natural contingencies. The feat of industrialism is the satisfaction of needs beyond scarcity through highly productive labour, allowing for the alteration of time economy in postcapitalist society: high productivity processes free time for pure human creativity (Marx 1978: 441).

Bittner, rejecting Marx's idealist conclusion, has argued that human autonomy is always intimately connected with human labour.

> The aspiration to autonomy is realised in people's struggles to live according to their creative understanding of necessity and purpose (Bittner 1983: 251).

While human labour cannot produce what runs counter to the forces of nature, it realizes freedom in harnessing those forces to design and construct processes and artefacts which do not occur and operate through "nature's spontaneous naturing" (Bittner 1983: 250).

> Human autonomy comes into being when humans cease to depend on the providence of nature for sustenance and begin to rely on the prudence of labour (Bittner 1983: 251–52).

The Marxist emphasis on material practice tends to ignore self-reflective inquiry into the symbolic dimension of social action. The analysis of human cognition and communication, however, permits the insight, that the effects of human participation in ecological systems are co-determined by the cognitive system through which human participation is signified. Language communication creates selective images which arbitrarily engage reality by creating situational definitions as a basis of human-action co-ordination.

Habermas (1971: 166) has argued that human actors are guided by a knowledge interest in "technical power of control over an environment," and thus perceive of nature and their participation in nature as a sphere of instrumental rationality. Scientific empirical knowledge provides for the efficacy of human labour and is conversely validated by the efficacy of human "intervention in nature." Accordingly, "instrumentally acting man" is required to conceive symbolically of nature as an instrumentally rational actor "who has habitualised the laws of nature as behavioural rules." Thus "man" projects "himself as an adversary of instrumentally acting nature" (Habermas 1971: 161). Mirroring Marx, Habermas claims that the evolution of the human species as a whole is characterized by the increasingly successful pursuit of the domination of nature, and that all human knowledge of nature is inevitably tied to the interest in technical control (Habermas 1971: 49). By contrast, the construction of social relations among human actors by means of "communicative action" is guided by the interest in "securing intersubjectivity of understanding in speech communication and in action within common norms" (Habermas 1971: 221; Habermas 1987). Social action is co-ordinated by "unconstrained consensus" achieved in discourses free of domination (Habermas 1983: 98–99; Habermas 1987).

Habermas's polarization of nature and society — assigning domination to human relations with nature and consensus to relations among human actors — is problematic. In their social relations, human actors draw on complex mixes of power modalities, including domination, manipulation and deception, deals and compromises, as well as consensus (Starkloff 1989). Meanwhile, non-human actors and phenomena may not only be experienced as adversaries, but also as allies, friends or gifts. The mutual adjustment of relations between human beings and nature through, for example, ritual negotiation with spirits, demons and deities, involves equally complex and differentiated symbolics of power as social action. If the scientific perspective was to judge these communications as imaginary and incapable of generating intersubjectivity, it will have to recognize that an imagined dialogue with non-human actors has the same symbolic status as imagining nature as an adversary. Furthermore, communicative exchanges with nature may establish an ethic of restraint.

The characterization of human labour as an instrumental act is a reductionist image which ignores the circumstance that labour is embedded in complex sociocultural and ecological matrices. Accordingly, instrumental rationality may be perceived as a central tenet of cultural lore in capitalist societies. It is the product of discursive practice. The construction of human relations with nature as a "functional sphere of instrumental action" (Habermas 1971: 166) rests on a more or less tacit agreement that non-human actors and phenomena may be treated as adversaries who are technologically apprehended in the interest of human control. This agreement does not only rest on the rationality, of validity claims, but also on the desire of a social group to conduct life within a context which defines nature as a functional sphere of instrumental action. It rests on a power claim. The commitment to instrumental rationality, as much as the belief in the existence of powerful deities, prove to be quite immune to criticism. Rational discourses are but one kind of manifestation of the conflictual negotiation of knowledge and power claims. The insight of the Thomas theorem may thus guide our understanding of the social and ecological consequences of socially accepted cognitive models.

> If men define situations as real, they are real in their consequences (quoted in Merton 1967: 421).

Still, the motivation for a commitment to an adversarial approach to nature appears not to be sufficiently explained by a power claim for technical control. Fellman (1989: 23) has pointed out the need for a theoretical framework which explains motivations for adversarial encounters in human conduct. From a psychoanalytic perspective, he argues that the explanatory value of theories, which focus on the consciously intended outcomes of adversary contact, is limited, as they ignore its emotional dimension, and may actually serve to rationalize repressed impulses (Fellman 1989: 2–3). Fellman explores the emotional commitment of human actors to the pursuit of triumph and submission, of victory and defeat, the central role of fear and the denial of empathy with others. The projection of repressed emotions onto an "other" in whom they may be ruthlessly defeated, is perceived as a central tool of unconscious ego defence.

The human experience of both society and nature is fundamentally ambivalent. The family, community, ethnic group, nation state etc., as well as our interactions with other species and natural phenomena, are sources of both nurture and threat. Social relations entail belonging and mutual care, but also exclusion, conflict and exploitation. Nature provides health and sustenance, but also inflicts disease, famine and catastrophes. The cognitive models and cosmologies of human cultures may serve as collective "projective systems" (Obeyesekere 1987: 428) in the attempt to find resolutions for existential ambivalence. The conflicting emotions of love and anger, of security and fear that we feel among our fellow human beings, as well as in the natural environments in which we participate, precipitate multiple instances of repression and projection. Projective systems are complex, often contradictory, and always culture specific.

The image of nature as an adversary, predominating in industrial modernity, is a projective resolution of existential ambivalence by means of dichotomization. Nature and society are split apart, and the threatening dimension of experience is projected towards nature, while society, by dominating nature, is proclaimed as the source of sustenance and security. Nature is accordingly constructed as an adversarial other, whose violence we fear, and with whom we deny empathy. Human violence against nature is legitimized as the liberation from the "fetters of nature" through technical progress. In the process of this emotional manipulation, we repress our connectedness, our existentially ineluctable interdependence with the diversity and complexity of life-forms in nature, and most importantly, our subjection to risk. In nature, all beings are living with the constant possibility of death at any moment. Nature is an interplay of nurture and destruction, a dance of life and death. Society too engenders multiple splits or layers of exclusion and inclusion. The other nation state, other ethnic group, other gender and so forth, are constructed as adversaries towards whom emotions of fear and anger are projected, engendering violent encounters and defensive strategies.

The preoccupation of modernity with security, and the systematic reduction of risk is, however, a "double edged phenomenon," as Giddens (1990: 7, 92–100, 124–34) has pointed out. The ontological security of reliable environments under human control is created by means of domination of, and violence against, the "forces of nature" and the "evil" forces of human "others," a circumstance which paradoxically places us under the threats of large-scale nuclear annihilation, protracted ethnic warfare and ecological destruction.

Nevertheless, the motivation for adversarial conduct cannot be reduced to collectively shared emotional deep structures. Each human actor is inevitably socialized into a symbolic and material universe already in place. Therefore, the motivational force of an "ensemble of economic, political and cultural practices, structured by the capitalist relations of production" (Berger, translated by R. Starkloff 1991) can hardly be escaped.

Historically, human beings have evolved as increasingly powerful actors, whose cognitive capacities and use of tools have radically extended human autonomy. The capacity to transform nature according to human design has, however, created a cognitive illusion, which assumes that human dependence on nature has been lessened or supplanted by reliance on human skill. Yet, the prudence of human labour continues to depend on the providence of nature. Human production depends on activities beyond its own capacity and reach. The production of food and energy requires solar radiation, photosynthetic activity, water and biomass flows, and fertile soil. While human beings can significantly impact the conditions of production of natural resources, they cannot substitute human labour for the productive activities of the beings and phenomena which constitute nature. They can only participate in this interactive network. Dependence remains an inevitable constraint of autonomy, and subjection to risk an ineluctable condition of life. We will have to come to terms with the fact that there is no prudence in labour which does not recognize the fundamental interdependence of nature's participants.

Eco-Marxists, in their recent ecological revision of Marxism, are therefore emphasizing the dependency of human actors on the maintenance of the "conditions of production" — i.e., labour power and material resources, comprising nature, as well as socially produced infrastructures. They discern a fundamental contradiction between the capitalist mode of production and its own conditions of production (O'Connor J. 1989: 16, 25). As capital does not reproduce its conditions of production, but treats them as externalities, the demands of capital accumulation on resource bases cause scarcities and the impairment of ecosystems. Scarcities and natural limits are consequently considered historically specific social phenomena, rather than given by nature (Collins 1992: 181, 186).

The insight that human practices in nature are always social practices, and as such, co-determined by historically mutable social structures, is crucial. Ecological crises can neither be understood nor resolved by analysing human land-use practices in isolation from the social relations structuring human action, and thus constraining ecological systems. However, as eco-Marxists continue to treat nature as a mere resource pool of human social production, they fail to recognize the implications of the mutual conditionality of co-dependent relations of production in nature. Their concept of production remains limited to human activity, and natural systems are viewed as conditions of production operating within human practices as mere environments. Eco-Marxists maintain the perspective which imagines and treats nature as an assemblage of fragmented objects of unidirectional appropriation. Thus, Brinda Rao (1989) has analysed resource scarcity as a consequence of social conflicts over the appropriation and management of a particular condition of social production — water — without analysing the conditions of production of water in the hydrological cycle.

The concepts of "production" and "conditions of production" indicate two concurrent aspects in the co-dependent relationships among the diverse actors and phenomena in nature. Nature may be characterized as a productive network in which its participants produce their existence, by cycling biomass, water and air, and thereby simultaneously provide conditions of production to each other. Each productive activity, by drawing on opportunities existing within the network, contributes to creating a set of opportunities for others. At the same time, productive activities also act as constraints limiting the opportunities of particular actors and phenomena. If constraints significantly outweigh opportunities, an actor's or phenomenon's ability to exist may be severely reduced or become impossible.

For example, in the case of the land phase of the hydrological cycle, water enables biomass production as it infiltrates into the soil. Conversely, a particular biomass structure (i.e., a network of plants, micro-organisms, animals, human actors, soils etc.) facilitates this infiltration process. If, however, this structure is altered and soil-water infiltration is sufficiently limited, water availability to biomass production will be reduced, and the system will experience a crisis and change its composition and patterns of activity (Falkenmark and Chapman 1989). Consequently, water scarcity may pose a serious threat to human livelihoods. Yet, not only crop cultivation and household water consumption would suffer from water scarcity. Forests may alter their species composition or die altogether, and aquatic life may decline or disappear. Other life forms dependent on these specific habitats could lose their sources of sustenance. This set of changes may, on the other hand, increase the opportunities for other species not affected by increasing water stress. These may then occur in such proportions as to threaten human interests by causing crop damage or diseases.

Each ecosystem component has therefore a set of requirements, its conditions of production, which must be met for this component to exist and reproduce itself, and act as a condition of production for others. These requirements constitute limits for the opportunities of other actors. They are intrinsic to nature rather than produced by human social action. Accordingly, the physical requirements of non-human ecosystem participants act as constraints of social systems and may contradict social practices. Two conflicting modes of economic action may be derived from this framework: first, human production which maintains links between ecosystem components by recognizing the limits inherent in the requirements of co-dependent actors and phenomena; and second, human production which ruptures these links and appropriates a maximum of productive output in ecosystems for selective human uses.

The analysis of ecological crises requires the study of both the productive activity and the conditions of production of critical ecosystem components affected by crisis. Given the powerful role human actors play in the making of ecological crises, environmental sociologists need to scrutinize the social relations and material practices which comprise the human component of an ecosystem. However, an analysis which reduces nature to a condition of production for human ends, may miss the point that social systems, by themselves, are not capable of reproducing their conditions of production. The participants of ecological systems, of which human actors may be but one component, are reproduced through the composite effects of their interactions.

Nature as Victim

As human actors in the 20th century strive to come to terms with the experience of ecological crises, functionalist-systems theory has successfully emerged as an alternative to the adversarial cognitive model. Scientific ecology perceives nature as a hierarchy of distinct, functionally integrated systems, rather than a realm of adversarial struggle. It informs a wide array of perspectives in Green discourses which view nature as a victim of human action.

The functionalist perspective represents participants of natural and social systems as comprising structures which are integrated by functional relations among constituent components. The delineation of a system is somewhat arbitrary — i.e., constructed according to an observer's interest and point of reference. Each system is integrated on a larger scale with others, which define its environment. This circumstance permits the observer to situate relevant phenomena outside his immediate focus in an environmental position. Located on a particular hierarchical level, a system is related horizontally to corresponding systems, and to the encompassing higher-level constructions through "input/output" relations of energy, matter and information. Internally, systems components maintain "throughput" relations (Odum 1971; Marten and Saltman 1986; Rambo and Sajise 1984). The structure of an ecosystem refers to its "component parts ... and how they are arranged in space and time" (Marten and Saltman 1986: 24). Function constitutes the dynamic element of interactions and relationships — i.e., "work" and "co-ordination" (Odum 1971: 293; Kucera 1973: 9). The flow (input/output/throughput) of energy, matter (nutrients, biomass) and information (language, genes) are subsumed under this concept. Functions are indicated by the achievement of such systemic goals and purposes such as productivity, stability, sustainability, self-maintenance, resilience to disturbance and regeneration. Homeostatic feedback provides "checks and balances" for system components, and thus promotes system integration.

Succession theory accounts for linear processes of change in ecological systems. It is assumed that ecosystems move through subsequent stages of species composition in an ordered fashion until they reach a state of climax: a self-perpetuating, stable, diversified and complex community in equilibrium with its environment. So-called "pioneer" species or "opportunists" "colonize" a habitat (grassland), and are gradually replaced by equilibrium species (old-growth forest), a state which may persist indefinitely. An important assumption of the idea of succession is that this sequence will only occur in the absence of "external disturbances" which may prevent the establishment of a "climax community" (Whittaker 1970: 73; Kucera 1973: 86–97; Emberlin 1984: 61).

Many writers view functional integration in ecological systems as comprising highly organized and orderly, even imperative, processes. Some project an organic metaphor onto ecosystems. Hence, a community of plants is perceived as "a complex organism," "a new kind of organic being" (Clements, quoted in Worster 1977: 211), and the planet as a "living organism" (Lovelock 1979; Lovelock 1988: 40) or a "single living system" (Goldsmith 1988: 161). Goldsmith (1988: 161–62, 168–69) explicitly posits an ecosystemic telos emanating from the global biosphere ("Gaia's strategy"), which is transmitted through its subsystems and components. This telos is binding for all constituent parts, which are assigned roles (via structure/function) and thereby contribute to "maximizing" systemic stability as "Gaia's overriding goal." The idea of ecological superorganisms has been dismissed by ecologists such as Tansley and Whittaker (Botkin 1990: 98–99; Worster 1977: 301; Catton 1994: 90). Botkin (1990: 41–47, 101–10) and Worster (1977: 291–315), however, note that the dominant metaphors of 20th-century scientific ecology reflect the dominance of industrial capitalism as a social system. Reduced to interactive sets of energy flows, exchanged between "producers" and "consumers," nature is depicted as a functionally integrated stable mechanism constituting a network of quasi-economic exchanges. Mechanistic ecology treats energy, calculated in units of calories, as the universally quantifiable means of exchange, analogous to money, in "nature's economy."

Catton and Dunlap have sought to assimilate the assumptions of modern functionalist ecology into environmental sociology. Their "New Ecological Paradigm" states that human actors are subject to the homeostatic checks and balances of nature, which eventually maintain species populations within carrying capacities and equilibrate ecological systems (Catton and Dunlap 1980: 34). They suggest that the interest in attaining a "sustainable society" which participates in a "climax community" in "steady state" falls within the proper subject matter of environmental sociology (Catton and Dunlap 1980: 36–37). Concerned primarily with the prescription of a new paradigm as the foundation of environmental sociology and human ecology, Catton and Dunlap fail to perceive that environmentalism is a highly complex and contradictory discourse, informed by powerful socially grounded metaphors. Environmental sociology would be enhanced by reflecting upon the sociological significance of the resurgence of organic metaphors and their often-confused co-existence with mechanical, economic, sociological and biological metaphors in contemporary Green discourses, including environmental sociology itself.

Organic and mechanical metaphors are appealing to human actors because they suggest the conceptualization of ecological and social systems as orderly, predictable and therefore manageable, processes. Bodies are invariant structures comprising of organs with physically necessary links. All life-forms are bounded by their bodies and act as whole bodies. Machines comprised of neatly co-ordinated parts designed to operate as integrated closed systems. As organs and machine parts do not act on their own, it may be warranted to speak of their functions and explain the parts in reference to the whole. While bodies are integrated and predictable systems, not all of their functions and processes are consciously controllable by human actors. Correspondingly, the organic metaphor implies that human beings lack encompassing control of social and ecological systems. By contrast, the mechanical metaphor presumes that social and ecological systems can be engineered like machines.

However, social systems are complex interactions among human individuals involved in variant social relations, and ecosystems are complex interactions among variant networks of individual bodies and geophysical phenomena. Life constitutes historically evolving practice rather than functional fit. The idea of a "function" in living systems is of limited explanatory value. Functionalism portrays relationships between actors and phenomena as necessary or inevitable, because functional links are constructed on the basis of observable interactions judged as useful. The historical emergence of specific relationships remains obscure. The utility of a phenomenon does not explain its being. Explanation requires specific historical analysis of activity among concrete actors and phenomena. Action theorists in the social sciences have long resisted the deceptive elegance of functionalist models, and their sociological critique is pertinent to ecological discourses:

> The contribution of the consequences of structures of action to the maintenance of a social system cannot by itself explain why these structures exist (McCarthy 1986: 203).

Therefore, Giddens has argued that function is an effect yet to be explained (1979: 113).

The composition of social and ecological systems and the links among their components are variant, flexible and subject to multiple concurrent factors of change. Each individual and phenomenon encounters the system as a changing set of conditions, and contributes to this changing character of the system by its actions. Action constitutes a "continuous flow of conduct" (Giddens 1979: 55) and events. Activity in society and in nature has systematic effects resulting from mixes of intended and unintended consequences of action (Giddens 1979: 56). These effects are regularized and networked by actors who are linked in time and space, and constitute composite conditions of action which are, to some degree, reliable: i.e., adaptations. The structure of a system is a set of resources (material, energy, information) upon which living actors draw in order to reproduce their bodies and their activities. Giddens's theory of structuration is relevant to ecological theory:

> The structural properties of social systems are both medium and outcome of the practices that constitute the system (Giddens 1979: 69).

By recursively participating in the continuous flow of conduct, actors render the system self-productive — i.e., autopoietic (Giddens 1979: 69). The reproduction and transformation of a system is the accomplishment of its participating components, not of the system.

Functionalism betrays an important logical inconsistency in dealing with "disequilibrating activity," in particular, the destructive effects of human action. Functionalist concepts are constructed from the vantage point of order and integration within undisturbed systems. The forces of conflict, rupture and disintegration are posited not within the system, but outside in its environment. They are defined as "external disturbances," such as "anthropogenic intervention" (Kucera 1973: 88; Emberlin 1983: 61; Goldsmith 1988: 174; Meyer 1989: 159–61, 172). Though functionalist ecology perceives nature as the victimized other, instead of as humanity's adversary, it continues to construct nature and society as discontinuous. Accordingly, human ecologists juxtapose two distinct hierarchies of natural and social systems which maintain input/output transfers (Marten and Saltman 1986; Gallopin, Gutman, and Maletta 1989; Rambo and Sajise 1984). Ecological traditionalists, such as Goldsmith (1988) or Shiva (1988), dichotomize social systems as, on the one hand, traditional social orders, grounded in naturalist religious ideologies and harmoniously integrated with nature, and, on the other, secular rationalist societies, whose technological capacities and economic systems obstruct nature's integrationist imperative. Meanwhile, scientific ecology identifies human population growth and the progress of technology as the causes of human disruption of ecosystems (Odum 1971: 510–16; Emberlin 1984: 262–81).

A striking, if sociologically unsophisticated, example of the rationalization of socially and ecologically destructive human behaviour as "functional," is Odum's discussion of "succession in human societies" (Odum 1989: 264–66). In a metaphorical confusion of organisms, societies and ecosystems, he transposes the succession model from ecosystems to human societies and assigns developmental trajectories leading from youth to adulthood, to both social and ecological systems. Thus, Odum depicts the genocide of native peoples, slavery and the industrial revolution as the actions of a conquering "pioneer society" which ruthlessly exploits both people and the environment. Odum is prepared to consider this "opportunistic" pioneer conduct necessary and justifiable, since high-productivity processes and the displacement of competitors are capable of rapidly producing economic wealth. As such, they mirror the high consumption levels of energy and nutrients, and the rapid production of biomass in ecological "pioneer" systems. However functional this "youthful" greed may have been, according to Odum, now it is time for human society to reach the mature stage, analogous to a climax ecosystem, in which environmental protection and human rights contribute to a new level of functional integration. In this scheme of things, even environmental destruction and tremendous human suffering can be integrated as functional.

Despite the emphatic claim of most ecologists that human actors are part of nature, to conceive of human action as functional for ecosystem integration meets with difficulties. If human actors are part of nature, social and ecological systems need to be placed on different levels of the same hierarchy. A social system is comprised of *intra*species relations among individuals of the human species, and an ecological system is comprised of *inter*species relations among individuals of a great number of different species. Therefore, social systems need to be situated *within* ecological systems, rather than in their environment, and analysed as to their "functions" in throughput relations of matter, energy and information exchange among systems components, to stay within the functionalist idiom. Faced with the destructiveness of human activity, this approach would, however, compel ecologists to recognize that maintaining an integrationist perspective is inconsistent. Instead, the complex interplay of destructive and integrative activity among ecosystem participants needs to be acknowledged as a "normal" feature, which may cause a system to develop towards a highly unsteady state. Human actors may prevent ecosystems from attaining climax equilibria and wear them down. This may or may not eventually prevent human beings from participating in particular ecosystems, or ultimately lead to human extinction. However, the time-scales involved may be too large, the multitude of concurrent factors of change too numerous, and the features of posthuman ecosystems too different, for us to ascertain the grip of "homeostatic checks" on human action, and to conceive of a process of systemic self-equilibration.

Sociologists concerned with ecological issues can learn much from the insights of "open-systems theory," which is increasingly influential in both ecology and the social sciences. This perspective proposes that ecological and social systems are "indeterminate" and volatile, exhibit "erratic, random qualities," are subject to chance and permanent change, and may persist in a state of non-equilibrium (O'Connor M. 1989: 49–51; O'Connor M. 1991: 111–17; Botkin 1990: 121–31; Bailey 1994: 123). If human actors can maintain a far-from-equilibrium state by importing resources from a system's environment, they must inevitably engage in resource conflict. For example, an agricultural community may engage in land-use practices and reach a level of population growth which causes the depletion of local water resources below the threshold of the community's viability. It may thus mobilize political and financial resources to build a public water-supply scheme which imports water from another area by means of a reservoir and pipes, and thereby continue its land-use practices. If human actors in the source area of the supply scheme fail to successfully contest the resource claims of the inhabitants of the target area, the source area may experience desiccation at a level which undermines local agricultural production and plant growth in riparian ecosystems. The outcomes of such conflicts are unpredictable, as they depend on a complex and dynamic set of actors who juggle the opportunities and constraints of their respective circumstances, and thereby affect processes of ongoing change (Starkloff 1994).

Ecosystems are but the relationships and interactions of their participants. They are temporally and spatially open, and thus volatile. They exhibit persistent change produced by multitudes of integrative and destructive effects. "Disequilibrating activity" is a matter of power relations. If a dominant group of actors asymmetrically controls and appropriates the resources of a co-dependent network, other participants may be constrained beyond their capacity to act, as conditions of production and the state of the system may significantly be altered. Yet, ecological systems evolve regardless of their specific composition, properties and patterns of activity, with and without human participation. Our current concern with sustainability signifies our worry about the human destruction of ecosystems which are capable of sustaining human livelihoods. Thus, human actors are motivated to form an image of necessary integration by which to control human activity.

The Problem of Control

Functionalist ecology shares with the adversarial cognitive model a fundamental concern with, and desire for, control. As, however, human beings, rather than nature, appear as aggressors who are "out-of-control," Green visions of the integration of human societies into ecological orders raise the issue of ecological totalitarianism. Eco-managerial and eco-traditionalist versions compete for the social recognition of a way of cognitively ordering reality and strive to impose this order on human actors.

The assumption that natural systems "design" themselves is central to functionalist-systems theory and reveals its practical purpose.

> The design of a natural ecosystem serves one function above all — the continued functioning of the ecosystem, its persistence (Marten and Saltman 1986: 34).

Tautological reasoning aside, the idea of design presupposes a designer, a being or a group of beings, who signify and select a particular pattern of how what is being designed ought to be. Design and its implementation necessitate power and co-ordination among social actors. These categories are inapplicable to the biosphere or other levels of ecological systems, which are complex systems of recursive participation of multiple actors of different species, unless one is prepared to assert that all these are under the control of an "almighty" being or a group of co-ordinated powerful beings, such as deities or ancestral spirits.

Eco-traditionalist Goldsmith explicitly proclaims a cosmology in which Gaia, the Greek nature goddess, appears as the supreme systems technologist. According to Goldsmith, traditional societies were guided by a knowledge and ritual practice which contains an unconscious cybernetic code of eco-functionalist instructions. These are grounded in the "great dawn" of ancestral origin and transmitted through myths, analogous to the genetically coded unconscious neurological control of organs in a body. Goldsmith has discerned "The Way," a cosmological principle and telos, binding for all life on earth, expressed in 67 "laws." Thereby, human actors would recreate "homeotelic society," a social order unified by its aspiration to integrate itself with nature, restore divine creation and serve their gods with zeal, lest they face the disastrous consequences of "maladjustment" (Goldsmith 1988: 163, 169, 183).

Be that as it may, the assumption of systemic self-design entails the projection of a capacity generic to actors to an authority beyond actors. Thus, a human cognitive construction is imbued with a suprahuman imperative force. However, as both the concept and the imperatives are human constructions, based on historically specific experiences and meanings, a design is limited in space and time. Designs are subject to conflict, to rupture and refusal. They are tools of power. A design is the anticipation of a co-ordinated effect, not the coming into being of a composite effect, a characteristic generic to systems. Thus, designs enter systems' evolution as factors. Neither do systems design themselves, nor can specific actors design the entire system, although they may desire to do so and invoke notions of systemic self-design on behalf of this desire.

This circumstance refers to the point of origin of systems theory. Bailey (1994: 127–30) has recently pointed out that its foremost rationale has always been the exercise of control. Giddens indicates that systems theory "has received a direct impetus from developments in modern technology":

> Systems technology does not just refer to computers, automated machines etc., but also to the incorporation of human beings, or their activities, within designed control systems (Giddens 1979: 74).

I should like to add that non-human actors and phenomena are incorporated into designed control systems as well. A scientific experiment, as much as an agricultural operation, can be imagined and implemented as a designed control system, where phenomena are brought within a designer's (scientist, farmer) cognitive and/or material control. In such pragmatic contexts, systems theory is a "potent ideological force" (Giddens 1979: 75). The perception of reality in terms of functionalist-systems theory is less of an analytical tool than a managerial device. Under conditions of socioecological crises and conflicts, an image of "necessary" integration is formed by which to motivate human actors to act according to its presumed imperatives. The forces of disorder, rupture and conflict contradicting integrative power are consequently designated as maladjusted, deviant, malignant, out of control and non-fit, and thus are to be reintegrated according to roles designed for them.

The "ecosystems approach to water management" of the United Nation's Economic Commission for Europe (ECE) is an instructive example of an eco-managerialist response to environmental crisis. The "Senior Advisors to ECE Governments on Environmental and Water Problems" assume that all forms of life and physical phenomena are "integrated into complex interactive systems" requiring "adaptability" and "long-term stability" in order to achieve "balance" or "equilibrium." Water is perceived as a central component of "one life-supporting system," into which human beings are "locked" (Tollan 1992: 29, 32; ECE 1989: 1, 4; ECE 1992: 4). "Holistic" "management from within" has the "function" of "maintaining ecological equilibrium in nature" (ECE 1989: 1, 6). Senior advisor Tollan expresses this framework in medical and organic metaphors:

> Our globe can be seen as an organism whose health depends on the health of all of its parts (Tollan 1992: 32).

The aspirations of ecosystems' management are formulated in the language of domination and control. Aided by a "centralised comprehensive data bank," the "surveillance" of ecosystems would be achieved by "monitoring systems" which trigger intervention through "corrective action" (ECE 1991: 8; ECE 1992: 7). The assignment of "ecosystems objectives" and the maintenance of "co-evolution conditions" for the purpose of "safeguarding the functional integrity of an aquatic ecosystem," necessarily rely on selective strategies of knowledge production. In a vast and volatile world, "uncertainty" is inevitable, and knowledge limited. Observation and measurement of "indicator species," which are broadly distributed in ecosystems and conveniently accessible to scientific treatment, as well as so-called "target variables" — i.e., cross-sections of species representing ecosystems — provide information for systems modelling, computer simulations, "simple overview models which capture the essence of complex systems" for daily management purposes, and ecological forecasting (ECE 1989: 14–20; ECE 1992: 6–11). Under these circumstances, the role of scientific experts is pivotal, as "expert judgement" (ECE 1992: 11) will have to bridge the gap between sampling procedures and management activities. The final word is the judgement of the same actors who have designed the system of control in the first place, and are thereby turning their cognitive model into a tool of social power.

The institutional set-up for the implementation of projects would be guided by national water master plans with an explicit ecosystems approach.

These translate into multipurpose water-management systems, integrating all institutions relevant to aquatic systems on regional and local levels (ECE 1989: 4, 8–12; ECE 1992: 13, 16). Mandatory environmental impact assessments prior to implementation of water-related projects are proposed (ECE 1989: 12–13; Tollan 1992: 31). The legal and educational systems of nation states and international law are to assimilate and promote the ecosystems approach (ECE 1989: 7, 21, 23; ECE 1991: 9; ECE 1992: 6–7).

The designs of expert culture, however, operate within the context of democratic polities in which social movements have initiated struggles over environmental concepts and practices. To integrate ecological movements, the ECE group has designed the requirement for "public participation" into its control system, which would mediate resource users and expert knowledge, and promote its translation into social practices through the formation of "voluntary organizations." From the perspective of the experts, their function is to provide "information on people's views and priorities" and to "engage the local population effectively in water-related environmental activities" (ECE 1989: 22). Public acceptance of the ecosystems' approach is fundamentally based on the inculcation of "the ecology frame of reference" and the translation of the complexities of ecosystems into "manageable facts" for public comprehension in order "to make valid judgements" (ECE 1989: 21; Tollan 1992: 29, 32). Cognitive and behavioural change among the public are considered crucial components in achieving "a state of equilibrium between economic activity and the pressure on the environment" (ECE 1989: 22).

The traditionalist version of totalitarian — i.e., all encompassing — design is merely communicated in a different idiom. The design of reality is assigned to powerful forces which oblige human actors to integrate themselves. Otherwise they will be punished, ultimately with annihilation, as these forces will sacrifice anything for their ultimate telos of order, integration and stability. Again, knowledge and proper action are difficult to discern. The communication with deities is not a common person's business. Therefore, the ritual expert and cosmologist is required to translate the "instructions" of the supreme beings to the commoners and design their behaviour according to her "intuition" and ritual tradition (Goldsmith 1988). Yet, the cosmologists are the ones who have designed the image of order (cosmology) in the first place.

The myths of the modern and traditional functionalists are self-legitimating symbolic systems. Their practical legitimacy would derive from their power to convince the human actors, whom they are intended to motivate, and from their capacity, to enable the actors operating within the horizon of the cognitive model, to achieve the objectives proclaimed. Faced with two competing totalitarian versions of coping with socioecological crises on our planet, the question arises: How much control is actually possible?

The Limits of Control

The "ecosystems' approach to water management" is an instance of what eco-Marxists have called the attempt of capital, mediated by the state, "to exercise more control or planning over production conditions" in order to deal with the self-induced impairment of the conditions of production (O'Connor J. 1989: 29, 23; O'Connor M. 1989: 35; O'Connor M. 1992: 1–2). By assigning commodity values and property status to fragments of nature (see Tollan 1992: 31–32; ECE 1992: 11; Panayotou 1992: 8–9), anything under the sun is potentially subject to the social control of capital. Martin O'Connor notes the symbolic manipulation involved. Since the natural conditions of capitalist production are not produced as commodities, they must be "codified as capital."

> "Exchange value" is instituted as a mode of coding and controlling social relations — the commodification and pricing in the market of all this hitherto external domain (O'Connor M. 1992: 2).

> Nature is conceived in the image of capital (O'Connor M. 1989: 35).

Planning and design are assigned the function of controlling the interface between nature and capital. The cognitive planning tool is functionalist-systems theory, by which the planners imagine to accomplish the transformation of 'disorder and conflict' into 'harmony and optimisation' (O'Connor M. 1992: 36). However, managerial control is constrained on three interrelated levels: the structural contradiction between industrial production and production in nature; social struggles over the appropriation and management of resources; and the indeterminacy of temporally- and spatially-open ecological systems.

Eco-Marxists argue that the "contradiction between capitalist production relations (and productive forces) and the conditions of production" (O'Connor J. 1989: 6) induces a "capitalist crisis of control" (O'Connor M. 1989: 51), which radically undercuts the symbolics of integration. As I indicated above, the eco-Marxist perspective subsumes production in nature under the capitalist mode of production as mere conditions of production. I suggest treating capitalistic production as a historically specific mode of human participation in productive activity in nature. I can thereby shift the analytic interest from a single-minded anthropocentric preoccupation with contradiction in capitalism, to the fundamental contradiction between capitalist production and production in nature. The concern for ultimately self-destructive tendencies of the capitalist mode of production is expanded to a concern with the destruction of the life-sustaining processes on Earth. Capitalist production depends on production in nature, but the rationales driving the two modes are irreconcilable.

The expanding appropriation and consumption of nature in capitalistic production processes, and the concomitant violation of relations of production in nature, result from structural pressures for increases in economic productivity. Competition in capitalistic markets and class conflict over the conditions of work (wages, length of the working day, organization of production) motivate ongoing productivity increases by means of technical reorganization of production processes (Marx 1973: 331–40). Rising productivity implies increases in commodity output per invested unit of variable and constant capital, or conversely lowers the production cost of a commodity. High productivity processes thereby increase the throughput of biomass (resource appropriation). At the same time, they motivate the externalization of costs in the form of socioecological damages (pollution, depletion, diseases, overproduction crises and unemployment). The internal logic of capital renders this process potentially ceaseless, as the socially valid standards of productivity level out, and new cycles of competition, class struggle and investment in new products and markets are set off. High productivity processes in the production of desired goods occur regardless of important social variations in free enterprise, state interventionist or state-managed systems.

For example, in agri- and silviculture, monocrops and high levels of agrochemical inputs afford an optimum output of commodities per units of time, space and investment at a given standard of productivity. High productivity systems select fast-growing early succession species, whose growth requires high levels of energy inputs, such as irrigation, fertilizer, fuel for machinery, pest control etc. (Odum 1971: 45, 267). Competition from undesirable life-forms, defined as pests and weeds, is limited or eliminated. Space, energy and water resources are dominated by a single species (crop) which is controlled and consumed by another single species (human beings). At the same time, these high-productivity processes have problematic environmental consequences, such as desiccation, nutrient depletion, soil erosion and destruction of genetic diversity. Agricultural production is, however, fundamentally dependent on the free availability of soils, water and genetic material. These cannot be produced by capital. They have been produced over long periods of time without concern for short-term optimum outputs. Ecologically diverse processes of production are not based on optimum productivity, but on flexible degrees of shared cycling of energy, nutrients and water in complex networks of interacting species and inorganic phenomena. No single species or group of individuals can appropriate an optimum of desired resources without simultaneously depriving others of their capacity to exist. Conversely, for a network of life-forms and phenomena to exist, their interactive requirements in terms of nutrition, energy, space and time must be met. If such a set of requirements is fundamental to the existence of the human species, the latter's optimum productivity processes can be operated only until they have consumed and impaired what ecological processes have built up in time spans and spaces not affected by capitalistic production. As soon as these resources are exhausted and their productive interactions eliminated, high-productivity-production processes are rendered defunct.

James O'Connor (1989: 23–24, 29) argues furthermore that the central contradiction between the relations and conditions of production is displaced into the spheres of the state and the public. The bureaucratization and politicization of the relationship between human actors and nature, on account of state intervention, induces the emergence of social movements contesting the "definition and use of production conditions." The relationship between the structural contradiction and social conflict requires to be theorized carefully. I suggest that the contradictory modes of capitalist production and production in nature are not symmetrically represented by particular social forces (i.e., capital and the state on one side, and social movements on the other). The fundamental contradiction is embodied by all social actors operating within the context of the capitalist mode of production. They may at the same time be victims and perpetrators of ecological destruction. While human actors are subject to the specific capitalist conditions of resource appropriation and conflict, they simultaneously require the continued viability of these resources. As long as livelihoods are dependent on the production of commodities, the economic survival of human actors and the survival of their bodies are in conflict. The contradiction culminates in a livelihood crisis when human economic activity destroys its ecological foundations.

The relationship between capital, the state and social movements cannot be inferred from structural theory. Ecology, as a concern of contemporary capitalist societies, is a contested terrain in which conflicting and shifting alliances of actors make knowledge claims and compete for the implementation of their projects. Capital does not act as "abstract capital," but constitutes a competitive social field in which the productive practices of one branch of production may cause the impairment of the conditions of production of another. The capitalist state is not a monolith, but another conflictual social field in which politicians, administrators and technical experts networked in national and global agencies, negotiate relations of relative autonomy and dependence among each other, as well as with capitalist producers and the public. The state as a producer of infrastructural commodities may cause the impairment of the conditions of production of other commodity producers. State regulation may therefore intensify, rather than displace, contradictions on the level of production. Social movements are not "class actors" unified by a common objective in a continuity of struggle. Social struggles may be confined to the particular spatial and temporal circumstances by which they are motivated, if they occur at all. The emergence of social movements in the event of a livelihood crises cannot be taken for granted. The preoccupation with struggles diverts the view from less extraordinary conflict situations among human actors. Spectacular confrontations may be a comparatively rare strategy. Empirical research would have to discern the actual constellations of conflict and co-operation among actors involved in ecological crises. Where conflict does occur, it infuses ecological systems with unpredictability which impacts the efforts of environmental managers, particularly if their designs are the subject of conflict.

Irrationality and failure of human knowledge and practices comprise a further social constraint of managerial control. The most careful design of control systems may be undermined by "crude and perfunctory reasoning, lack of discipline, incompetence and corruption" (Widstrand 1980: 65–72, 81–82) among administrators, experts, politicians, capitalists, environmental activists, farmers, workers etc. Recalcitrance, incompetence and parasitic behaviour may be calculated and intentional, and co-constitute the ecological system in which eco-managers intend to intervene. Rationally motivated resolution of crises and mediation of conflicts may not be attainable. Conflicts and failures may remain unregulated if the juggling of the status quo proves to be immune to recurrent waves of reform initiated by seemingly enlightened experts.

Furthermore, the importance of purely symbolic handling of ecological crises cannot be underestimated. An abundance of rhetoric, pretence, plans, designs and strategies, without necessarily producing practical consequences, supports the livelihoods of experts, administrators and activists participating in advisory boards, conferences, development agencies, government departments and non-governmental organizations. These actors may institutionalize inaction by creating mere talk.

Martin O'Connor (1989: 45, 52) has emphasized the systemic limits of human control on account of the volatility and complexity of interactions in spatially and temporally open ecosystems. Recurrence and order interlace with rupture and conflict. Time is both cyclical and linear, and space is possibly infinite. There are too many actors and phenomena at work for total system integration to be feasible. The systemic consequences of eco-managerial 'inputs' are not necessarily predictable in systems characterized by "indeterminacy" (O'Connor M. 1989: 49–51). Knowledge production and managerial intervention remain necessarily reductionist and partial. In the course of contradictory, contested and complex realities, ecosystems management can be expected to be patchy and discontinuous, a set of interactions among a multiplicity of others. Instead of assuming that nature and society can be controlled to achieve an ultimate state of integration, we may want to acknowledge that risk is an inevitable existential condition, for participation in nature is indeed a dance of life and death.

Yet, planning and design remain an inescapable responsibility, however partial and local the actual effects may be. Given the extent to which human actors are capable of transforming and impairing the biosphere, and the fact that somebody always does some planning, it is of great importance to understand who has access to discourses on knowledge and design, to decision-making processes, and to the implementation of human projects. Furthermore, we need to understand how the requirements for the survival of non-human actors and phenomena, who cannot participate in human discourses about their fates, are represented or repressed in human knowledge and planning. These are indeed political matters, because the institutions and actors involved in these tasks exert mechanisms of social closure, which contestants for participation may seek to breach. Democratization of technocratic processes has therefore been a central demand of ecological-social movements. However, democratization only offers a chance for better knowledge and action. It allows for access. There is no guarantee that democratized planning and project implementation would in fact realize ecologically sustainable human practices.

DISCUSSION QUESTIONS

1. Do you think nature can be adequately characterized as either an adversary or a victim of human actors?
2. What are the material and emotional motivations for perceiving and treating nature as an adversary?
3. Why is functionalist-systems theory attractive to eco-managerialist and eco-traditionalist perspectives?
4. Discuss the significance and problems of the use of metaphors in ecological discourses.
5. Which social and ecological factors constrain the managerial control of nature?
6. Discuss how systems' theoretical and eco-Marxist approaches (in particular the concepts of "open systems" and "conditions of production") are used to suggest a new synthesis in the analysis of the nature/society interface, which avoids the shortcomings of adversarial and integrationist cognitive models.

References

Bailey, Kenneth D. 1994. *Sociology and the New Systems Theory: Toward a Theoretical Synthesis.* Albany, N.Y.: SUNY Press.

Berger, Johannes. 1991. The Linguistification of the Sacred and the Delinguistification of the Economy. In *Communicative Action: Essays on Jürgen Habermas's The Theory of Communicative Action,* ed. Axel Honneth and H. Joas, 165–80. Cambridge, Mass: MIT Press.

Bittner, Egon. 1983. Technique and the Conduct of Life. *Social Problems* 30: 249–61.

Botkin, Daniel B. 1990. *Discordant Harmonies: A New Ecology for the Twenty-First Century.* New York: Oxford University Press.

Catton, W.R. 1994. Foundations of Human Ecology. *Sociological Perspectives* 37: 75–95.
Catton, W.R. and R.E. Dunlap. 1980. A New Ecological Paradigm for Post-Exuberant Sociology. *American Behavioral Scientist* 24: 15–47
Collins, Jane. 1992. Marxism Confronts the Environment: Labor, Ecology and Environmental Change. In *Understanding Economic Process*, ed. by S. Ortiz and S. Lees, 179–88. Lanham, Md.: University Press of America.
Economic Commission For Europe (ECE). 1989. *Ecosystems Approach to Water Management*. ENVWA/WP.3/R.7/Rev.1
———. 1991. *Report of the Seminar on Ecosystems Approach to Water Management*. ENVWA/SEM.5/3.
———. 1992. *Guidelines on the Ecosystems Approach in Water Management* (Preliminary draft). ENVWA/WP.3/R.28.
Emberlin, J.C. 1984. *Introduction to Ecology*. Estover, Eng.: MacDonald and Evans.
Falkenmark, M. and T. Chapman, eds. 1989. *Comparative Hydrology*. Paris: UNESCO.
Fellman, Gordon. 1989. *Peace in the World or the World in Pieces*. Jerusalem: The Hebrew University.
Gallopin G., P. Gutman and H. Maletta. 1989. Global Impoverishment, Sustainable Development and the Environment: A Conceptual Approach. *International Social Science Journal* 121: 375–97.
Giddens, Anthony. 1979. *Central Problems in Social Theory*. Berkeley: University of California Press.
———. 1990. *The Consequences of Modernity*. Stanford, Calif.: Stanford University Press.
Goldsmith, Edward. 1988. The Way: An ecological World-View. *The Ecologist* 18: 160–85.
Habermas, Jürgen. 1971. *Erkenntnis und Interesse*. Frankfurt am Main: Suhrkamp.
———. 1983. *Moralbewusstsein und kommunikatives Handeln*. Frankfurt am Main: Suhrkamp.
———. 1987. *Theorie des kommunikativen Handelns*. 2 vols. Frankfurt am Main: Suhrkamp.
Kucera, Clair L. 1973. *The Challenge of Ecology*. Saint Louis, Mo.: Mosby.
Lovelock, James. 1979. *Gaia: A New Look at Life on Earth*. Oxford: Oxford University Press.
———. 1988. *The Ages of Gaia: A Biography of Our Living Planet*. New York: Norton.
Marten, G.G. and D.M. Saltman. 1986. The Human Ecology Perspective. In *Traditional Agriculture in Southeast Asia: A Human Ecology Perspective*, ed. G.G. Marten, 20–53. Boulder, Colo.: Westview Press.
Marx, Karl. 1973. *Das Kapital*. Vol. 1. Berlin: Dietz Verlag.
———. 1978. Capital. Vol. 3. In *The Marx-Engels Reader*, ed. R.C. Tucker, 439–42. New York: Norton.
Marx, K. and F. Engels. 1969. *Die deutsche Ideologie*. MEW 3. Berlin: Dietz Verlag.
McCarthy, Thomas. 1986. Komplexität und Demokratie — die Versuchungen der Systemtheorie. In *Kommunikatives Handeln*, ed. A. Honneth and H. Joas. Frankfurt: Suhrkamp.
Merton, Robert K. 1967. *Social Theory and Social Structure*. New York: Free Press.
Meyer, Peter. 1989. *Waldfragmente im Uva-Becken/Sri Lanka: Eine Analyse der Artenzusammensetzung und Bestandesstrukturen sowie ihrer Bestimmungsgründe*. Diplomarbeit der Forstwirtschaftlichen Fakultät der Georg-August-Universität Göttingen.
O'Connor, James. 1989. Capitalism, Nature, Socialism: A Theoretical Introduction. *Capitalism: Nature: Socialism* 1: 11–38.
O'Connor, Martin. 1989. Co-dependency and Indeterminacy: A Critique of the Theory of Production. *Capitalism: Nature: Socialism* 3: 33–57.
———. 1991. Entropy, Structure, and Organisational Change. *Ecological Economics* 3: 95–122.
———. 1992. *The System of Capitalised Nature*. Unpublished manuscript.
Obeyesekere, Gananath. 1987. *The Cult of the Goddess Pattini*. Delhi: Motilal Banarsidass.
Odum, Eugene. 1971. *Fundamentals of Ecology*. Philadelphia: Saunders.
———. 1989. *Ecology and Our Endangered Life-Support Systems*. Sunderland: Sinauer Associates.

Panayotou, Theodore. 1992. *The Integration of Economy and Ecology in Sustainable Development and Its Implications for Policy*. Keynote address for the conference on The role of environmental economics in national development, Colombo.

Rambo, A.T. and P.E. Sajise. 1984. Introduction: Human Ecology Research in Tropical Agriculture in Southeast Asia. In *An Introduction to Human Ecology Research on Agricultural Systems in Southeast Asia*, ed. A.T. Rambo and P.E. Sajise, 1–24. Laguna: University of the Philippines.

Rao, Brinda. 1989. Struggling for Production Conditions and Producing Conditions for Emancipation: Women and Water in Rural Maharashtra. *Capitalism: Nature: Socialism* 2: 65–82.

Schmidt, Alfred. 1978. *Der Begriff der Natur in der Lehre von Marx*. Frankfurt: Europäische Verlagsanstalt.

Shiva, Vandana. 1988. *Staying Alive: Women, Ecology and Development*. London: Zed.

Starkloff, Ralf. 1989. *Eco-power Communications*. Unpublished manuscript, Brandeis University.

———. 1994. 'No water!' Socio-ecological Transformations in an Uva Basin Village in Sri Lanka. Ph.D. thesis, Brandeis University.

Tollan, Arne. 1992. The Ecosystem Approach to Water Management. *World Meteorological Organisation Bulletin* 41: 28–34.

Whittaker, Robert H. 1970. *Communities and Ecosystems*. London: Macmillan.

Widstrand, Carl. 1980. *Water Conflicts and Research Priorities*. Oxford: Pergamon Press.

Worster, Donald. 1977. *Nature's Economy: The Roots of Ecology*. San Francisco: Sierra Club Books.

TAKING STOCK OF TWO DECADES OF RESEARCH ON THE SOCIAL BASES OF ENVIRONMENTAL CONCERN

Allan Greenbaum

York University

INTRODUCTION

This chapter sets out to describe and analyse the results of a body of research concerned with the relationship between individuals' social characteristics (understood in terms of discrete socio-demographic variables) and their level of environmental concern (understood in terms of a variety of attitudinal and behavioural measures). This research is interested in finding the answers to the following questions. First, how does environmental concern vary depending on factors such as age, gender, race, class, socio-economic status, community size, political orientation, religious beliefs and so on? Second, how does the answer to the first question help us to explain why people think and feel as they do about "environmental issues?" What does it tell us about the nature and likely direction of environmentalism (and anti-environmentalism) as social movements? The means employed to address these questions in the literature under consideration here are those characteristic of post-World-War-II mainstream American social science: chiefly, the statistical analysis of large-sample survey data.

The literature of the socio-demographic research on "environmental" attitudes and behaviours is important for several reasons. In the first place, this work represents the application, to topics of interest to environmental sociologists and others who study environmental issues, of powerful and well-developed research methods. Although these methods — like any others

— have limitations, they do shed light on various aspects of the complex and multifaceted problem of understanding environmental concern and environmental conflict. Their application, moreover, constitutes a very significant current in the gathering stream of environmental sociology during the period under consideration. In terms of the sheer volume of work produced, it is one of the major contributions of sociologists, social psychologists and empirical political scientists to the environmental field. It is also a major contribution because of its inherent interest, and because it allows us to explore questions that naturally arise in the course of sociological study of environmental issues. Some of these questions can be explored in no other way. Consider, for example, the perennially vexed question of whether (and to what extent) the values, attitudes and priorities of the largely well-educated, upper-middle-class leadership of "mainstream" environmental organizations are reflective of those of the "general public" (Morrison and Dunlap 1986; Bosso 1994).

Before turning to a review of the results of research on the social bases of environmental concern, it is necessary to provide some brief comments on the theoretical context of this research. A detailed discussion of these theoretical issues is beyond the scope of this chapter; my intention here is only to provide the reader with sufficient background to understand some of the more salient questions and assumptions informing the research discussed.

THEORIZING THE SOCIAL BASES OF ENVIRONMENTAL CONCERN

Environmental sociology has yet to reach the stage where it would be possible to point to a fixed set of widely recognized competing theories regarding the social bases of environmental concern. This is not because there has been a shortage of activity in the area; indeed, an adequate analysis of that activity is beyond the ambit of this chapter. Rather, the activity has been diffused both in terms of the aspects of "social basis" or "environmental concern" under consideration, and in terms of the level of theoretical generality. While some theorists have attempted to develop high-level accounts of the origins and constituencies of environmentalism and other "new social movements," others have concerned themselves with lower-level hypotheses about the connections between particular social variables and particular environmental concerns. Theoretical issues relevant to a particular variable will be discussed later, together with the empirical findings regarding that particular variable. In this section, I will confine my comments to a few issues that cut across specific variables.

The explanatory question that faces us is that of why (some) people come to perceive and be concerned about "environmental problems" (leaving aside for now the question of how the latter is to be defined). Possible answers to this question can be arrayed along a continuum of kinds of explanations, with what might be called a "naïve environmental" position at one end, and what might be called an "extreme relativist" position at the other end. Towards the environmental side of the spectrum, environmental concern would be seen as an effectively unmediated response to objectively given environmental harms and risks. In an extreme form of this perspective, concern would be seen as a direct and automatic consequence of learning about these objectively given problems, either through personal experience, the media or through formal education.

As we move towards the other end of the continuum, there is an increasing emphasis on the ways in which knowledge is mediated by the knower's attributes or is constructed by the knower. There is a focus on the ways that beliefs and concerns about the environment are affected by expectations, biases, acquired and innate dispositions, schemas, paradigms, ideologies, world-views and so on. At the relativist end, we may distinguish between subjectivist approaches that focus on the psychological attributes of individuals, and what are referred to as "social-constructionist" approaches that focus on the collective creation of ideas about the world. I will confine my attention to the latter, as it is the version associated with sociology. Social-constructionist approaches have deep roots in the sociological tradition, drawing on the sociology of knowledge and ultimately on the "classical" sociological schools: from the Marxian analysis of class consciousness and ideology, from the Weberian notion of "elective affinities" between world-views and ways of life, and from the Durkheimian exploration of the social basis of concepts and classifications.

Intermediate between environmental and social-constructionist approaches are what we may call "interest-based" ones. According to interest-based approaches, how concerned one will be about a particular environmental problem, or whether one will perceive the environmental effect as a "problem" at all, will depend on where one stands with respect to the benefits and the costs or risks created by the activity in question. For example, Raymond Murphy (1994: chap. 8), drawing on the work of Perrow (1984), distinguishes several "environmental classes" on the basis of whether their members are chiefly beneficiaries of environmentally harmful activities, victims who also benefit as producers or consumers, and victims who do not benefit at all. Like the constructionist approach, this analysis assumes that perception and concern is mediated and socially situated. Like the environmental approach, it assumes that risks and benefits are objectively given and at least partially knowable.

By contrast, social constructionists like Mary Douglas and Aaron Wildavsky (1982; Douglas 1992) suggest that while risks may be real, they are impossible to assess objectively. They suggest that the risks we choose to emphasize are "selected" in accordance with our "cultural bias." To put the comparison crudely, for the interest approach, whether I think industrial pollution (say) is a serious problem depends on whether or not I am a victim; for (at least some versions of) the constructionist approach, whether I perceive myself to be a victim depends on whether I am socially and culturally disposed to see industrial pollution as a serious problem.

It is convenient, then, to contrast "interest based" theories of environmental concern with "cultural" theories. Though I have up until now treated these theories as resting on mutually incompatible premises regarding the social construction of environmental harm, they may also be regarded as complimentary.

That is, some differences in environmental concerns may be better explained by differences in interests, while others may be better explained by differences in culture, and the issue between them is an empirical one.

A relatively simple example of a cultural theory, and one that has been most influential in the literature under consideration in this chapter, is that developed by the American political scientist Ronald Inglehart. Inglehart's theory draws heavily on the thought of the psychologist Abraham Maslow, who described human needs as forming a hierarchy with basic survival and security needs at the bottom, and the need for self-realization at the apex. Higher-order needs emerge as lower-order needs are satisfied. Inglehart (1990) translates this model from the arena of personal wants to that of political priorities and cultural values. Inglehart, and those who make use of his method, ask survey respondents to rate or rank the priority of various social goals. Some of the goals, such as economic growth, national defence and crime control are intended to reflect lower-level material security needs, and are referred to as "materialist" goals. Others, such as free speech, having more say in government and at work, or a society where ideas are more important than money, are intended to reflect higher-level needs, and are referred to as "postmaterialist" goals. Depending on the kind of goals to which they tend to assign greatest importance, individuals may be categorized as being "materialist" or "postmaterialist" in their value structures. Inglehart suggests that individuals raised in circumstances of material security are more likely to become postmaterialist.

Survey data show postmaterialism to be associated with support for environmentalism in both Western Europe and North America (Cotgrove 1982: 29; Inglehart 1990: 383–88; Rohrschneider 1990; Tindall 1994). It appears that concerns about the balance of nature, the limits to growth and the domination of nature, are more strongly linked to degree of postmaterialism among Americans than they are among Canadians (Steger et al. 1989) or Japanese (Pierce et al. 1986, 1987a, 1987b). It is also likely that not all *kinds* of environmental concerns are related to postmaterialism — concerns about direct environmental threats to one's health or material well-being are very much "material" concerns in Inglehart's terms.

Research Findings

The results of recent empirical research on the social bases of environmental concern is reviewed below. The discussion is organized in terms of the main socio-demographic variables by means of which "social bases" have been characterized: age, education, gender, race, socio-economic status, occupation and sector, community size, religion. Although in one sense this is arbitrary, it is convenient because relatively little attention has yet been paid to the effects of interactions among these variables on environmental concern.

Age

With a very few exceptions (Derksen and Gartrell 1993; Woodrum and Hoban 1994), virtually all studies have found that environmental concern varies with age. As one major study concluded, "age is the strongest and most consistent predictor of environmental concern" (Mohai and Twight 1987). On the whole, younger people express higher levels of environmental concern than older people do (Van Liere and Dunlap 1980, 1981; Hamilton 1985a, 1985b; Rohrschneider 1990; Inglehart 1990; Olsen, Lodwick, and Dunlap 1992; Filson 1993). Jones and Dunlap (1992) find this tendency to hold in the United States in every year between 1973 and 1990, though the effect is strongest from 1977 to 1982, and gets weaker in the late 1980s. American studies have tended to find that, in particular, younger people are less likely to believe that humans ought to dominate and exploit non-human nature and non-human beings (Kellert 1984a; Kanagy and Willits 1993; Nibert 1994).

Why should environmental concern be related to age? Before addressing this question, it is necessary to say a bit more about how sociologists address the effect of age in general. Let us start by distinguishing between what are called "life-cycle effects" and "cohort effects." Life-cycle effects are the effects of being at a certain stage in one's life; cohort effects are the effects of belonging to a certain generation. Suppose, for example, a survey found a certain view to be characteristic of middle-aged people. We would not be able to tell on the basis of that survey alone whether this view comes from having reached mid-life, or whether it is characteristic of the particular generation that happens to be middle-aged at the time the survey was conducted. Life-cycle and cohort effects can only be separated by means of a series of studies over a period of decades.

Note that in terms of both life-cycle and cohort effects, age is best understood as the qualitative variable "belonging to a certain age bracket," rather than as the quantitative variable "being older or younger." Unfortunately, many studies are designed or analysed so as to provide information only on the effect of being older or younger, not on the effect of belonging to a certain age group. This is unfortunate because such studies cannot, for example, tell us if environmental concern rises to a peak at a certain age and then declines. We can, however, draw inferences from a series of such studies. The pattern reported by Jones and Dunlap (1992), mentioned above, might suggest that in addition to a constant life-cycle effect producing greater concern in the young, a particularly concerned cohort entered young adulthood in the late '70s and early '80s. Subsequent analysis of these data by Kanagy and his colleagues (1994) indicates that the increase in support for environmental protection during the 1980s was due *both* to the replacement in the population of older by younger, more environmentally concerned cohorts, and an increase of expressed support *among* the members of each cohort.

We are now ready to consider some theories about why the age effect exists, before returning our attention to the empirical details. Some theories focus on life-cycle effects. In Murphy's (1994: 185) interest-based theory, the young are a relatively victimized environmental class, as they constitute "the closest approximation" of future generations — the ultimate innocent victims of environmental despoliation. Other life-cycle theories take a social-constructionist tack: "Youth ... has become a symbolic definition.... [One is young because] one assumes cultural characteristics of variability and temporariness proper to youth" (Melucci 1985: 811). In the context of a European analysis that sees the ecology movement as part of a larger resistance to the domination of life by technocratic rationality, the social meaning of youth as freedom and spontaneity would come to be linked with that of ecology. Kanagy et al. (1994) conjectured that young people would have little to lose from increased spending on social problems generally because they typically have little income to tax (a life-cycle effect); they inferred that because the effect of age on support for environmental spending remained strong even when holding overall support for spending constant, the age effect is due in large part to cohort effects specific to environmental concern, and not merely to youthful generosity with public funds.

The best developed and most influential theory about age and environmentalism is Inglehart's (1990) cohort theory. Recall that Inglehart argues that environmentalism is linked to "postmaterialist" values, and that these values tend to result from early adolescence spent in circumstances of economic security. As the standard of living has increased over the past century, every generation has been more postmaterialist than the last. If his explanation is sound, however, we might expect a downturn in postmaterialist values in that cohort, born after the early 1960s, whose early adolescence coincided with the period of increased economic instability, increased unemployment and declining real wages that commenced in the mid to late 1970s. Consistent with this hypothesis, those respondents to a 1988 Canadian national survey born after 1964 were the *least* likely of any age group to say that environmental protection is more important than job creation (Bakvis and Nevitte 1992).

The available empirical literature suggests that, at least in North America and Australia, expressed environmental concern (as distinct from environmentally concerned behaviour) seems to peak in the cohorts born between 1942 and 1964 (George and Southwell 1986; Mohai and Twight 1987; MacDermid and Stevenson 1991; Blaikie 1992; Bakvis and Nevitte 1992; Simmons and Stark 1993). This is consistent with the prediction made above. Unlike reported concern, reported personal behaviour such as recycling, tends either to increase with age, or to peak at a later age (Van Liere and Dunlap 1981; MacDermid and Stevenson 1991; Kanagy and Willits 1993; Derksen and Gartrell 1993). It may be that these behaviours are influenced by frugal "waste-not" values formed under conditions of relative scarcity, and would therefore not be associated with postmaterialism. The relationship between age and activism or participation in the environmental movement is more complicated. Different kinds of involvement seem to be associated with different age profiles. For example, the membership of nature-preservation organizations tends to be disproportionately drawn from pre-World-War-II cohorts, while members of less conservative environmental groups, and the most active nature preservationists, tend to be younger (Cotgrove 1982; Tindall 1994). Since environmental activists tend to be individuals with resources to spend on activism (see section on socio-economic status), one might expect them to be disproportionately drawn from those in mid-life (Mohai and Twight 1987).

Though Inglehart (1990; see also Rohrschneider 1990) drew much of his data from Europe, there is some indication in the literature that the youth effect is less marked in England and Germany than it is in North America. Milbrath (1984) found that in the U.S., those who favoured technological solutions to environment problems tended to be older than those who favoured social solutions. On the other hand, he found that older Germans were more likely than younger ones to believe in limits to growth (see also Schahn and Holzer 1990), and that environmental concerns were more or less evenly distributed across age groups in England. Eiser et al. (1988) found that the oldest residents were most opposed, and the youngest least opposed, to proposed local nuclear power plants in England; this pattern is unlike that found in the U.S. regarding nuclear power (George and Southwell 1986; Dunlap and Olsen 1984).

Education

The empirical literature on the relation of education to environmental concern is ambiguous and inconclusive. A majority of studies find that expressions or indications of environmental concern increase with education (Van Liere and Dunlap 1980; Inglehart 1990; Arcury and Christianson 1990; Schahn and Holzer 1990; Mohai 1990; MacDermid and Stevenson 1991; Olsen, Lodwick, and Dunlap 1992; Kowalewski and Porter 1992; Kanagy et al. 1994; Tindall 1994). Bakviis and Nevitte (1992) report that, controlling for other variables, education is the best demographic predictor of preference for environmental protection over job creation among Canadians; they also found that better-educated people were more likely to name the environment as the most important political issue. George and Southwell (1986) found that both general education and specific knowledge about the issue were positively related to opposition to nuclear power plants. Jones and Dunlap (1992) report that in the late 1980s the relation between education and support for spending on environmental problems became stronger than it had previously been.

There are, however, contradictory indications. Van Liere and Dunlap (1981) found that education was related to level of concern, but not to level of personal behaviour; Derksen and Gartrell (1993), who studied recycling in Alberta, found just the opposite. Kellert (1984a) found that those with lower levels of education had more negative attitudes towards animals, and tended to take greater satisfaction in mastery and control, than did those with higher levels of education; those with less education were also less likely to indicate affection for individual animals, or to regard cruelty to animals as morally wrong. On the other hand, Nibert (1994) found that educational attainment had no effect on support for animal rights, and Woodrum and Hoban (1994) found it made no difference to belief in human mastery over nature. In some cases, statistical analyses suggest that education has only an indirect effect on environmental attitudes, by way of its effect on socio-economic status or religious beliefs (Hamilton 1985a; Kanagy and Willits 1993).

It is possible that the effect of education is not a simple linear one. Milbrath (1984), who analysed the results of two American surveys, an English and a German one, found that in all four samples, both the "vanguard" (those who thought environmental problems were serious, who favoured social solutions, and who believed there are limits to growth) and the "rearguard" (those who thought environmental problems were not serious, who favour technological solutions and who did not believe there are limits to growth) tended to be better educated, than those who subscribed to other permutations of these beliefs. Similarly, Olsen, Lodwick, and Dunlap (1992) found that the proportion both of those who fully accepted, and those who fully rejected the attitude cluster they labelled the "technological paradigm" (which involves faith in technology and support for the domination of nature), was highest among the best educated of those surveyed. This polarization may be a manifestation of a more general tendency of the better educated to display greater ideological consistency in their responses to survey questions (see the discussion of "constraint" in Pierce et al. 1987b).

Gender

Although social scientists have amassed a sizeable body of empirical data on gender differences in attitudes about topics connected with environmental concern, there has been surprisingly little specifically sociological theorizing on whether and why any such differences exist. In the broader literature of social and environmental thought, several currents have developed since the 1970s under the rubric of ecofeminism, but these theoretical explorations have tended to be concerned with the cultural, historical and philosophical links between patriarchy and the destructive exploitation of nature, and thus the links between the feminist and ecology movements. Ecofeminist thought has been, on the whole, only indirectly relevant to questions relating to the empirical relation between gender and environmentalism.

Drawing on these indirect implications, it is possible to generate several theories of gender difference in environmental concern, reflecting corresponding tendencies in ecofeminist thought. These theories may be grouped into three categories, which may be labelled as "biological," "psycho-cultural" and "social," respectively. On the biological view, men's and women's experiences differ because their bodies differ biologically, and their attitudes reflect these different experiences. On one version of this approach, the experience of gestation gives mothers an embodied sense of connection to future generations, while menstruation gives all women a similar sense of connection to the cyclicity of nature. By contrast, "the male bodily experiences is quite lacking in experiences [of weakness, limit or intrinsic connection to natural processes] which would help him forge an adequate ... relation to nature or concept of limits" (Gray 1979: 110). One might predict, on the basis of such a theory, that men would be less concerned about disruption of natural processes, less concerned about future generations, and less likely to recognize limits to growth than women would, and that this difference would hold across differences in culture, class and context.

On one influential version of the psycho-cultural approach, gender differences arise not out of innate bodily experience, but out of the formation of masculine and feminine consciousness under patriarchy. Theorists differ on the details of how this difference comes about, but tend to agree that in patriarchal cultures, masculine identity is bound up with the ideals of mastery, domination, detachment and (at least in some such cultures) with those of abstract, instrumental and universalizing modes of reasoning (Gray 1979; Gaard 1993). The corresponding feminine identity is bound up with particularity, connection and an "ethic of care" (as opposed to a putatively masculine ethic of justice and individual rights) (Gilligan 1982). Such an approach might predict that men would be more sanguine than women about science, technology and bureaucratic or technocratic management; it might suggest that women would be more concerned than men about environmental issues involving suffering or harm inflicted on particular human or non-human others, though perhaps not more concerned about environmental issues that are harder to conceive in such terms.

The social approach tends to attribute the differences between men's and women's knowledges, beliefs, attitudes and values regarding the natural environment to the different places assigned to them in the social division of productive and reproductive labour. The different attitudes of men and women would reflect the different experiences, competencies, interests and dispositions that come from performing (and being socialized to perform) these different roles (see, for example, Mies and Shiva 1993: chap. 11). This approach tends to give relatively greater emphasis to broadly "economic" factors, than do the other two. Thus, it would tend to predict that as these economic relations vary from one country, region, class and time period to another, so gender differences in environmental concern would vary accordingly.

The pattern of findings in the empirical literature is complex and by no means unequivocal. Studies from the 1970s reviewed by Van Liere and Dunlap (1980) did not consistently find major differences between men and women in their level of environmental concern. In a number of subsequent studies, women were shown to have higher levels of environmental concern than men (Van Liere and Dunlap 1981; McStay and Dunlap 1983; Fortmann and Kusel 1990; Schahn and Holzer 1990; Blaikie 1992). Milbrath (1984) found the environmentalistic "vanguard" to be among the most female dominated — and the anti-environmentalistic "rearguard" consistently the most male dominated — of all environmental attitude categories in each of the three countries studied. Blaikie's (1992) results suggest that the gap is smallest in the cohort in which overall environmental concern peaks — those born between 1954 and 1964 — but is wider among those born before and after that period.

Not all research from the 1980s and 1990s tends to support the simple formula that "women care more about the environment," however. The pattern is more complex and ambiguous than that. Derksen and Gartrell (1993) found no difference between male and female Alberta respondents in their responses to a very general and abstract question about their level of concern for the

Earth's environment. The gender gap also appears to be small or non-existent on measures of what might be called "ecological" concern. Surveys in Kentucky and Pennsylvania found no difference between male and female scores, on a scale intended to capture beliefs and concerns about the balance of nature, limits to growth, and human domination and exploitation of non-human nature (Arcury and Christianson 1990; Kanagy and Willits 1993). Gender made no difference among Washington state survey respondents with respect to scores on "ecological belief" and "ecological value," scales which together capture much the same area of concern (Olsen, Lodwick, and Dunlap 1992).

Little (Blaikie 1992) or no (Woodrum and Hoban 1994) gender difference seems to exist about whether humans were meant to rule over nature. There appears to be little or no gender difference regarding the ecological valuation of wildlife or its protection (Kellert 1984a, 1990; but compare Kuitunen and Tormala 1994). Though Fortmann and Kusel (1990) found women to be more "pro-environmental" and less "pro-commodity" on National Forest management issues in the U.S. southwest, Tindall (1994) found men and women equally represented in the Vancouver Island rain forest preservation movement (because some male-dominated occupations were overrepresented, and some female-dominated ones underrepresented among the membership; however, women in non-traditional occupations may have been overrepresented). Outside of the environmental movement, Washington State women seemed to be less concerned about overpopulation than men (McStay and Dunlap 1983).

There is a much more marked difference between men and women in their levels of concern about pollution and support for environmental regulation (McStay and Dunlap 1983). These differences may obtain particularly with regard to local issues (Blocker and Eckberg 1989). As Krauss (1993) points out, the leadership of grass-roots activism against toxic wastes tends to be predominantly female. Among Ontario farmers, women are less likely to favour the use of chemical fertilizers and pesticides (Filson 1993). What would account for these differences? The pattern of evidence suggests that they may be the result of a combination of two factors. The first has to do with gendered differences in the relative evaluation of environmental health and economic prosperity. The second has to do with attitudes towards science and technology. Let us start with the first factor.

Looking only at aggregate data, the gender gap appears to be modest on the issue of the trade-off between economic and environmental values. In Australia, Blaikie (1992) found the smallest gender differences on these issues. Washington State studies (Van Liere and Dunlap 1981; McStay and Dunlap 1983) found some of the smallest differences on environmental spending items, and in annual U.S. national polls, women have been more supportive than men of spending on environmental problems in fewer than half the years between 1973 and 1990 (Jones and Dunlap 1992). Averaged over the 1980s, there was virtually no direct effect of gender on support for spending on environmental problems (Kanagy et al. 1994). Among respondents to a 1988 Canadian national survey, men were somewhat more likely than women to agree that environmental protection is more important than job creation (Bakvis and Nevitte 1992). However, there is — as will be discussed below — reason to believe that such aggregate data conceals as much as it reveals.

Stout-Wiegand and Trent (1983) surveyed residents of a coal-producing region in West Virginia about their attitudes toward environmentally controversial proposed coal-related energy developments. They found that there was no difference in support for the developments between male and female respondents who were generally unconcerned about the environment, and among those who did not work in the energy sector. Being environmentally concerned made women, but not men, *less* likely to support the developments; being employed in the energy sector made men, but not women, *more* likely to support the developments. Why might environmental concern affect female but not male attitudes towards development? Two possibilities come to mind. The first is that the male respondents were less likely to believe that the developments posed a serious environmental threat. There is some evidence to the effect that male-female differences in environmental concern are related more to differences in perception of environmental threat, than to differences in value structures (Stern, Dietz, and Kalof 1993). The other possibility is that the men weighed direct economic concerns more heavily than the women did. The latter possibility seems consistent with the finding that employment in the energy sector increased male but not female support for the developments.

More light is thrown on this issue by studies that examine the effects of parenthood on men's and women's environmental concerns. Hamilton (1985a, 1985b), who studied toxic contamination issues in three New England towns, consistently found that mothers of young children exhibited the highest levels of concern and action. Fathers tended to be more active and concerned than other males, but much less so than mothers. Results of some other American studies suggest that not only are mothers less inclined than other women to support nuclear energy, or to regard environmental protection as an economic burden, fathers are *more* inclined than other males to support nuclear energy, and regard environmental protection as an economic burden (George and Southwell 1986; Blocker and Eckberg 1989).

Though more research needs to be conducted on the interaction of parenthood, occupation and gender on environmental concern, the results of these studies are suggestive. A possible explanation is as follows. Mothers are socially assigned primary responsibility for the health and physical well-being — and fathers, the economic well-being — of their families. Insofar as an environmental conflict is constructed in terms of health risk versus economic risk, mothers' and fathers' anxieties about the discharge of their respective parental responsibilities will drive them in opposite directions with respect to the issue. On this account, mothers and fathers do not necessarily experience different degrees of connection with, and responsibility for, their offspring; rather, the socially constructed nature of the connection differs.

The second factor that appears to shape male-female differences in environmental concern has to do with differing attitudes toward science and technology. Surveys consistently reveal that women are less confident in, and more fearful and critical of, science and technology than are men — see the findings of Blaikie (1992) in Australia; Olsen, Lodwick, and Dunlap (1992) in the U.S.; and MacDermid and Stevenson (1991) in Canada. Milbrath (1984) found that, especially in England and the U.S., those who preferred technological to social solutions for environmental problems, were disproportionately male. There is some evidence that males tend to have somewhat higher levels of scientific knowledge about wildlife (Kellert 1984b, 1990) and about environmental issues generally (Schahn and Holzer 1990); however, the latter study found that the level of general scientific knowledge — as opposed to concrete practical knowledge — about environmental matters was unrelated to environmentally responsible behaviour.

Fox and Firebaugh's (1992) analysis of U.S. national opinion-poll data reveals that the difference between male and female levels of confidence in "the scientific community" was greater than that for any other major institution (including the military, medicine and big business), that this gap grew steadily wider between 1973 and 1989, and that the gap was unaffected by statistical controls for confounding effects of education, occupational prestige, religiosity, attitude to the military, and support for environmental spending. On no issue are gender differences in environmental risk perception and in support for technology more striking than on that of nuclear power, "where a strong divergence turned up in every poll, in every country, at every time" (Weart 1988: 367; see also Brody 1984; Dunlap and Olsen 1984).

Another issue about which gender differences are marked and consistent is that of the welfare and rights of individual (especially, but not only domestic) animals. Women and girls are much more likely to hold humane attitudes and support animal rights, and much less likely to hold "dominionistic" attitudes towards animals, than are men and boys (Kellert 1984a, 1984b, 1990; Nibert 1994). Attitudes to science and animals intersect on the issue of the use of animals in painful scientific and medical experiments. In a comparison of survey results from 15 Western countries, Pifer et al. (1994) found that in every country, women were more opposed to animal experimentation than were men. The size of the gender gap, however, varied considerably between countries, as did the overall level of opposition to animal experimentation among survey respondents. In some countries, people who had lower levels of general scientific knowledge tended to be more opposed to animal experiments than those who had more knowledge, but in other countries the reverse was found. The gender gap was found to be largest in those countries (such as the Netherlands and the U.S.) where overall levels of opposition are low to moderate, and where opposition tends to decrease with level of scientific knowledge. The gender gap was found to be smallest in countries (such as Germany, Belgium and Ireland) where the overall level of opposition was high, and where opposition tends to *increase* with level of scientific knowledge. In both groups of countries (but not in all countries), those who expressed opposition to animal experimentation tended to express interest in environmental pollution issues.

It is not obvious what to make of these intriguing findings. For a start, one would want to know the partial effects of each of gender and scientific knowledge, controlling for thc other; unfortunately this analysis is not provided. One might speculate that, assuming one's level of scientific knowledge to be at least in part a function of one's interest in and affinity with science, these results suggest that in some countries, attitudes to both science and animals are less "gender laden" than in others. Where the gender difference is smaller, attitudes to animal experimentation and to science *per se* may be less tightly coupled, or vice versa.

To conclude, I would suggest that the complex pattern of findings reviewed in this section tend not to support biological theories of gender difference. The evidence of interplay between gender, occupation and parenthood seems to point in the direction of "social" explanation, but there is a need for more research specifically designed to develop and test hypotheses about the connection between gender and division of labour in shaping environmental concerns. These studies also give qualified support to psycho-cultural theories, particularly regarding attitudes to science, technology and the domination of non-humans.

Race

Very little research has been undertaken on the relationship between ethnicity and environmental concern. Historical investigations into the origins of the ecology movement have, for example, traced the connections between some strands of environmentalism and specifically Northern European forms of Romanticism and organicism (Bramwell 1989). In Canada, francophones were much more likely than anglophones to name the environment as the most important issue in the 1988 federal election, but were less likely to agree that environmental protection was more important than job creation (Bakvis and Nevitte 1992). Currently in North America, a burgeoning literature on environmental racism has documented the disproportionate infliction of environmental (especially toxic) hazards on native Canadian and American,

African-American and U.S. Hispanic communities (Bullard 1990, 1993; Hofrichter 1993). Given the social and cultural contexts illuminated by both these areas of investigation, it should be possible to bring our theoretical framework into play.

Interest-based theories such as Murphy's (1994) would tend to predict that those minorities who have borne the brunt of environmental contamination would, as disproportionate victims, be more concerned than others about environmental issues of the kind that bear on the health of victimized communities. On the other hand, culture-based constructionist theories and those influenced by work such as Inglehart's, might predict that concern for ecological integrity, protection of non-human beings and harmony with nature would be highest among relatively affluent people (especially of Northern European descent). Literature on the American environmental movements tends to suggest that the "grass-roots" movement may differ from the "mainstream" and "radical-ecology" movements in just some such demographic respects (Dunlap and Mertig 1992; Bosso 1994), but only survey research can shed light on the question of the distribution of kinds of concern in the "general public."

Available survey research only allows us to compare American whites and American blacks. Consistent with the predictions of the interest-based theory, Cutter (1981) found that among Chicago residents, blacks were more concerned about pollution than were whites, and this held, even controlling for the levels of pollution to which respondents were exposed in their neighbourhoods. The literature reviewed by Taylor (1989), however, much of it concerned with nature-preservation issues, found a "concern gap" between blacks and whites, with blacks expressing lower levels of concern. These findings suggested cultural differences, and such differences might be expected to show up most clearly in attitudes towards animals. Kellert (1984a, 1984b) found that among both adults and children, blacks expressed more negative and exploitative attitudes about animals than did whites. The attitudinal differences between blacks and whites were similar to those found between low- and high-income respondents, and between rural and urban ones; since these other variables were not controlled for, the effect of race may be a spurious one. Nibert (1994) found no difference between whites and non-whites (race not specified) in support for animal rights.

Any notion that American environmental concerns differ significantly by race is called into question by the most thorough study on the subject. Mohai (1990) analysed the results of a nation-wide U.S. Department of Agriculture survey that employed a variety of measures of concern, and covered a wide array of environmental issues, ranging from toxic wastes, through soil conservation, to the preservation of wildlife and wildlife habitat. Controlling for income, education and occupational status, Mohai found virtually identical levels of concern among blacks and whites across all issues. Although, other variables being equal, blacks were just as environmentally concerned as whites and, in general, as politically active, they were much less likely to be politically active in regard to environmental issues.

These results suggest that while African-Americans may be alienated from mainstream American environmentalism, this is not because their environmental concerns are necessarily either less, or different, than those of other racial groups. However, it is important to note that the survey upon which this study was based was conducted in 1980, during a period in which non-white support for spending on environmental protection tended to be high relative to white support, and that by 1990, non-white support was lower than White support. Moreover, while support for spending on environmental protection tended to be somewhat higher on average among blacks than among whites during the 1970s and 1980s, whites during this period tended to place a higher priority on environmental protection than did blacks, relative to other social objectives such as "improving the conditions of blacks" and "improving the education system" (Jones and Carter 1994). Thus, even during periods when overall environmental concern is significantly higher among blacks than whites, the attention of black activists may be expected to be occupied by issues regarded as still more pressing.

Socio-economic Status

Socio-economic status (SES), which is measured in terms of income and occupational prestige, is obviously closely related to occupation, the next social variable to be considered. Because it is unidimensional, it is a simpler concept, and our discussion will be brief. Interest in the relation of SES and environmentalism stems mainly from the issue of what Morrison and Dunlap (1986) call "compositional élitism" — that is, the question of whether environmental concern and activism is a function of social privilege.

The most consistent finding in the literature is that higher SES is positively related to both environmental activism (Cotgrove 1982; Mohai 1985; Mohai and Twight 1987; Kowelewski and Porter 1992) and to personal behaviours such as recycling (MacDermid and Stevenson 1991; Derksen and Gartrell 1993; Kanagy and Willits 1993). When it comes to environmental concern as determined by survey responses, there appears to be no such consistent effect. Though some studies report that higher-income people are more likely to subscribe to an ecological worldview (Arcury and Christianson 1990; Kanagy and Willits 1993), analyses of U.S. nationwide survey data revealed no effect of either income or occupational prestige on environmental concern (Mohai 1985; Mohai and Twight 1987; Kanagy et al. 1994). Studies cited by Van Liere and Dunlap (1980) tended to find a slight positive effect for prestige only; on the other hand, Hamilton (1985a) found a positive effect for income only, and Mohai (1990) found that, controlling for income and education, those with higher prestige were slightly *less* concerned. Inglehart (1990) suggested that those with postmaterialist values tend to have occupations that are relatively prestigious for the amount they pay, but this is an effect of education. Consistent with Inglehart's view are data suggesting that those with higher-prestige occupations and those with lower incomes were more likely to regard the environment as the most important issue in the 1988 Canadian federal election (Bakvis and Nevitte 1992).

There is evidence that high income is related to positive attitudes towards high technology. Olsen, Lodwick, and Dunlap (1992) found that those attitudinal measures that included items tapping faith in technology, tended to be positively associated with high income. Members of pro-nuclear energy groups tended to have higher incomes than members of pro-solar and energy-conservation groups (Dunlap and Olsen 1984), and one study found high income to be the best demographic predictor of support for a local nuclear powerplant (George and Southwell 1986). On the other hand, those with high incomes and high-prestige occupations, who presumably tend to enjoy a relatively high degree of economic security, may be much more willing than others to favour environmental protection at the expense of job creation (Bakvis and Nevitte 1992). Both these tendencies are explicable in terms of the interests of lesser-skilled workers most vulnerable to unemployment. Such workers would be disinclined either to sacrifice job creation, or to support environmentally dubious capital-intensive high-technology development paths, that may offer them little in the way of steady employment opportunities.

It appears that the relationship between income and attitudes connected with environmental concern or environmentalism, may not be a linear one. Specifically, environmentalist attitudes may be higher among middle-income people than among either high- or low-income people (MacDermid and Stevenson 1991; Olsen, Lodwick, and Dunlap 1992). Kellert (1984a) found that middle-income people were most likely to express humane attitudes toward animals, and that high- and low-income people were most likely to express "dominionistic" attitudes. Milbrath (1984) found that in Germany, England and the U.S., those who thought that environmental problems were minor, posing no limits to growth, and amenable to technological solutions, tended to have high incomes. In England and Germany, those most opposed to such views tended to have low incomes, but in the U.S., they tended to have "medium-high" incomes. Some common and powerful statistical methods are not designed to detect this kind of pattern, and would misleadingly indicate "no relationship" between income and environmental concern if applied to these data. This illustrates how quantitative methods, if not used carefully, can underestimate certain sociological relationships.

Occupation and Sector

Of all the variables reviewed in this chapter, those related to occupation, sector and class are the most complex, and have received the most theoretical attention among sociologists. It is commonly suggested that the main constituency of "new social movements" in general, and of environmentalism in particular, is the "new middle class" — well-educated professional and semi-professional service and knowledge workers, especially those employed in the non-market sector (Offe 1985; Buttell 1992). The definitions of "new middle class" tend to vary in content and emphasis, and the precise set of occupations that make up the class would vary accordingly, but by all definitions, it would include academics, teachers, social workers and the various other professionals that staff public agencies.

Connected to the issue of definition, but more important for our purposes, is the question of the nature and explanation of the connection between environmentalism and the new middle class. Assuming some such connection exists, there are (at least) four possible kinds of answers to the latter question. The first possibility is that while those who are active in the environmental movement are disproportionately drawn from the new middle class, this is not because environmental concerns or environmentalist attitudes are more prevalent in this class, but because members of this class tend to possess the knowledge, skills, time, money and other social resources that are useful for activists (Bagguley 1992; see also Mohai 1985).

The remaining three answers assume that new middle-class members are indeed more likely to hold pro-environmentalist views than are members of other classes, but differ on why this is so. One possibility is that members of this class do not come by their environmental views because of their occupations, but rather that both their views and their choice of occupations arise from values and opportunities shaped by the circumstances of their formative years. This theory is most associated with the work of Inglehart (1990), who suggests that people raised in circumstances of security and affluence are more likely both to obtain education necessary for professional occupations, and to acquire postmaterialist values that make them disinclined to seek careers in private business. Cotgrove (1982: 45) provides some evidence that postmaterialist values (themselves associated with environmental concerns) affect both the programme choices and career aspirations of university undergraduates.

On the other hand, the views of new middle-class members may be (at least in part) affected by the occupations they hold. The nature of the connection may be explained in terms of interest-based or cultural theories. Our third answer is interest based: members of the new middle class disproportionally benefit from environmentalism. For those employed directly or indirectly by the regulatory apparatus of the state, any social forces contributing to a demand for increased regulation of business will tend to increase their own power and better their career prospects. In addition, professionals employed outside of the industrial-production sector benefit from the protection of environmental amenities (which they in particular have the time, money and education to fully appreciate), while the cost of such protection is disproportionately borne by investors in, and employees of, the industrial production sector. Finally, a fourth approach would be to look for distinctive cultural biases arising out of the nature and organizational structure of new middle-class occupations (Douglas and Wildavsky 1982: 159–60).

The typical class, sector and occupation of those *opposed* to environmentalism have been a matter of somewhat less discussion, but anti-environmentalist attitudes have been associated with two groups. The first group consists of "industrialists," that is, private-sector executives, managers or professionals, especially those in resource extraction or manufacturing industries (Cotgrove 1982). The second group consists of workers and self-employed people whose jobs are directly related to resource extraction or pollution-causing activities. Interest-based theories have obvious applicability here, but there is evidence of cultural factors in play, as well (e.g., see Dunks 1994).

Turning to the empirical literature, we find that, as expected, members of environmental, radical-ecology and "soft-energy" advocacy groups are disproportionately employed in the non-market sector (Cotgrove 1982; Dunlap and Olsen 1984; Marangudakis 1991). Tindall (1994) found that among Vancouver Island forest preservationists, the most overrepresented occupations were those in natural science, teaching, administration, medicine and health; 42 per cent of the members of the organizations studied had new middle-class occupations. Social scientists and artists, so prominent in new social movements generally, were not significantly overrepresented, however.

It is difficult, on the basis of existing studies, to answer the question of whether environmental concerns are more prevalent among the new middle class than among other groups, because these studies have typically not been designed to provide precisely this information. One study that explicitly addressed the question of whether, in Western Europe, new middle-class members were more pro-environmentalist than members of other classes, found that they were not (Rohrschneider 1990). Like Inglehart (1990), Rohrschneider found environmentalism to be related to postmaterialism; unlike Inglehart, he found that, depending on the country, the association between postmaterialism and new middle-class location was either weak or absent.

Few American studies provide much detail on occupation. In their study of a toxic-waste controversy in Indiana, Simmons and Stark (1993) report that the occupational groups, whose members were most likely to name toxic waste as the most important local election issue, were students, academics, professionals and civil servants.

Kellert's (1984a) survey data indicate that professionals tend to value wildlife and oppose cruelty to animals; they are the least likely to value domination of animals. Olsen, Lodwick, and Dunlap (1992) report that those in "high white-collar" occupations were most likely to subscribe to ecological beliefs and values, though generally the relationships between their occupational categories and any of their attitudinal variables tended to be weak.

Milbrath (1984) found that in England and Germany, as well as in the U.S., fewer than a third of those who thought environmental problems were small and who favoured technological solutions, were employed in the "service sector," while over half of those who thought environmental problems were large and who favoured social solutions, were so employed. Of those Americans in the latter group, who also believed there were limits to growth, up to three-quarters were in the "service sector." The association between service-sector occupation and support for environmental spending was weak between 1973 and 1989, but much stronger in 1990 (Jones and Dunlap 1992). On the whole, these findings provide some support (albeit weak, ambiguous and qualified) that members of new middle-class occupations are not only more likely to participate in the environmental movement, but to hold attitudes associated with at least some aspects of that movement.

Findings are less ambiguous in the case of industrialists. Cotgrove (1982) and Milbrath (1984) found industrialists to be, on practically every attitudinal measure, diametrically opposed to environmentalists, and certainly farther from the views of environmentalists than was the general public. This was particularly the case in the U.S., where Milbrath found that almost half of business leaders (as compared with about one-fifth of the general public) fell into the least environmentalist of his attitude categories. European business leaders were more likely to recognize limits to growth, and less likely to endorse technological approaches to solving environmental problems. Note, however, that these studies were conducted during a period (late 1970s and early 1980s) when in the U.S. the environment was a low-profile issue associated with the political left, and environmental concern was relatively concentrated among the less privileged (Jones and Dunlap 1992). After a decade of unprecedented media attention (one that also saw members of environmentally concerned postwar cohorts begin to enter positions of power), not only had overall support for environmental protection greatly increased, but it had disproportionately gained in relative strength among males, whites, the highly educated, the highly paid and conservatives (Dunlap 1992; Jones and Dunlap 1992; Mazur and Lee 1993). We might therefore expect to find that the attitudinal gulf between corporate élites and environmentalists to have narrowed, whether or not actual corporate behaviour has become any "Greener."

Findings regarding occupations in agriculture, resource extraction and polluting industries is inconclusive. Lowe and Pinhey (1982) found that those who worked in agriculture, or whose father worked in agriculture, had lower levels of environmental concern than those in other occupations, but that workers in mining or polluting industries and their offspring did not. Freudenburg (1991) found just the reverse of this. Filson (1993) found that among Ontario farmers, those with small farms and those with off-farm income tended to have the highest environmental orientation and were most opposed to use of chemicals; capitalist farmers (owners with at least two employees) were most opposed to conservation regulations and least opposed to chemicals. Stout-Wiegand and Trent (1983) found, as we have seen, that coal-related employment was associated with reduced opposition to coal development among men but not women. Fortmann and Kusel's (1990) data suggest than experience in forestry occupations was related to pro-commodity attitudes regarding forest management primarily among forest service employees (a case where public-sector employment was *not* associated with pro-environment views). Effects on environmental attitudes of living in, or coming from, rural areas is discussed further in the next section.

Community Size

Though the earlier studies reviewed in Van Liere and Dunlap (1980) did not consistently find that community size made a difference to people's levels of environmental concern, results of several more recent studies suggest that urban residents are more "environmentalist" in their views than are rural or small-town residents (Dunlap and Olsen 1984; Rohrschneider 1990; Arcury and Christianson 1990; but cf. Fortmann and Kusel 1990; see also Kellert 1984a and Nibert 1994 regarding animals). Other studies suggest that rural people are not on the whole less concerned about the environment, but are less favourably disposed towards land-and resource-use regulation (Van Liere and Dunlap 1981; Hawkins 1984: ch. 6; Freudenburg 1991). Still other studies have also found environmentalism to be an urban phenomenon, but that having grown up in a city has a greater effect than currently residing in one, on whether one is likely to adopt environmentalist views (Lowe and Pinhey 1982; Kellert 1984a; Jones and Dunlap 1992).

Why is it that urban residents, or those of urban origin, might express higher levels of environmental concern? In the first place, we must ask whether the urban-origin effect is a spurious one. Some researchers have found that while the urban-origin effect is a strong one, it is much reduced or else eliminated when controlling for other variables (Mohai and Twight 1987; Jones and Dunlap 1992). It might, for instance, be an effect of age:

younger people are more environmentally concerned, and because of increasing urbanization, younger people are more likely to have been raised in a city. However, Americans who have grown up on farms appear to be significantly less likely to support increased spending on environmental problems, even when the effects of age, region and overall willingness to spend on social problems are controlled for (Kanagy et al. 1994). If the urban-origin effect is not spurious, it is evidence for the proposition that preadult socialization is more important than current circumstances in shaping environmental attitudes, which would in turn tend to support cultural over interest-based theories of environmental concern. This still leaves open the question of just what it is about the differences in social relationships and relationships with nature characteristic of rural versus urban environments, that generate the observed attitudinal differences. The question is a fascinating one, but we would have to look elsewhere than survey data for the answer.

Religion

In a famous and controversial essay, the historian Lynn White (1967) argued that the historical roots of the environmental crisis can be traced to Western Christianity's interpretation of passages in the book of Genesis in which God gives Adam dominion over non-human creatures and orders him to subdue the Earth. Although White's thesis concerned the influence of religious ideas on the formation of the Western world-view in the Middle Ages, and not the present-day association of religious beliefs and environmental attitudes, a number of American social scientists have investigated the latter topic. Most (but not all: Woodrum and Hoban 1994) studies have found some relationship between religious variables and environmental concern, but the pattern is not a simple one.

Eckberg and Blocker (1989) found that, controlling for other socio-demographic variables, those who believe the Bible to be literally true, tend to have lower levels of environmental concern than those who do not. Olsen, Lodwick, and Dunlap (1992) also detected a general association between religious conservatism and views favourable to the exploitation of nature for human benefit, but found that those with no religion were the most likely to either fully reject or fully endorse such exploitation. Kellert (1984a) and Kanagy and Willits (1993) found church attendance to be positively associated with attitudes favouring the domination of animals and nature. Hand and Van Liere (1984), however, found that while high church attendance was associated with lower levels of environmental concern among members of the more conservative, fundamentalist Protestant denominations, the reverse was true among members of more liberal Protestant denominations. Among people with the same environmental attitudes, those who attend worship services more often are more likely to engage in environmentally responsible behaviours; but since those who attend more often are less likely to have environmentalist attitudes, the two effects cancel themselves out (Kanagy and Willits 1993).

Greeley (1993) conducted a detailed analysis of U.S. national survey results in an effort to determine just what aspects of religious beliefs had an effect on support for spending on environmental problems. Especially among Catholics, those who regarded God as a mother or friend were more supportive of environmental spending, than those who regarded God as a father or judge. Greeley found that Biblical literalism *per se* was not negatively associated with support for environmental spending, but was often accompanied by political conservatism, moral rigidity and authoritarian attitudes; these latter variables explained the apparent association between fundamentalism and low-environmental concern. It is also the case that in the U.S., religious fundamentalism is often an aspect of a more general cultural conservatism, that also has implications for attitudes regarding the relationship between humans and non-humans: Nibert (1994) found that opposition to abortion, gun control and neighbourhood racial integration correlated strongly with opposition to animal rights.

Conclusions

The findings of the literature reviewed here cannot be summarized in a few simple propositions. Moreover, statistical analyses have found standard socio-demographic variables to "explain" only about 10 per cent of the variation in measurable environmental concern among members of the public; this increases to 15 or 20 per cent if political and religious variables are added to the model (Lowe and Pinhey 1982; Hand and Van Liere 1984). It would be tempting to conclude from these results that the researches reported here have come up empty-handed, or that they have only the negative value of showing that environmental concerns are distributed almost at random in the population. I would suggest a contrary conclusion.

The lesson learned from this research is that the social bases of environmental concern are complex and subtle. Environmental concern is not one thing. The category of the "environmental" is a social construction, whether or not the content of that category is. That issues as diverse as species extinction, the thinning of the ozone layer, and the contamination of local drinking water by toxic chemicals are categorized together as "environmental," may not be arbitrary, but it is certainly socially and historically contingent; and those contingencies are variably reflected in the social bases of concern about those issues. Further, the effects of socio-demographic variables are often of kinds that are easily overlooked or underestimated if data is collected or analysed on the basis of overly simplistic assumptions. We have found interactions, non-linear relationships and so on whenever the data have been presented in a way that allow us to look for them. My objective in reviewing these results has been to juxtapose diverse findings in such a way as to suggest new insights and hypotheses for future investigation, whether that investigation be quantitative, qualitative or theoretical.

DISCUSSION QUESTIONS

1. How, if at all, are theories or ideas about the social bases of environmental concern related to theories or ideas about environment and environmental problems themselves? Do those who subscribe to realist, interest-based and social constructionist explanations of environmental concern necessarily differ in their views regarding the nature and reality of environmental problems? Discuss.
2. What are some of the various ways that environmental concern is measured or operationalized in the research reviewed in his chapter? Do these various ways all measure the same underlying thing? If not, does it make sense to compare the results of studies that employ different measures of environmental concern? What might we learn from such comparisons?
3. For any of the social variables discussed in this chapter (e.g., age, gender, occupation etc.), or any set of these variables, formulate an "interest-based" theory and a cultural theory about the relationship between the variable(s) you have chosen and attitudes and/or behaviours concerning the environment. How do your two theories differ? What does your interest-based theory predict the effect of the variable(s) in question to be, and why? What does your cultural theory predict the effect of the variable(s) in question to be, and why?
4. Do the theories make similar predictions? If not, which is best supported by the results of the research reviewed in this chapter? What further research would you need to undertake, and what specific hypotheses would you need to test in order to help you to decide with greater confidence which of the two theories is better? What would be the advantages and disadvantages of using the survey method to undertake this research?

References

Arcury, T. and E. Christianson. 1990. Environmental Worldview in Response to Environmental Problems: Kentucky 1984 and 1988 Compared. *Environment and Behavior* 22: 387–407.

Bagguley, P. 1992. Social Change, the Middle Class and the Emergence of 'New Social Movements': A Critical Analysis. *The Sociological Review* 40: 26–48.

Bakvis, H. and N. Nevitte. 1992. The Greening of the Canadian Electorate: Environmentalism, Ideology and Partisanship. In *Canadian Environmental Policy: Ecosystems, Politics and Process,* ed. R. Boardman. Toronto: Oxford University Press.

Blaikie, N. 1992. The Nature and Origins of Ecological World Views: An Australian Study. *Social Science Quarterly* 73: 144–65.

Blocker, T.J. and D.L. Eckberg. 1989. Environmental Issues as Women's Issues: General Concerns and Local Hazards. *Social Science Quarterly* 70: 586–93.

Bosso, C. 1994. After the Movement: Environmental Activism in the 1990s. In *Environmental Policy in the 1990s*, 2d ed., ed. N. Vig and M. Kraft. Washington: Congressional Quarterly Press.

Bramwell, A. 1989. *Ecology in the Twentieth Century: A History*. New Haven: Yale University Press.

Brody, C. 1984. Differences by Sex in Support for Nuclear Power. *Social Forces* 63: 209–28.

Bullard, R. 1990. *Dumping in Dixie: Race, Class and Environmental Quality*. Boulder, Colo.: Westview.

———. 1993. *Confronting Environmental Racism: Voices from the Grassroots*. Boston: South End.

Buttel, F. 1992. Environmentalization: Origins Processes, and Implications for Rural Social Change. *Rural Sociology* 57: 1–27.

Cotgrove, S. 1982. *Catastrophe or Cornucopia: The Environment, Politics and the Future*. Chichester, Eng.: J. Wiley and Sons.

Cutter, S.C. 1981. Community Concern for Pollution: Social and Environmental Influences. *Environment and Behavior* 13: 105–24.

Derksen, L. and J. Gartrell. 1993. The Social Context of Recycling. *American Sociological Review* 58: 434–42.

Douglas, M. 1992. *Risk and Blame: Essays in Cultural Theory*. London: Routledge.

Douglas, M. and A. Wildavsky. 1982. *Risk and Culture: An Essay on the Selection of Technological and Environmental Dangers*. Berkeley: University of California Press.

Dunks, T. 1994. Talking About Trees: Environment and Society in Forest Workers' Culture. *Canadian Review of Sociology and Anthropology* 31: 14–34.

Dunlap, R. 1992. Trends in Public Opinion Toward Environmental Issues: 1965–1990. In *American Environmentalism: The US Environmental Movement, 1970–1990*, ed. R. Dunlap and A. Mertig. New York: Taylor and Francis.

Dunlap, R. and A. Mertig. ed. 1992. *American Environmentalism: The US Environmental Movement, 1970–1990*. New York: Taylor and Francis.

Dunlap, R. and M. Olsen. 1984. Hard-path versus Soft-path Advocates: A Study of Energy Advocates. *Policy Studies Journal* 13: 413–28.

Eckberg, D.L. and T.J. Blocker. 1989. Varieties of Religious Involvement and Environmental Concern: Testing the Lynn White Thesis. *Journal for the Scientific Study of Religion* 28: 509–17.

Eiser, J.R., J. Van Der Pligt and R. Spears. 1988. Local Opposition to the Construction of a Nuclear Power Station: Risk and Rationality. In *Environmental Social Psychology*, ed. D. Canter, J. Correia Jesuino, L Soczk and G. Stephenson. Dordrecht, Germany: Kluwer.

Filson, G. 1993. Comparative Differences in Ontario Farmers' Environmental Attitudes. *Journal of Agricultural and Environmental Ethics* 6: 165–84.

Fortmann, L. and J Kusel. 1990. New Voices, Old Beliefs: Forest Environmentalism Among New and Long-standing Rural Residents. *Rural Sociology* 55: 214–32.

Fox, M.F. and G. Firebaugh. 1992. Confidence in Science: The Gender Gap. *Social Science Quarterly* 73: 101–13.

Freudenburg, W. 1991. Rural-urban Differences in Environmental Concerns: A Closer Look. *Sociological Inquiry* 61: 186–98.

Gaard, G., ed. 1993. *Ecofeminism: Women, Animals and Nature*. Philadelphia: Temple University Press.

George, D. and P. Southwell. 1986. Opinion on the Diablo Canyon Nuclear Power Plant: The Effects of Situation and Socialization. *Social Science Quarterly* 67: 722–35.

Gilligan, C. 1982. *In a Different Voice*. Cambridge, Mass.: Harvard University Press.

Gray, E.D. 1979. *Green Paradise Lost*. Wellesley, Mass: Roundtable Press.

Greeley. A. 1993. Religion and Attitudes Toward the Environment. *Journal for the Scientific Study of Religion* 32: 19–28.

Hamilton, L. 1985a. Who Cares About Water Pollution? Opinions in a Small-town Crisis. *Sociological Inquiry* 55: 170–81.

———. 1985b. Concern About Toxic Wastes: Three Demographic Predictors. *Sociological Perspectives* 28: 463–86.

Hand, C. and K. Van Liere. 1984. Religion, Mastery-over-nature, and Environmental Concern. *Social Forces* 63: 555–70.

Hawkins, K. 1984. *Environment and Enforcement: Regulation and the Social Definition of Pollution*. Oxford: Clarendon Press.

Hofrichter, R., ed. 1993. *Toxic Struggles: The Theory and Practice of Environmental Justice*. Philadelphia: New Society.

Inglehart, R. 1990. *Culture Shift in Advanced Industrial Society*. Princeton: Princeton University Press.

Jones, R.E. and R. Dunlap. 1992. The Social Bases of Environmental Concern: Have They Changed Over Time? *Rural Sociology* 57: 28–47.

Jones, R.E. and L. Carter. 1994. Concern for the Environment Among Black Americans: An Assessment of Common Assumptions. *Social Science Quarterly* 75: 560–79.

Kanagy, C. and F. Willits. 1993. A "Greening" of Religion? Some Evidence from a Pennsylvania Sample. *Social Science Quarterly* 74: 674–83.

Kanagy, C., C. Humphrey and G. Firebaugh. 1994. Surging Environmentalism: Changing Public Opinion or Changing Publics? *Social Sciences Quarterly* 75: 804–19.

Kellert, S. 1984a. American Attitudes Toward and Knowledge of Animals: An Update. *Advances in Animal Welfare Science 1984*, ed. M. Fox and L. Mickley. Boston: Martinus Nijhoff.

———. 1984b. Attitudes Towards Animals: Age-related Development Among Children. *Advances in Animal Welfare Science 1984*, ed. M. Fox and L. Mickley. Boston: Martinus Nijhoff.

———. 1990. *Canadian Perceptions of Marine Mammal Conservation and Management in the Northwest Atlantic*. International Marine Mammal Association Technical Report.

Kowalewski, D. and K. Porter. 1992. Ecoprotest: Alienation, Deprivation or Resources? *Social Science Quarterly* 73: 523–34.

Krause, C. 1993. Blue-collar Women and Toxic-waste Protests. In *Toxic Struggles: The Theory and Practice of Environmental Justice*, ed. R. Hofrichter. Philadelphia: New Society Publishers.

Kuitunen, M. and T. Tormala. 1994. Willingness of Students to Favour the Protection of Endangered Species in a Trade-off Conflict in Finland. *Journal of Environmental Management* 42: 111–118.

Lowe, G. and T. Pinhey. 1982. Rural-urban Differences in Support for Environmental Protection. *Rural Sociology* 47: 114–28.

MacDermid, R. and M. Stevenson. 1991. *Identification with New Social Movements: The Structure of Public Opinion on Environmental Issues*. North York, Ont.: York University Institute for Social Research.

Marangudakis, M. 1991. Emerging Ideologies in the Environmental Movement: The North American Case of "Deep" and "Social Ecology." Master's thesis, McGill University.

Mazur, A. and J. Lee. 1993. Sounding the Alarm: Environmental Issues in the US National News. *Social Studies in Science* 23: 681–720.

McStay, J. and R. Dunlap. 1983. Male-female Differences in Concern for Environmental Quality. *International Journal of Women's Studies* 6: 291–300.
Melucci, A, 1985. The Symbolic Challenge of Contemporary Movements. *Social Research* 54: 789–816.
Mies, M. and V. Shiva. 1993. *Ecofeminism*. Halifax: Fernwood/London: Zed.
Milbrath, L. 1984. *Environmentalists: Vanguard for a New Society*. Albany: State University of New York Press.
Mohai, P. 1985. Public Concern and Elite Involvement in Environmental-conservation Issues. *Social Science Quarterly* 66: 820–38.
———. 1990. Black Environmentalism. *Social Science Quarterly* 71: 744–65.
Mohai, P. and B. Twight. 1987. Age and Environmentalism: An Elaboration of the Buttel Model Using National Survey Evidence. *Social Science Quarterly* 68: 798–815.
Morrison, D. and R. Dunlap. 1986. Environmentalism and Elitism: A Conceptual and Empirical Analysis. *Environmental Management* 10: 581–89.
Murphy, R. 1994. *Rationality and Nature: A Sociological Inquiry into a Changing Relationship*. Boulder, Colo.: Westview Press.
Nibert, D. 1994. Animal Rights and Human Social Issues. *Society and Animals* 2: 115–24.
Offe, C. 1985. New Social Movements: Challenging Boundaries of Institutional Politics. *Social Research* 52: 817–66.
Olsen, M., D. Lodwick and R. Dunlap. 1992. *Viewing the World Ecologically*. Boulder, Colo.: Westview Press.
Perrow, C. 1984. *Normal Accidents: Living with High-Risk Technologies*. New York: Basic Books.
Pierce, J., T. Tsurutani, N. Lovrich Jr. and T. Abe. 1986. Vanguards and Rearguards in Environmental Politics: A Comparison of Activists in Japan and the United States. *Comparative Political Studies* 18: 419–47.
Pierce, J., N. Lovrich Jr., T. Tsurutani and T. Abe. 1987a. Culture, Politics And Mass Publics: Traditional and Modern Supporters of the New Environmental Paradigm in Japan and the United States. *Journal of Politics* 49: 54–79.
———. 1987b. Environmental Belief Systems Among Japanese and American Elites And Publics. *Political Behavior* 9: 139–59.
Pifer, L., K. Shimizu and R. Pifer. 1994. Public Attitudes Toward Animal Research: Some International Comparisons. *Society and Animals* 2: 95–113.
Rohrschneider, R. 1990. The Roots of Public Opinion Toward New Social Movements: An Empirical Test of Competing Explanations. *American Journal of Political Science* 34: 1–30.
Schahn, J. and E. Holzer. 1990. Studies of Individual Environmental Concern: The Role of Knowledge, Gender and Background Variables. *Environment and Behavior* 22: 767–86.
Simmons, J. and N. Stark. 1993. Backyard Protest: Emergence, Expansion and Persistence of a Local Hazardous Waste Controversy. *Policy Studies Journal* 21: 470–91.
Steger, M.A., J. Pierce, B. Steel and N. Lovrich. 1989. Political Culture, Postmaterial Values and the New Environmental Paradigm: A Comparative Analysis of Canada and the United States. *Political Behavior* 11: 233–54.
Stern, P., T. Dietz and L. Kalof. 1993. Value Orientations, Gender and Environmental Concern. *Environment and Behavior* 25: 322–48.
Stout-Wiegand, N. and R. Trent. 1983. Sex Differences in Attitudes Toward New Energy Resource Developments. *Rural Sociology* 48: 637–46.
Taylor, D. 1989. Blacks and the Environment: Explaining the Concern and Action Gap Between Blacks and Whites. *Environment and Behavior* 21: 175–205.
Tindall, D. 1994. *Collective Action in the Rainforest: Personal Networks, Identity and Participation in the Vancouver Island Wilderness Preservation Movement*. Ph.D. Dissertation, Department of Sociology, University of Toronto.

Van Liere, K. and R. Dunlap. 1980. The Social Bases of Environmental Concern: A Review of Hypotheses, Explanations and Empirical Evidence. *Public Opinion Quarterly* 44: 181–97.
———. 1981. Environmental Concern: Does It Make a Difference How It's Measured? *Environment and Behavior* 13: 651–76.
Weart, S. 1988. *Nuclear Fear: A History of Images*. Cambridge, Mass.: Harvard University Press.
White, L. 1967. The Historical Roots of Our Environmental Crisis. *Science* 155: 1203–7
Woodrum, E. and T. Hoban. 1994. Theology and Religiosity Effects on Environmentalism. *Review of Religious Research* 35: 193–206.

ENVIRONMENTAL STUDIES IN CONTEXT: KNOWLEDGE, LANGUAGE, HISTORY AND THE SELF

Éric Darier

Brock University

The emergence and institutionalization of Environmental Studies in the education curriculum is a confirmation of the prevalence and the worsening of the "environmental crisis" over the past 30 years or so. The concern for the environment is moving away from the discursive monopoly of the natural sciences towards a broader concern which includes the various social interactions shaping the environment. Therefore, an "environmental sociology" is a necessary and welcome development. Before embarking on an environmental sociology, it is important to map out the theoretical landscape from which an environmental sociology is sprouting. The conventional approach is to take for granted the environmental crisis. However a sociological study must observe not only sociological "objects," but also the sociological conditions for the emergence of objects in themselves. Rather than establishing the ground for an epistemology of environmental sociology, I would invite the reader on a historical journey about the conditions of the emergence of the various discourses and social practices around the "environment." More specifically, I would argue that the intellectual context of the late 20th century has to be taken into consideration in order to make sense of an environmental sociology.

The central point of my argument centres around four contextual premises about the possibilities of undertaking Environmental Studies in our historical context: (1) the conditions for knowledge; (2) the role of "language" in the shaping of the perception of "reality" of the world; (3) the plurality of histories; and (4) the theories about the "self." These four premises are not merely philosophical and/or epistemological. They are social practices which have been long established in, what I call, the "North-West" (North America and Western Europe). I am fully aware of the "North-Western" culturo-centric bent I have adopted. In response to possible accusations of an "inverted form of Eurocentrism," I would argue that the so-called environmental crisis which the world currently appears to be facing is largely the result of the historical dominance of European and North-American patterns of development for the past 300 years (Turner 1990). Obviously, this doesn't mean that other cultures and civilizations are, or were, more ecologically friendly. However, the cultural awareness of something called an "environmental crisis" is specific to the North-Western context. Therefore, we should try to understand the specific conditions for the emergence of discursive practices around this theme called the "environment" (Darier 1991).

As there are already excellent historical studies on this topic, I shall focus on the 20th century (Worster 1988, 1994a). I leave it to someone else to demonstrate that the core North-Western views concerning the environment are in fact strongly influenced by the "contact" which occurred during the European colonization of the world. Edward Said's challenging thesis outlined in *Orientalism*, and more recently again in *Culture and Imperialism*, might indeed extend to the North-Westerners' views about the environment (Said 1978, 1994). If Edward Said is correct, this would mean that the conventional sharp separation between North-Western views about the environment, and the heterogeneity of views from the South, would be questioned.

Finally, I would like to mention that, obviously, I am not the first person to make this kind of argument. Sociologists of science, philosophers, social and political theorists and other writers have made these points before me (Bird 1987). I want to bring together the discursive practices of the sociologists of science, the philosophers, the social and political theorists and so on, to help us undertake a critical and practical investigation of Environmental Studies as an emerging new discipline and as a new set of social practices.

The Conditions for Environmental Knowledge in Our Historical Context

> So here the sentence "I know...." expresses the readiness to believe certain things (Wittgenstein 1969: section 330).

North-Western knowledge about "knowledge" — and the conditions for knowledge — are radically different in the late 20th century than they were before. Until recently, knowledge, and more precisely scientific knowledge, was about the discovery, the revealing, the unmasking of the world as it was. As

Michel Foucault (1926–1984) has shown, there was a rupture in scientific discourse in the 17th and 18th centuries, away from representational knowledge and towards the new-life sciences, which focused on the process of life rather than any static and visual similarities. For example, the botanical inventories and visual displays of Carl von Linné (1707–1778) were displaced by the new biological sciences (Foucault 1973). In both cases, the world was presumed knowable, and proper scientific methods would eventually result in the revealing of the truth about the world. René Descartes (1596–1650) established the faith in the possibility for humans to become detached and "free" observers of the world, and to apply a rationality based on reproducible experimentation to gain comprehensive knowledge about "nature." Cartesian assumptions have been successfully challenged (Taylor 1987), and the traditional relationship between natural sciences and "nature" has also been questioned. For example, two sociologists of science, Bruno Latour and Steve Woolgar, have shown that "natural facts" are the products of power relations within a scientific community. Furthermore, "facts" — as "natural objects" — are sometimes the creation of the scientific community itself through its research practices such as laboratory research. By isolating specific elements under laboratory conditions, scientist are creating new natural "facts" which could not have existed previously without their intervention (Latour 1983; Latour and Woolgar 1979).[1]

Today, even most natural scientists would agree that what is called "science" is one among many cultural artefacts and constructions (Simmons 1993; Evernden 1992). Of course, this does not mean that "science" has no privileged position in legitimating "modern" knowledge. In fact, it might be its prime function: to create a community of believers. Science — a peculiar cultural artefact — is therefore the result of "social construction." Today, most social scientists would agree, too, that our knowledge in general — including knowledge about the environment — is the result of social construction. Elizabeth Ann Bird is correct when she points out the contradiction between the obvious social construction of science, as a social practice, and the object of science, which is supposed to be nature itself. In this context, the concept of "environmental problems" assumes that there is a fundamental natural order which is being disrupted, usually by humans. It implies a truth-claim about nature based on an accumulative knowledge and an objective certainty about the working of the "natural" world. This does not mean that environmental problems are less real or simply don't exist. The arguments against social construction are similar to the ones against relativism: "relativism is relative" (Slezak 1994). As a way out of this tension, Bird suggests that environmental problems should be examined "through a variety of health, survival, moral, empathetic, aesthetic, political, economic, and cultural interests" (Bird 1987: 260). This would enable us to get away from the self-imposed imperative to prove the existence of environmental problems under criteria of the natural sciences, which cannot be certain to be accurate knowledge about nature in the first place. Similar to Charles Taylor's hermeneutics, Bird's solution is an "inter-subjective, socially negotiated moral truth achieved in the interests of (environmental) justice" (Bird 1987: 261).

Finally, I want to focus on the issue of the use — and consequently absolute misuse — of naturalist justification (Shimony and Nails 1987; Shimony 1993). Because of the probable and practical impossibility of knowing nature in itself, truth-claims about nature have often been used to justify conflicting political projects and also social and political knowledge (Mackenzie 1979). For example, naturalistic arguments were called on to justify both the origins of absolutist power of the monarchy and the natural rights of the king's subjects to withhold the ultimate political consent. A more recent example is the recourse to naturalistic arguments to condemn "homosexuality" — "it is against the natural reproductive function, and therefore it should be repressed" — or in favour of greater social acceptance — "there are many examples of homosexual behaviour in the natural world, therefore it is natural." In both cases, the recourse to naturalist arguments is a dead end because nature itself cannot be called to the witness stand. Furthermore, the issue of homosexuality is more a matter of social, cultural, religious and political belief, than a sharply defined natural category. Naturalistic arguments should also be viewed with great suspicion, even when the principal object of attention is the "natural world." For example, pest-control techniques create new conditions which cannot be measured against an absolute natural scale. John Perkins has shown that various insecticides can only be measured in relation to the cultural and social norms at a given time (Perkins 1982).[2] If, as Wittgenstein suggested, knowledge is a matter of believing, then there is an even more radical challenge laying at the doorsteps of Environmental Studies: "the difficulty is to realize the groundlessness of our believing" (Wittgenstein 1969: section 166).

The prevalence of the social-construction thesis reflects, in part, the intellectual context of late 20th century in the North-West which no one can avoid, namely "postmodernism" in the social sciences and the humanities, and "post-Kuhnian" approaches in the natural sciences. Obviously, I am not suggesting that every social scientist is postmodern or that every natural scientist is post-Kuhnian. However, the prevalence of the social-construction thesis reflects the growing importance of the terminology of postmodernism in the social sciences. My point here is not to come down in defence of either modernity or postmodernity *per se*. More simply, I want to point out that we live in a world in which there is an intense debate between modernity and postmodernity, which has also ramifications for Environmental Studies (Bennett and Chaloupka 1993). Because of this debate, students in

Environmental Studies can no longer relate to themes like "truth," "progress" and even the "environment" in the same way that it was possible before postmodernism emerged. Our discursive practices have changed and therefore the objects of these practices, like the environment, have changed too. In this context, Environmental Studies has become more than the mere study of the natural world from the point of view of a modernist, and more precisely, from the view of the natural scientist. The environment is not only the aggregate of rocks, trees or rivers, or any interaction between them, it is also the various ways which humans have come to perceive and interpret those "rocks," "trees" and "rivers," and their "interactions." The environment is not merely a set of biological interactions, it is also cultural, social, political, economical, ideological, artistic and so on, which take place at the level of what Wittgenstein would call "language-games."

One positive contribution of postmodernism to the study of the environment has been to promote a plurality of interpretations about the environment which could be called interdisciplinary. In part, it is the debate around postmodernity — and post-Kuhnian approaches like those of Paul Feyerabend (1924–1994) — which have made it possible for the emergence of an interdisciplinary approach to the environment (Feyerabend 1975, 1987). If I am correct in my understanding of interdisciplinarity — i.e., a knowledge attitude based on an heterogeneity of interpretations which cannot be synthesized easily in a coherent and simple theory — it also implies that no single interpretation is superior to the others. They are just different. Therefore, I believe that the main purpose of an interdisciplinary approach to the environment is to map out the broad spectrum of the various types of knowledge around the environment, and the specific conditions for their emergence. I should point out that interdisciplinarity is not merely the sum of the knowledge of each individual disciplines. It is a totally new knowledge based on the understanding of the intrinsic complexity of the world and of the heterogeneity of interpretations. It is also an attitude towards knowledge which is based on theoretical scepticism and humility about the limits of that knowledge, and on practical restraints on acting in the world. However, Environmental Studies is not simply the result of a postmodern attitude towards knowledge. It is also embedded in at least three other major shifts which took place around language, history and the self.

Language and the Environment

Italo Calvino, the well-known Italian writer, illustrates very well the problem surrounding language and knowledge about the world. Calvino's book, *Mr. Palomar*, is a series of short stories about the reflection of an aging man (in fact Calvino himself).

> If man were to invest in whistling everything he normally entrusts to words, and if the blackbird were to modulate into his whistling all the unspoken truth of his natural condition, then the first step would be taken toward bridging the gap between ... between what and what? Nature and culture? Silence and speech? Mr. Palomar always hopes that silence contains something more than language can say. But what if language were really the goal toward which everything in existence tends? Or what if everything that exists were language, and has been since the beginning of time? Here Mr. Palomar is again gripped by anguish (Calvino 1994: 27).

I believe that Mr. Palomar's anguish is a cultural anguish that is specific to our late 20th century context in the North-West. It is an anguish about the growing certainty about the context-specificity of language and the world in which humans live.[3] Let's examine the context about language.

Around the turn of this century, language *per se*, became a legitimate object of human investigation. Until then, the study of language was limited to the study of the general rules governing the proper use of a language — the grammar. It was then presumed that language was only the medium used to express the true nature of the world as humans were slowly but surely discovering it. In some cases, just the uttering of words called up the magical and/or religious worlds. In general, language was literally the "representation" of the world as it was. There was a direct relationship between a "word" and the object it named. More exactly, the relationship between the word and the object it named was not yet problematized. Because there was no doubt about the existence of the object (or God or a magical world), the word was presumed to represent the object in question. However, the existence of many languages and therefore, many different words for the same object, was causing problems for the unproblematized relationship between language and the world.

It was Ferdinand De Saussure (1857–1913) who broke the direct relationship between language and the world by introducing the "signifiant" and the "*signifié*" [signifier and signified] (De Saussure [1922] 1983). This rupture in the direct relationship between language and the world challenged another accepted assumption: the primacy of the positivist view of reality, that is to say, the undoubted existence of objects in themselves.

Ludwig Wittgenstein explored the philosophical implications of this issue and turned the relationship between language and reality upside down (Heaton and Groves 1994; Monk 1990). For Wittgenstein (mainly after 1929), it was language which constituted our reality, not the other way round. He held that there were words in language which were irreducible to rational explanation. "You must bear in mind that the language-game is to say something unpredictable. I mean: it is not based on grounds. It is not reasonable (or unreasonable). It is there — like our life" (Wittgenstein 1969: section 559). In order to explain his conception about language, Wittgenstein used the analogy of a rhizome. Traditional philosophical conceptions perceive language like a tree. There is a clear origin — the roots — and a linear and rational progression towards the branches. For Wittgenstein, the image of the tree — and its liner roots — doesn't render the full complexity of language. On the contrary, language should be perceived as a rhizome — bulbs and tubers — to illustrate that every points of the language is simultaneously linked to all other parts. Unlike the roots of a tree, it is not possible to disentangle a rhizome. In the same way, there are parts of language which cannot be disentangled from the language itself: "it is there — like our life." One cannot make sense of a rhizome by making sense of the separate elements forming a rhizome. It is the articulation between the various points in the rhizome which gives sense to the whole. Wittgenstein's prime concern is not about gaining absolute knowledge — i.e., "truth" — but the clarity about statements about truth-claims. "The *truth* of my statements is the text of my understanding of these statements" (Wittgenstein 1969: section 80).

This assumption about language has important implications for Environmental Studies. To start with, the word "environment" itself has to be contextualized within the conditions of its emergence. The word "environment" was probably used for the first time in an English-speaking scientific publication in the 1930s, and became a popular term in the late 1950s and 1960s. The institutions created in this period adopted this word "environment" to describe the object of their attention: for example, the Ministry/Department of the Environment. Following Wittgenstein and Foucault, one could therefore assume that the emergence — and the widespread adoption — of the word "environment" reflected a perception of the world — of the reality — in a specific historical context, in a specific configuration of the "field of power/knowledge."[4] The "environment" is not simply a new term to describe some ahistoric and universal reality, but it is the "constitution" — the "construction" — of a new reality. Therefore, "Environmental Studies" is about the study of the conditions of the emergence of this new category — i.e., the "environment" — including all the "contextual" narratives ranging from biology to literature (Jordanova 1986).

This suggested "language" approach for Environmental Studies is not new. It has extensively been practised in the field of Cultural Studies, its purpose being to breakdown the existing barriers between different types of human experience as they related to knowledge. Raymond Williams was the chief proponent of this holistic and interdisciplinary approach (Williams 1963). It was an attempt to resist hyperspecialization in the various human practices around knowledge. It was also about the recognition of the importance of language-games as constituting our sense of reality and our identification to that reality.

The Plurality of Environmental Histories

The general context surrounding postmodernity and language-as-constitutive leads us to challenge a third concept, the one of "history." For some historians, history is still the somewhat objective account of how nature and people were in the past. The assumption behind this concept of history is the possibility of finding an order, a purpose, a unity in direction and a general temporal coherence in the world of the past. The main challenge facing historians, and which can't be avoided, is to project contemporary concerns back into the past. "Presentism" — as it is called — is always at play behind the backs of all narrators, even narrators of natural history. For example, it is now well-established that Charles Darwin's natural evolutionist theory was strongly influenced by Darwin's exposure to late 19th-century capitalist social ruthlessness (Tort 1992; Crook 1994).

Donald Worster suggested a way out of the tendency of historians to impose universal rules and order on the workings of history. Worster points to the various rates of change in the natural world and in human histories which might direct us toward a multiplicity of histories. Indeed, "(h)istorians can no longer claim that there has been a single universal narrative of change that all species, all communities, all places have conformed to. 'History' has given way to 'histories'" (Worster 1994b: 14).[5]

In this context, history should now abandon its pretension of unearthing the definitive and the final account of the past. The diversity of the historical narratives written by historians over, let's say, the past 200 years, should convince us that, far from discovering how things and people were in the past in a definitive fashion, the practice of historians has been to offer seducing interpretations of the past, from the vantage point of the concerns of the present, from a specific social, cultural or gender position. "Traditional" historians would argue that their mission is to reveal elements of a past which have a definite certainty about them. This position is flying against the practice of the traditional historians themselves, and against the purpose of history as an academic discipline. The practice of history has been precisely to challenge previous interpretations of historical narratives, sometimes to turn them upside down. Therefore, history is facing a dilemma between adopting a truth-claim or a narrative-contingent position. I don't think there is a way out of this dilemma which might reflect our postmodern condition. We just have to live with it. One way of dealing with it, in an academic context, would be to avoid imposing one single history. On the contrary, one should promote the plurality of histories and let the marketplace of ideas determine which ones are the most convincing narratives for the time being, while certainly subject to subsequent revisions.

"Environmental history" does not escape the above-mentioned dilemma. What environmental history is about, remains the central issue. Some, mainly in the natural sciences, would focus on a grand narrative of the history of the natural world (Ponting 1991). However, even within the context of the natural sciences, there are many histories of the natural world, not systematically contradicting or confirming each other. At least, since the publication of Thomas Kuhn's *Structure of Scientific Revolution*, we know that scientific truth is validated within the boundaries of a scientific paradigm (Kuhn 1962). In the context of the natural sciences, there are many examples of disciplines or scientific approaches which have disappeared. Therefore, a more promising environmental history would focus on the study of the various interpretations offered by the diverse schools in the natural sciences, rather than directly on the object of the natural sciences. As I have already pointed out in the previous section, it is language which constitutes the boundaries of reality. In this case, it is, in large part, the language of the natural sciences which constitutes the imageries of our world.

However, in the context of environmental history, there is another tension which is more pernicious than the previous one. When writing narratives about environmental history and the interactions of humans with the natural world, one could be tempted to adopt a naturalist position. Many environmentalists and ecologists extract — from natural history narratives — fundamental and normative principles guiding human actions in history. The epistemological problem is that these fundamental principles emerged from interpretations of natural-history narratives, which themselves are the products of cultural construction. As nature does not talk to us directly, one cannot rely on the presumed superiority of "naturalness" to ground one's investigations and historical research. The histories of the use and misuse of naturalist arguments to justify and legitimize conflicting social, cultural and political projects, should be a convincing reminder of the risks involved in bringing in naturalist arguments (Phelan 1993). There is also a danger of reintroducing a universalist narrative by the means of "environmental principles" which are themselves historically contingent.

For example, Martin W. Lewis, in his attempt to debunk one of the environmental myths about a lost time when "humankind lived in easy harmony with the rest of the natural world," argues in favour of a single global history. For Lewis, "environmental history's prime challenge now is to outline and explain the long-term development of human pressures on ecosystems around the globe" (Lewis 1994: A56). The only problem with this approach is that it attempts to reduce and universalize human historical experiences to a single category (i.e., the ecosystem). In support of his argument, Lewis cites Paul S. Martin's research about the destruction of 80 per cent of large mammal species in North and South America by late-Pleistocene Paleo-Indian migrants some 10,000 years ago (Martin 1967, 1984). Although this might be correct, I doubt very much that the Pleistocene Paleo-Indians migrants would have seen what they were doing as an attack on the ecosystem, or even as an environmental crisis as such. Ecosystem and environmental crisis can only be understood by people sharing a common type of knowledge, and using the same referential language-games. This closely-knit cultural web between knowledge, language and historical narratives sets the conditions for the constitution of specific identities adopted by humans. Therefore, the issue of the "self" becomes an intrinsic part of environmental issues.

The Theories of the Self and the Construction of an Environmental Subject

> We live our lives inscrutably included within the streaming mutual life of the universe (Buber 1970).

Finally, it is important to look at human subjectivity from a perspective which takes into consideration the three previous premises. If one takes for granted that, indeed, (1) there is a plurality of knowledge; (2) language is constitutive of reality; and (3) there are histories rather than one "history," it should be possible to approach human subjectivity from a similarly reflexive, critical and sceptical perspective. This argument is highly relevant for Environmental Studies, as one of the central challenges is to explain the process by which individuals in the North-West have problematized their relationship to nature. In this sense, the environmental crisis is largely a cultural crisis in which the self-definition of humans is being problematized. The search for a new self is on. The governing of individual and collective environmental "conducts" (Tully 1988) implies the construction of an "environmental self" — of an environmental subject.

Each school of ecology has articulated a project which includes a radical transformation of the self, and sometimes the construction of a totally new self. For example, John Dryzek and Raymond Murphy, following the footsteps of Jürgen Habermas, argue in favour of an environmentally rational self, as the next expansion of modernist "rationality" (Dryzek 1987; Habermas 1979; Murphy 1994). On the other hand, "deep ecology" promotes an "ecological self," which is a total transformation of the individual's awareness towards other humans and non-humans. For "social ecology," the new self must be the fulfilment of human self-consciousness through which nature expresses itself.

"Ecofeminism" attempts to break down the traditional gender boundaries as a way of creating a new self, which would recognize the mutuality and the necessary integration of both male and female in a way which would be mutually enhancing (Pearce 1990). Beyond these four normative projects which focus on the self, there is at least one issue which should be addressed. The emergence of discourses about the self is as relevant as the substance and qualities that the self ought to adopt. Charles Taylor has remarkably illustrated the heterogeneity of the various conceptions of the self within the Judeo-Christian world (Taylor 1989).

There are two broad categories of approaches about the self: the "positivist" self and the "self-reflective" self.

The positivist self presupposes the possibility for humans to detach themselves — at least methodologically — from the world in order to gain a neutral, and superior, observation point. This positivist position assumes that there is a sharp epistemological difference, and hierarchy, between the object and its observer. The positivist self sees itself as above the non-human world.

On the contrary, the self-reflexive self questions the place of humans in relation to the world. However, there are at least three competing interpretations. The first one — the "humanist self" — assumes that the self possesses an ontological essence which is permanent, universal and transhistoric. The purpose of the study of the humanist self is to discover, behind all cultural and historical differences, a deep and universal self. This type of approach is what individuals like Hans Georg Gadamer or Charles Taylor would call the "fusion of horizon" (Gadamer 1975, 1984; Taylor 1985).

On the opposite spectrum, the "social constructionist self" — à la Jacques Derrida — would argue that the search for the "real" self is as problematic and futile as the search for "truth" (Derrida 1981; Bennington and Derrida 1993). For this group, as no one can prove that the self possesses a universal, transhistoric, ontological essence, the self becomes the mere result of the social and historical processes.

You could ask me what the theories of modern identity have to do with Environmental Studies, whose focus of attention is supposed to be the natural environment. I would argue that the category "environment," and the related themes such as "pollution" and "deforestation," are more about human activities — i.e., results of human activities — than merely environmental problems. In this context, if the central environmental issue is human activities, it is important to find out what drives the modern subjects to adopt such damaging environmental practices. What are the elements in the modern identity which make it possible to engage in damaging environmental practices. This focus of the self-construction of an environmental self can lead to the attempt to find alternatives: ecofeminism, ecospirituality, "environmental citizenship" (which is an overtly humanist strategy promoted by Environment Canada), Green ethics, discovery or rediscovery of indigenous world-views and so on.

In order to illustrate the relevance of the self to Environmental Studies, let me mention briefly the "discovery" of the self by quantum physics, which happens to be very similar to Buddhism (Zohar 1990). Einstein's discovery of quantum physics confirmed the necessarily simultaneous and complimentary existence of particles and waves. This constituted a radical challenge to established Newtonian's physics, which perceives the world as only matter (i.e., particles). Consequently, quantum physics questions all the assumptions of Newtonian physics such as time, space and causality. For example, "Max Planck proved that all energy is radiated in individual packets called 'quanta' [hence 'quantum leaps'] rather than in flowing streams over a continuous spectrum" (Zohar 1990: 14; Planck 1922). The quantum view of the world tends to be based on "indeterminency maze of probabilities," rather than on the material certainty of the Newtonian physics (Zohar 1990: 12). The particle/wave complementarity of the quantum physics better explains the behaviour of correlated photons. Experiments have shown that "the photons' behaviour patterns are so eerily linked across any spatial separation — it could be a few centimetres, it could be all the way across the universe — that it appears there is no space between them" (Zohar 1990: 20). The consequences of this discovery challenges the current North-Western paradigm of total separation between matter and consciousness.

The scientific view of quantum physics is similar to the spiritual Buddhist world-view of energy flow. Quantum consciousness is not about an awareness of an external material reality, grasped by thinking subjects equipped with a special human faculty, called consciousness. On the contrary, consciousness becomes the manifestation of the phenomenon of the particle/wave complementarity. The consequences of quantum physics for the conceptualization of the self are profound. The self is not any longer the attribute of atomized and self-contained individuals, but the attribute of all life-forms, human and non-human. The self becomes an "extended self, a much larger part of which is I-and-you" (Zohar 1990: 147; Zohar and Marshall 1994). Secondly, the quantum self establishes our sense of freedom and re-establishes our responsibility which had been undermined by modern science. "In Newton's dead and silent universe, we conscious human beings have neither any part to play nor have we any power to still the march of blind and immutable forces" (Zohar 1990: 156). A quantum self can offer not only a radically "new" view of the world, but also a sense of belonging to the world as responsible "co-authors of the world" (Zohar 1990: 170) and a "re-enchantment of the world" (Berman 1981, 1989).

Regardless of the scientific validity of quantum physics, it remains an unexpected attempt, originating from the core of modern sciences, to reincorporate humans in the "oneness of being." It is ironic that modern sciences — or at least some elements of them — which always attempted to keep human subjectivity away from scientific observation, is now discovering that human subjectivity is an integral and active part of the world. The solutions to the environmental crisis probably don't lie in the promises of technological innovation, but in the necessity of constructing a self as part of the world. Some Marxists, committed to "historical materialism" as a revolutionary telos, are now articulating a radical social project which explicitly includes the construction of a new self: a self which consciously would puts limits to its own self (Gorz 1993; Bahro 1994).

Conclusion

What I have attempted to do in this presentation, was to review the conditions under which Environmental Studies have emerged in the context of our late 20th century. It should be obvious that these conditions do not constitute a kind of manifesto on my part. I wanted only to contextualize the possibilities for such an approach. Based on the four premises I outlined in this article (knowledge, language, history and the self), I believe that Environmental Studies should reflect and promote the heterogeneity of the various environmental types of knowledge, and not be limited to academic disciplines and, I might add, to specific parts of the world.

Based on the historical experience of the world in the 20th century — two world wars; one cold war; dozens of conflicts across the world; the Nazi regime, and the systematic and organized genocide of entire populations, the bombing of Hiroshima; the threat of nuclear winter; the re-emergence of nationalism in Eastern Europe, and the possibility of endless civil wars as in the former Yugoslavia; the re-emergence of fundamentalism (including neo-Christian American fundamentalism); and the genocide in Rwanda — it easy to understand the scepticism, and even overt pessimism, which seems to prevail in the end of this century. Even if the world managed to survive all the risks listed previously, the emerging ecological crisis in its global, regional, local and social dimensions, is still facing us. Karl Marx said that the role of intellectual endeavour was not to interpret the world, but to change it. Now that humans have changed the planet beyond recognition, and in a way which probably threatens the long-term biological viability for humans, it might be more appropriate to "save" the world rather than to "change" it. Despite his 19th century optimism, I would argue that Marx is still correct today in the sense that the change he was calling for, was largely a change within and between humans — a "social revolution." In a similar way, the environmental issue is cultural, political, social, ethical and so forth. In this context, an interdisciplinary approach to Environmental Studies is a new form of knowledge and, as Foucault taught us, it also offers possibilities, and risks, in the broader field of power relations.

Finally, my concluding remarks seem to shift away, from a generally favourable attitude towards postmodernism, towards a more conventional "modernist/humanist" conclusion. You might rightly ask me to explain this apparent contradiction. For me, "postmodernism" is largely a descriptive category of the present. The prescriptive agenda of postmodernism, such as the celebration of differences, the plurality of meaning, of knowledge, of histories, of languages and so on, is very similar in its intentions to the humanist and the subsequent modernist project. The humanist concerns for "religious tolerance" as articulated, for example, by Spinoza (1632–1677), was an early form of the recognition, if not the celebration, of differences (Spinoza 1989). The liberal themes of "equality" or "rights," articulated and institutionalized during the French Revolution, and the theme of "human emancipation" announced by Marx, were both attempts to resist and overturn, respectively, a rigid social order and the unidimensional economic order of capitalism. In the realm of knowledge, the Cartesian doubt — doubt about nearly every single accepted convention — was the basic premise for scientific knowledge. Doubt is still the basic premise for most postmodernists.

In this context, an interdisciplinary approach to Environmental Studies should pursue the humanist/modernist project, not as an ontological category (the "subject" or "knowledge"), but as a strategic category, or more exactly, as an attitude to be constructed. I would suggest that this attitude should have three guiding axes: (1) modesty about the extent of knowledge and the possibility of knowledge; (2) extreme prudence in applying technological innovations; and (3) critical reflection about the self leading towards a human community relying less on technological solution. To paraphrase Wittgenstein, one could say that Environmental Studies "... is not a body of doctrine but an activity" (Wittgenstein 1961: section 4.112).

DISCUSSION QUESTIONS

1. Choose one ecological principle (like "ecosystem," "sustainable development" and so on), and find out when and how this concept emerged.
2. Choose a picture or a painting of a "natural" landscape you know personally. Ask your colleagues in your class to identify what seems to be the "natural" features. Tell them the history behind the "scene." For example: "this wonderful picture of the Scottish Highlands, with its rolling bleak hills, is the result of an ecological trauma caused by the cutting of the old-growth forest centuries ago."
3. Find one thing, or a phenomenon, which is usually considered "natural" (for example: milk, soil erosion, lawn etc.). Do some research to find out some of the arguments which suggest that it is "natural," and those arguments which suggest that it is not "natural." What conclusions can you draw?

Endnotes

1. See also the British-based journal *Science as Culture*.
2. The first sentence of the book sets the tone: "Science and technology are cultural phenomena."
3. Wittgenstein's belief about absolute relativism is encapsulated in the following sentence: "If you tried to doubt everything you would not get as far as doubting anything. The game of doubting itself presupposes certainty." In *On Certainty*, section 115.
4. In response to G.E. Moore's "Proof of an External World," Wittgenstein wrote: "So when Moore sat in front of a tree and said: 'I know that that's a tree,' he was simply stating the truth about his state at the time." In *On Certainty*, section 532.
5. However, Worster rejects a relativist position ("radical historicist") and still clings to the naturalist concept of "interdependency" without giving us the basis for epistemological grounding. I would argue that "interdependency" is also a narrative which I am willing to believe for the time being, but refuse to give it any special transhistorical and universal grounding.

References

Bahro, Rudolf. 1994. *Avoiding Social and Ecological Disaster*. Bath, Eng.: Gateway Books.

Bennett, Jane and William Chaloupka, eds. 1993. *In the Nature of Things — Language, Politics, and the Environment*. Minneapolis: The University of Minnesota Press.

Bennington, Geoffrey and Jacques Derrida. 1993. *Jacques Derrida*. Chicago: Chicago University Press.

Berman, Morris. 1981. *The Reenchantment of the World*. Ithaca, N.Y.: Cornell University Press.

———. 1989. *Coming to Our Senses: Body and Spirit in the Hidden History of the West*. New York: Simon & Schuster.

Bird, Elizabeth Ann R. 1987. The Social Construction of Nature: Theoretical Approaches to the History of Environmental Problems. *Environmental Review*, 255–64.

Buber, Martin. 1970. *I and Thou*. New York: Scribners.

Calvino, Italo. 1994. The Blackbird's Whistle. In *Mr. Palomar* (Italo Calvino). Toronto: Lester & Orpen Dennys.

Crook, Paul. 1994. Naturalistic Fallacies and Noble Ends. In *Darwinism, War, and History to the Debate over the Biology of War from the "Origin of the Species" to the First World War*, ed. Paul Crook, 176–91. Cambridge: Cambridge University Press.

Darier, Éric. 1991. Une généalogie du «conservationnisme» et de l'écologie au Canada. In *To See Ourselves / To Save Ourselves — Ecology and Culture in Canada*, 37–49. Association for Canadian Studies, Montreal.

De Saussure, Ferdinand. [1922] 1983. *Course in General Linguistics*, ed. Charles Bally and Albert Sechehaye. London: Duckworth.

Derrida, Jacques. 1981. *Dissemination*. Chicago: Chicago University Press.

Dryzek, John A. 1987. *Rational Ecology — Environment and Political Economy*. New York: Basil Blackwell.

Evernden, Neil. 1992. *The Social Creation of Nature*. Baltimore: The Johns Hopkins University Press.

Feyerabend, Paul. 1975. *Against Method; Outline of an Anarchist Theory of Knowledge*. London: NLB.

———. 1987. *Farewell to Reason*. London: Verso.

Foucault, Michel. 1973. *Order of Things: An Archaeology of the Human Sciences*. New York: Vintage Books.

Gadamer, Hans Georg. 1975. *Truth and Method*. New York: Crossroad.

———. 1984. *Reason in the Age of Science*. Cambridge: MIT Press.

Gorz, André. 1993. Political Ecology: Expertocracy versus Self-Limitation. *New Left Review* 202: 55–67.

Habermas, Jürgen. 1979. *The Theory of Communicative Action*. Boston: Beacon Press.

Heaton, John and Judy Groves, eds. 1994. *Wittgenstein for Beginners*. Cambridge: Icon Books.

Jordanova, Ludmilla J., ed. 1986. *Languages of Nature — Critical Essays on Science and Literature*. London: Free Association Books.

Kuhn, Thomas. 1962. *Structure of Scientific Revolution*. Chicago: University of Chicago Press.

Latour, Bruno. 1983. Give me a Laboratory and I Will Raise the World. In *Science Observed: Perspectives on the Social Study of Science*, ed. Karin D. Knorr-Cetina and Michael Mulkay, 141–70. Berverley Hills, Calif.: Sage Publications.

Latour, Bruno and Steve Woolgar. 1979. *Laboratory Life: The Social Construction of Scientific Facts*. Beverly Hills, Calif.: Sage Publications.

Lewis, Martin W. 1994. Environmental History Challenges the Myth of a Primordial Eden. *The Chronicle of Higher Education* 40(35): A56.

Mackenzie, W.J.M. 1979. *Biological Ideas in Politics — An Essay on Political Adaptivity*. Manchester: Manchester University Press.

Martin, Paul S. 1967. *Pleistocene Extinctions: The Search for a Cause*. New Haven: Yale University Press.

———. 1984. *Quaternary Extinctions: A Prehistoric Revolution*. Tucson, Ariz.: University of Arizona Press.

Monk, Ray. 1990. *Ludwig Wittgenstein: The Duty of Genius*. London: J. Cape.

Murphy, Raymond. 1994. *Rationality and Nature*. Boulder, Colo.: Westview Press.

Pearce, Jacqueline. 1990. Three Visions of An Ecological Self. *Undercurrents — A Journal of Critical Environmental Studies* 2: 28–34.

Perkins, John H. 1982. *Insects, Experts, and the Insecticide Crisis: the Quest for New Pest Management Strategies*. New York: Plenum Press.

Phelan, Shane. 1993. Intimate Distance: The Dislocation of Nature in Modernity. In *In the Nature of Things — Language, Politics, and the Environment*, ed. Jane Bennett and William Chaloupka, 44–62. Minneapolis: The University of Minnesota Press.

Planck, Max. 1922. *The Origin and Development of the Quantum Theory*. Translated by H. T. Clarke and L. Silberstein. Oxford: Clarendon Press.

Ponting, Clive. 1991. *A Green History of the World*. London: Sinclair-Stevenson.

Said, Edward W. 1978. *Orientalism*. London: Routledge & Kegan Paul.

———. 1994. *Culture and Imperialism*. New York: Vintage Books.

Shimony, Abner and Debra Nails. 1987. *Naturalistic Epistemology — A Symposium of Two Decades*. Boston: D. Reidel Publishing Company.

Shimony, Abner. 1993. *Search for A Naturalistic World View*. Cambridge: Cambridge University Press.

Simmons, Ian G. 1993. *Interpreting Nature — Cultural Constructions of the Environment*. London: Routledge.

Slezak, Peter. 1994. The Social Construction of Social Constructionism. *Inquiry — An Interdisciplinary Journal of Philosophy* 37(2): 139–57.

Spinoza, Carl von [Benedictus de]. 1989. *Tractatus Theologio-Politicus*. Translated by Samuel Shirley. Leide, N.Y.: E.J. Brill.

Taylor, Charles. 1985. *Human Agency and Language*. Cambridge: Cambridge University Press.

———. 1987. Overcoming Epistemology. In *After Philosophy: End or Transformation*, ed. Kenneth Baynes, 464–88. Cambridge, Mass.: MIT Press.

———. 1989. *Sources of the Self: The Making of the Modern Identity*. Cambridge, Mass.: Harvard University Press.

Tort, Patrick, ed. 1992. *Darwinisme et société*. Paris: Presses universitaires de France.

Tully, James. 1988. Governing Conduct. In *Conscience and Casuistry in Early Modern Europe*, ed. Edmund Leites, 12–71. Cambridge, Cambridge University Press.

Turner, B.L. et al., eds. 1990. *The Earth as Transformed by Human Action—Global and Regional Changes over the Past 300 Years*. Cambridge: Cambridge University Press.

Williams, Raymond. 1963. *Culture and Society 1780–1950*. London: Harmondsworth.

Wittgenstein, Ludwig. 1961. *Tractatus Logico-Philosophicus*. Translated by D.F. Pears and B.F. McGuiness. London: Routledge & Kegan Paul.

———. 1969. *On Certainty*. Edited by G.E.M. Anscombe and G. H. von Wright. New York: Harper Torch Book.

Worster, Donald. 1988. *The Ends of the Earth: Perspectives on Modern Environmental History*. Cambridge: Cambridge University Press.

———. 1994a. *Nature's Economy: A History of Ecological Ideas*. New York: Cambridge University Press.

———. 1994b. Nature and the Disorder of History. *Environmental History Review* 18(2): 1–15.

Zohar, Danah. 1990. *The Quantum Self—A Revolutionary View of Human Nature and Consciousness Rooted in the New Physics*. London: Bloomsbury.

Zohar, Danah and Marshall, Ian. 1994. *The Quantum Society: Mind, Physics and the New Social Vision*. New York: William Morrow and Co., Inc.

ON THE PROBLEMS AND POSSIBILITIES OF GREEN SOCIAL THEORY

Joseph G. Moore

McMaster University

The recent history of sociology owes much to its relation with social movements — a relation that has two facets. On the one hand, social movements which have been the focus of social theory, have been subjected to a sociological gaze that has sought to interpret, explain and understand the social, political and cultural significance of these movements. On the other hand, sociologists have sought out the forms of alternative knowledge (re)produced by these movements, as solutions to the epistemological, theoretical and methodological problems that face the discipline. This dual relation between sociology and social movements, acknowledges that social movements are both the subject of social theory and the authors of new ways of knowing (including innovations in social theory).

The relation of ecological movements and sociology is a particularly interesting example of this movement/theory connection. First, environmentalism has pushed ecological destruction and resource depletion onto the agenda of sociology. It has forced sociologists to reconsider the human/nature relation and to contemplate the idea of "natural limits" and biological "laws." Here, the relation is complicated by the often intimate connection between the "natural" sciences and environmental movements, for the knowledge produced in many of these movements is weighed with the knowledge and epistemology of biology, physics and chemistry.

Secondly, environmental, political ecology, animal rights and deep ecology movements have become the subject of much empirical and theoretical work, and the study of these movements and the political and cultural space that they produce, has forced social theorists to confront notions such as the "environment" and "nature" as contested concepts.

The following will evaluate Green social theory with a recognition of its place at the nexus of politics, science and ethics. It will be argued that this theory tends towards a dependency on essentialist/naturalistic explanation, traceable to the borrowing and reproduction of natural science frameworks by environmental movements. However, it will also be acknowledged that the study of the wider political and cultural practice of social movements comes to breach these certainties, and suggests that a coherent sociology of knowledge (which understands the discontinuities and limits of knowledge) must become the central component of a critical Green social theory.

A New Ecological Paradigm: Bringing Biology (and Chemistry and Physics) Back In

When I refer to Green social theory, I speak of a particular strand of environmental social theory that criticizes mainstream social thought as being based on untenable distinctions between the natural and social worlds, views these distinctions as the foundation of anthropocentric and unecological social theory that neglects important "natural limits," and seeks to reconstruct social theory through an appeal to an overarching naturalistic theory and method that, nonetheless, eschews reductionism. This direction in social theory is loosely equatable with that which Buttel refers to as the "New Human Ecology" (Buttel 1987).

Much contemporary work in Green social theory may be traced to that of Catton and Dunlap in the late 1970s. In a series of papers at the turn of the decade, Catton and Dunlap made a concerted effort to bring the knowledge of ecosystems into mainstream sociology, not as an analogy for social systems, but rather, as the inseparable context of these social systems. In the authors' words, this project sought to go beyond a sociology of environmental issues, and move towards an environmental sociology that focused on the "relationship between human society and the biophysical environment" (Catton and Dunlap 1980: 32). This move, developed mostly as a metatheoretical critique of existing social theory, argued that existing social theories, despite their diversity, could be placed within a broad paradigm that had at its centre a shared anthropocentricism — a world-view that insisted on an arbitrary division between humans and nature (Catton and Dunlap 1978: 42). In response, the authors called for, and traced the beginnings of, a new paradigm of social thought that (1) placed humans within a global ecosystem, (2) emphasized the interaction (interdependence) of social/cultural and natural systems, and (3) (re)introduced the understanding of natural limits, biological constraints and unalterable ecological laws.[1]

Even as their calls for a rejection of all previous social theory became more tempered, Catton and Dunlap's critique remained one which, in moving to theoretical and methodological practice, would require a thorough rethinking of the boundary between the social and natural sciences. At the least, it would require that social theory find room for the consideration of natural limits and bio-physical "laws."

The goals that Catton and Dunlap identify as the basis for a New Environmental Paradigm are far too broad to identify with one particular theoretical stance. Indeed, these basic assumptions are most often shared by those found across the wide spectrum of environmental sociology (and by many who are quite reluctant to be labelled as environmental sociologists). However, it is arguable that the heart of this project — the thorough integration of biophysical laws into social theory — has largely met with silence. As Buttel (1987) concludes, the New Human Ecology has had a minimal effect on the discipline of sociology as a whole, and has largely been relegated to a subsection of environmental sociology.

Now, there are some significant reasons for this "neglect," not the least of which is the spectre of reductionist sociobiology, where biophysical laws are taken seriously indeed.[2] Nonetheless, certain aspects of Green social theory have been adopted by a diverse range of environmental sociologists that explicitly reject biological reductionism. Most directly, the last few years have seen authors come to call for a reconstruction of the link between the natural and social sciences.

A Cautious Development

Ted Benton's (1991) call for the "cautious welcoming" of biology by social theory, attempts to move Green social theory forward by suggesting a "realignment" of the social and natural sciences. Benton claims that contemporary biology and the social sciences (especially those concerned with health, the body and the environment) have developed interests and understandings that both insist on this realignment, and have already carved out the theoretical and methodological space in which this realignment is to take place.

Now Benton's choice of the word realignment is a careful one.[3] Here he suggests that the nature of this realignment is best seen as the elaboration of an "already occupied" (although sparsely) space between nature/culture dualism and biological reductionism. In a manner very similar to that of Catton and Dunlap, Benton claims that if one is to work within this theoretical space, certain anthropocentric aspects of "traditional" social theory must be cleared away.[4]

Of the many such obstacles, it is the nature/culture and body/mind dualisms that follow from the distinction between human autonomy and natural necessity that come to be the focus of Benton's work (Hayward 1992: 5). Benton's project becomes that of transcending these dualisms which, echoing Catton and Dunlap, are said to structure most mainstream social scientific discourse, condemning such sociology to an inescapable anthropocentrism. According to this reasoning, avoiding, or transcending, these dualisms requires that sociology reposition itself vis-à-vis the natural sciences (Benton 1991).

This transcendence of human/nature dualism requires that Green social theory start from an understanding of similarity rather than difference. This requires that we focus on an explanatory strategy that:

> begins with the common predicament of natural beings and moves from that basis to render intelligible their specific differences in constitution, structure and modes of life (Benton 1988: 12).

It is Benton's position, following from that of Catton and Dunlap, and echoed in other recent attempts at Green social theory such as Dickens (1992), that such explanatory strategies must be based in a renewed naturalistic understanding of social life. A naturalistic theory, methodology and epistemology is viewed as the key to a repositioning of social theory such that natural limits, ecological laws and the similarity of natural beings can be provided adequate conceptual room within the social sciences.

The adoption of a naturalistic framework and the strategy of moving from similarity to difference (scarcity to possibility) suggests that this repositioning of the natural and social sciences is to take the form of an integration. Although Benton is careful in the use of this term, Dickens is quite brazen in his position that the goal is to develop "a set of methods which allow the natural and social worlds to be constructed and examined in a similar way" (Dickens 1992: 193). Following from the transcendence of human/nature dualism in social theory, it is expected that a more unified understanding of the natural-social world will be developed.

In a much reduced form then, Green social theory can be said to: (1) begin with the understanding that an anthropocentric bias in mainstream social thought makes it difficult, if not impossible, to incorporate an understanding of natural limits into social theory, (2) claim that this anthropocentrism is rooted in an arbitrary distinction between human autonomy and natural necessity, which is expressed in human/nature, body/mind (etc.) dualisms, (3) argue that these dualisms must be pushed aside in favour of an explanatory approach which starts with the understanding of similarities and only then moves towards an understanding of difference, and (4) contends that a naturalistic theory can best serve these goals and may eventually yield an explanatory framework that can bridge the natural and social worlds.

From Critique to Practice: A Critical Approach to Green Social Theory

The recognition that social theory has suffered from (been defined by?) a lack of serious concern for the place of physical (built and "natural") environments, is an important starting point for Green social theory. Also, it is evident that human/nature and other dichotomies, are intimately connected with the way in which social theory has explained the natural world. These dichotomies have made it very difficult for social theory to develop concepts that capture human/environment interaction and interdependence, without falling into a sociobiology that reduces human activity to biological need or a "Prometheanism" that reduces environmental limits to technological or organizational problems to be overcome.

While this critique is necessary and instructive, the promise of Green social theory rests on its ability to offer a coherent alternative. To remain useful, Green social theory must prove that it is possible to develop a social theory that: (1) is based on naturalism, (2) avoids anthropocentrism, and (3) remains within the theoretical space between nature/society dualism and biological reductionism. One of the most promising attempts at constructing a theory based upon this foundation, starts with the elaboration of a critical realism and moves towards a Marxist methodology and theory. Yet, we will see that the debate over Marx's work, highlights the difficulty in developing a non-reductionist approach while avoiding anthropocentrism. It will be argued that the balancing of each of these requirements is only possible if one makes unwieldy assumptions about what is and is not knowable.

Bringing Marx Back In: Green Social Theory and Critical Realism

Although the practice of Green social theory is less developed than its critique of anthropocentric social theory, Dickens has suggested that the way forward rests in an elaboration of a "critical realism" that avoids both the eclecticism and hopeless relativity of idealism, and the mechanistic, essentialist pitfalls of enlightenment science. The title "critical realism" covers a wide range of theory and method, and in this context no overview is possible. However, as this approach represents some of the brightest possibilities and most stubborn problems of Green social theory, a basic understanding is crucial.[5]

In a nutshell, critical realism seeks to replace the search for a linear relation between cause and effect with a scientific study of their dialectic relation. In a sympathetic reading of Marxian epistemology (most explicitly developed in the 1857 *Introduction* but also in the *Grundisse*), Jessop (1990:

162–66) provides one of the clearest summaries, reproduced in an even further reduced form below.

First, Marx's method rests on a set of ontological assumptions that together assert that the real world consists of the entirety of contingently realized natural necessities. Thus labour power is for Marx a real power, however its realization rests contingently upon a struggle between capital and labour (Jessop 1990: 163). Continuing with Marxist terminology then, the task of theory comes to that of moving from the "real concrete" to the "concrete in thought," all the while recognizing that there is no simple (i.e., linear) translation of the former to the latter. As Marx most effectively demonstrates in *The 18th Brumaire of Louis Bonaparte* and in *Class Struggles in France*, a realist methodology must ground itself in the real historical moments of complex multiple determination.

Contemporary realist theorists have further clarified the terminology of "real concrete" and "concrete in thought" by emphasizing that although they do assume the existence of a real world outside of thought, there are important discontinuities that foreclose on the direct access to this reality. It is recognized that knowledge of the world is "never theoretically innocent," that "the starting point of any enquiry is discursively constructed" (Jessop 1990: 163). The goal of theory moves from the testing of final theory, to the development of concepts capturing different levels of reality. One does not move from the concrete to the abstract, or from the abstract to the concrete, but rather dialectically between these two in a process in which the articulation of each level of concept introduces changes to the other (Jessop 1990: 164). As we will see, there is a trend in Green social theory, that biases it towards an non-dialectical movement from the concrete to the abstract, suggesting a dangerous move to "final theory."

On Needs and Powers: Green Social Theory and the Debate over Marx

For Green social theory, the central Marxist concepts, raised to the level of real phenomena by Marx, are the natural powers and needs of humans and of species. In short, Marx proposed that humans share with other species certain natural powers and needs (including both those requisites of survival such as food and those of companionship, hunting and even the appropriation of objects in their environment). These natural needs could only be met through interaction with an environment which makes any approach that includes the analytical separation of the organism from its environment, misguided.

Species powers and needs are, for Marx, those specific to a species. Here, some Green social theorists have turned to the *Economic and*

Philosophical Manuscripts of 1844 to echo, a young Marx's argument, that before the advent of private property the relationship of humans to nature was one which affirmed their species-being.[6] Production under capitalism, by separating the producer from the product "tears away from him his [sic] species-life," such that "his inorganic body, nature, is taken away from him" (Marx 1844, cited in Dickens 1992: 68).

Therefore, while there exists for Marx a real essence to humans' species-being, the goal of Marxist theory is to enter the complex reality of a historical moment of the capitalist mode of production, and elaborate the contingent specificity of the condition of human species-life. The concept of alienation is but an attempt to capture the specific, distorted form of species-life in capitalist relations of production.

Now, while it is quite fair to say that Green social theory positions critical realism as the vehicle that moves social theory away from dualism (as equated with anthropocentricism and reductionism), there is considerable debate over how successful Marx was at developing a theory and method that met these goals. This debate is informative, as it sheds light on an underlying tension (contradiction is too strong here) between the attempt to move beyond dualism and the move away from reductionism. Benton insists that, in developing the essentially parallel arguments of humans species powers, and the natural powers shared by all species,[7] Marx sustains a human/animal dualism that fails the project of Green social theory (Benton 1988).

In short, Benton claims that a non-dualistic reconstruction requires that the distinction between natural and species powers be dissolved. In its place, Benton suggests that all species have their own characteristic species-life and can be said to "flourish only when the conditions for the living of the mode of life characteristic of the species are met" (Benton 1988: 13). A more suitable approach, it is claimed, starts not from how human needs are different than non-human, but rather from how these needs are essentially similar. The important difference between species is to be found in the way they fulfil their needs, not in the needs themselves.

Benton claims that this reconstruction of early Marxism provides the framework by which dualistic, anthropocentric understanding is replaced by an eco-centric naturalism. In so doing, Benton positions Green social theory some distance from the method of theoretical inquiry suggested by critical realism. While Dickens suggests that Benton is overly hostile towards Marxist theory, which Dickens claims can be read as rather close to Benton's reconstruction, he nonetheless supports the move towards the primacy of similarity. What I wish to argue, is that by insisting that theory move from similarity to difference, Green social theory, at the best, loses critical potential and, at the worst, may end in biological reductionism.

Back to Nature?

As we have seen, Benton's "reconstruction" of Marxism rejects the notion that human and non-human needs are substantively different. Both human and non-human species have needs for survival and for flourishing that are substantively similar. The difference between species is merely the way in which each goes about meeting these needs. The goal of Green social theory becomes the development of a conceptual framework that explains what a species needs to flourish and how it can go about meeting these needs. Given the relative similarity of the needs of human and non-human species, it is expected that Green social Theory will transcend the tradition boundary between the biological and the social sciences. As Dickens rightly, although somewhat uncritically, suggests, Green social theory lies on the verge of Marx's notion of "one science" (Dickens 1992: 152).

As Hayward (1992) has noted though, this project falls on its extension of the decidable of natural science into the undecidable of normative action and belief. While not denying that non-human species can flourish, human knowledge of this flourishing is highly questionable. At the worst, claims about the flourishing of non-human species may, in fact, be a version of the anthropocentrism that Benton seeks to reject (Hayward 1992: 10).

In attempting to develop an understanding of the relation between natural limits and human "needs," while avoiding the human/animal dualism, Benton ends up overextending the limits of naturalistic knowledge. In this case, the inability to accept that the knowledge we can have of human and non-human needs is qualitatively different (which is traceable to the insistence on similarity and a fear of reconstructing a human/non-human dualism), leads Green social theory to insist that we can distinguish the difference between a turtle that is surviving and one that is flourishing. This type of fallacy suggests that it is quite possible that the transcendence of dualism, and the goal of an overarching naturalism, is overambitious and misguided.

Now, I don't claim here, that Green social theory seeks out such an overambitious project. Indeed, the theorists mentioned here differentiate themselves from reductionist sociobiology by imposing strict limits on determinist explanation. However, it does seem that, in practice, Green social theory has a problem limiting its work to the theoretical space it claims. This problem, I would argue may be attributed to a project that seeks to transcend, rather than interpret, human/nature dualistic understandings, and is exacerbated by the methodological approach of moving from similarity to difference. Together these factors move Green social theory away from the non-reductionist, dialectic relation between society and the environment, that the critical realist project demands (the space between a dualism that ignores the environment and a reductionism which ignores the social).

If Green social theory is to recapture the critical possibilities of this project — a better understanding of the social movement — social theory relation is crucial.

Social Movements and Green Social Theory

Many of the problems of Green social theory may be traced to the rather uncritical stance that these theories have taken towards environmental movements. Early in the development of environmental sociology, Catton and Dunlap noted that the increased salience of ecological constraints (particularly as embodied in the concept of "ecological scarcity"), "encouraged environmental sociologists to more boldly challenge existing social theory" (Catton and Dunlap 1980: 32). What was not stressed however, was that the "increased salience of ecological constraints" and concepts such as "ecological scarcity" were, in the late 1970s, being (re)constructed in the dialogue between environmental movements and natural scientists. At the time, many environmental movements were rather straightforwardly reproducing the knowledge of systems biology (Eyerman and Jamison 1989). This reproduction of scientific knowledge was smuggled into early environmental sociology in the form of the essentialist notion of "ecological limits."

Later Green social theorists have been more explicit about the role of ecological and other social movements in the development of social theory. Benton makes his debt to social movements explicit — it is the interpretive and explanatory concepts created within contemporary social movements, especially feminist, gay and lesbian and animal-rights movements that enable, and perhaps demand, an integration of biological and social theory, and the development of a non-reductionist, but naturalistic, social theory. Benton credits recent feminism with problematizing the sex/gender relation, environmental movements with exploring the interconnectedness of humans and ecosystems and the animal-rights movement with blurring the distinction between the human and non-human.[8] Once again, though, the appropriation of the knowledge (re)produced by social movements is done in a rather uncritical fashion.

As in the work of earlier Green social theorists, Benton commandeers the language of these movements (with a special emphasis on animal-rights movements) and inserts it within a naturalistic social theory without reflecting on the political struggle from which these truths have arisen. This time, the failure is compounded by the fact that struggles that problematize dichotomies such as the human/nature relation, also problematize projects that seek to transcend them through the appeal to an overarching naturalistic theory. The study of these social movements forces one to recognize the social, historical and linguistic construction of the human/environment relation, and indeed the concepts themselves.

The critical understanding of the practice of social movements, one that recognizes the historical contingency of the human/nature relation, leads in one of two directions, both of which stress a retreat from the notion of natural limits and a renewed emphasis on the limits of knowledge.[9]

The first of these directions, and I would expect the least palatable to theorists in the Catton and Dunlap/Benton/Dickens tradition, is the investigation of the discursive construction of nature that is at the centre of poststructuralist/postmodernist theory. This approach, informed by literary theory, insists on displacing the "logic of nature" for the "rhetoric of nature" (Bennett and Chaloupka 1993: xii). This approach maintains that language does not, and cannot, simply refer to a "nature" that is external to itself. From this vantage, the quest of Green social theory for an overarching naturalistic theory and method, must be viewed as one among many linguistic claims, or more sinisterly, as "rhetoric-hiding-as-scientific-fact" (Bennett and Chaloupka 1993: xii). Such work finds an affinity with Green social theory in that it recognizes the problematic foundations of human/nature and culture/nature dichotomies. This approach differs, however, in limiting its scope to the deeper understanding of how such dichotomies come to be linguistically reproduced and/or challenged. What these theorists avoid, is the turn back to an essentialist social theory, such as the naturalistic understandings offered by Green social theory.

This is not to say that discourse analysis offers a simple substitute for the essentialism of Green social theory. The radical relativism of discourse theory is much better at deconstruction than it is at reconstruction. At the worst, it can be argued that this approach, recognizing that there can be no direct way to capture the "reality" of nature, comes to deny any consequence of the material — nature, society and politics are reduced to text. While the debate will not be settled here, it will suffice to say that poststructuralism challenges Green social theory to critically examine its own political foundations. Such reflection need not lead to a rejection of Green social theory, but it does suggest a move away from Benton's naturalism, and a renewed emphasis on the political and contingent nature of "nature."

The second direction that Green social theory may head, one arguably compatible with an understanding of the role of language in environmental politics is towards a radical political ecology. Such an approach, would attempt to recapture the dialectical possibility of the of critical realist epistemology, and seek to understand the relation between ecology and politics without subsuming one under the other.

Green social theorists have traced the beginning of such a political ecology. Dickens's insightful account of Marxian understanding of the human-nature relation is an important step. Yet, what is crucial to emphasize, is that having learned from social movements that nature is a contested concept, and from political economy of the co-determinant relation between humans and their environment, many of the goals of Green social theory must be forsaken. The lesson to be learned above all, is that overarching understandings that draw continuous explanations (from the natural to the cultural or vice versa) are problematic.

Having been thus informed (humbled?), Green social theory must turn its back on the project of transcending human/nature dualisms and settle for an approach that recognizes and explains these dualistic understandings as social, cultural and political constructions. Focus must be shifted to the better understanding of how nature/human dualistic understandings come to be transformed into relations of domination. This is the direction suggested by Dickens when he insists that Green social theory address questions of the "real, material practices involved in contemporary alienated relations with nature" (1992: 195).

In many regards, I think that this project must open itself to the charge of anthropocentrism, if by that we mean that dualistic understandings are not to be explained away, but rather to be viewed as unresolvable discontinuities in knowledge. A critical Green social theory must recognize that such discontinuities preclude the notion that we can (or should) develop definitive understandings of either the natural or the social. The way forward, from the understanding of such discontinuity, is not to develop grand explanatory schema but to understand dialectically the relations between the natural and the social. This approach does not move from similarity to difference, from the natural to the social, but rather dialectically between these constructions in a real, particular historical moment.

Conclusions

For all its difficulties, Green social theory remains a relevant and vital area of social thought. It is imperative that humans come to more adequately understand the co-determinacy of human activity and environmental context. The history of social theory is one of profound neglect of this relation, a neglect that has paralleled the development of sociocultural systems destructive and exploitive of the human and non-human world. Yet, nothing is gained by underestimating the complexity of these relations. The knowledge we gain from studying the practice of social movements comes to warn us that there is no simple way to reintroduce biophysical laws into social theory. We come to acknowledge the historical contingency knowledge which turns our attention away from the search for a historical understanding and back towards the real limits of human knowledge and practice.

DISCUSSION QUESTIONS

1. What role have contemporary social movements (e.g. feminism, gay/lesbian, animal-rights, etc.) played in the renewed interest in the place of natural and built environments in social theory? In what ways do these movements raise questions about this project?
2. The charge of anthropocentrism has spurred some social theorists to demand a fundamental change in the way we think about the human/environment relation. Others have suggested that a form of anthropocentrism may be an unavoidable part of social theory. What are the bases for each of these arguments? What do they suggest for social theory?
3. Some have claimed that Green social theory may be viewed as a move toward the goal of achieving "one science." How is this goal both informed and criticized by Marxist theory and method?
4. It has been argued that we must seek a dialectical understanding of the relations of domination in the social world (e.g., sexism, racism, class domination, etc.) and human/nature relations. Using specific examples, speculate what such an understanding might look like. What methods might one use to examine these relations?

Endnotes

1. The particulars of the "paradigmatic shift" called for by Catton and Dunlap have been adequately addressed elsewhere (see Catton and Dunlap 1978; 1980; Buttel 1987).
2. Here the relation between social movements and social theory seems to be of some import, as many radical sociologists, drawn to environmental movements would also be those most opposed to conservative sociobiology. While this may have originally kept some from exploring the field, Benton (1991) places hope that this same caution will help Green social theory avoid crude reductionism. However, as has been well argued, there is no necessary connection between political position and the attraction of environmental politics (Yearly 1991).
3. See Benton (1991).
4. Much of Benton's 1991 article consists of an attempt to show that the obstacle of biological reductionism has already been, or is soon to be, cleared by advances in the natural sciences. This type of analysis of the movement of natural science away from determinist models, has become commonplace in both Green social theory and those seeking to develop a sociology of knowledge that spans the natural and social sciences. See Dickens (1992) for an example of the former, and Aronowitz (1988) or Wright (1992), for the latter. I would argue with Wright (1992) that the extent to which "advances" in theoretical physics and biology have undermined the practice of science, or the hegemony of "traditional" science in our technological culture, are somewhat overemphasized. Here, I avoid this debate and focus on the demands that Green social theory places on social theory itself, rather than the hopes it places on the natural sciences.
5. See Dickens chap. 8.
6. See Dickens pp. 60–89.
7. Benton notes that Marx echoes this dualism in the theory of alienation which differentiates between " 'crude', 'physical' or 'animal' need, on the one hand, and 'human' need on the other" (Benton 1988: 8).

8. Benton, in acknowledging his debt to movements that offer radical critiques of scientific knowledge, also claims that the knowledge produced by these groups is also that most hostile to biological reductionism, which bodes well for Green social theorists, who must occupy a space between such reduction, and the mainstream position of human/nature dualism.
9. This terminology is from Hayward (1994).

References

Aronowitz, Stanley. 1988. *Science as Power: Discourse and Ideology in Modern Society*. Minneapolis: University of Minnesota Press.

Bennett, Jane and William Chaloupka, eds. 1993. *In the Nature of Things: Language, Politics and the Environment*. Minneapolis: University of Minnesota Press.

Benton, Ted. 1988. Humanism = Speciesism. Marx on Humans and Animals. *Radical Philosophy* 50: 4–18.

———. 1991. Biology and Social Science: Why the Return of the Repressed Should be Given a (Cautious) Welcome. *Sociology* 25(1): 1–29.

Buttel, Frederick H. 1987. New Directions in Environmental Sociology. *Annual Review of Sociology* 13: 465–88.

Catton, William R. and Riley E. Dunlap. 1978. Environmental Sociology: A New Paradigm. *The American Sociologist* 13: 41–49.

———. 1980. A New Ecological Paradigm for a Post-Exuberant Sociology. *American Behavioral Scientist* 24(1): 15–47.

Dickens, Peter. 1992. *Society and Nature: Towards a Green Social Theory*. Philadelphia: Temple University Press.

Eyerman, Ron and Andrew Jamison. 1989. *Social Movements: A Cognitive Approach*. University Park: Pennsylvania State University Press.

Hayward, Tim. 1994. The Meaning of Political Ecology. *Radical Philosophy* 66: 11–20.

———. 1992. Ecology and Human Emancipation. *Radical Ecology* 62: 3–13.

Jessop, Bob. 1990. Regulation Theories in Retrospect and Prospect. *Economy and Society* 19(2): 153–208.

Marx, Karl. 1990. *The 18th Brumaire of Louis Bonaparte*. New York: International Publishers.

Wright, Will. 1992. *Wild Knowledge: Science, Language and Social Life in a Fragile Environment*. Minneapolis: University of Minnesota Press.

Yearly, Steven. 1991. *The Green Case*. London: HarperCollins.

ENVIRONMENTAL RISK: A MACROSOCIOLOGICAL PERSPECTIVE

Michael Dinshaw Mehta

York University

Sociological conceptions of environmental risk are rapidly changing the role of sociology as a policy science. For example, there are studies on organizational aspects of risk (Perrow 1984), analyses of risk distribution among classes and populations (Beck 1992), studies on epistemology or legitimation of risk knowledge (Jasanoff 1986) and feminist perspectives on risk (Welch 1990). The conflict between sociologically informed concepts of risk and the more traditional, probabilistic calculations of risk, represents a contest of competing social philosophies and visions about the future development of natural resources, the relationship between economic growth and environmental protection, the role of government, and projections and visions about our common future. A sociologically informed understanding of risk illustrates the interconnectedness of population, environment, organization, technology and social life. From this perspective, risk is more than a calculation of possible costs and benefits; it is a theoretical mechanism for weighing different sets of values and political orientations.

Introduction

Throughout history, risks have presented a challenge to survival. An ever-changing environment, incurable diseases, famine and human-made threats such as war, have always presented risks to both individual and collective survival. Arguably, the most insidious risk facing both the individual and the collective is the danger from the steady decline in the quality of the natural environment. Several examples, such as the wanton use of non-renewable resources, air and water pollution, careless management of natural areas which support the survival of plant and animal species, disposal of toxic and nuclear waste, pesticide residues in food and the thinning of the ozone layer should make it abundantly clear that environmental problems involve risk. Ulrich Beck (1992) believes that we are nearing the end of an era concerned with building an industrial society, and moving into a postindustrial "risk distributing" society, concerned chiefly with controlling environmental risks created by modern technology. As such, risk is not simply the lack of safety, but rather, the possible damage that may result from one's decisions.

In seeking to master these pervasive risks, democratic societies are committed to preserving certain basic values, such as the citizen's right to understand, and to participate in, governmental decision making. In this context, the relationship between the average citizen and expert, who possess specialized knowledge about risk, has become a matter of acute concern. Designing effective participatory mechanisms for the lay public is increasingly difficult in a decision-making environment heavily dominated by technical expertise. By adopting technical definitions of issues, the language of political discussion becomes specialized.

In essence, the kind, magnitude and distribution of risks borne by industrialized societies are determined by regulations and the effectiveness of their implementation and enforcement. Perhaps the most dramatic shift in policy on both environmental and health risks has been the acceptance of quantitative risk assessment as an analytic instrument for regulatory decision making. Because risk assessment involves conflicts or trade-offs between values, it is inherently political. If there are a number of different actors in the societal decision-making game, then conflicts are likely to emerge. Understanding the relationship between regulation and the political conflict over risk is critical when one heeds the veiled warning of Plough and Krimsky (1987): "Those who control the discourse on risk, will most likely control the political battles as well" (p. 4).

The *Oxford Dictionary of the English Language* (1991) defines "risk" as the "chance or possibility of loss or bad consequence." Those who study risk have definitions of this concept which range in complexity. Mandl and Lathrop (1983: 173) quote definitions of risk that are common in the literature:

1. Risk is the expected number of fatalities per year resulting from the consequences of an accidental event.
2. Risk is the probability of an injurious or destructive event generated by a hazard, over a specified period of time.
3. Risk is defined as the frequency at which certain numbers of acute fatalities are expected from accidents.

These definitions suggest that risk arises from a hazard and some uncertainty about its effects.

Krimsky and Plough (1988) define "hazard" as a source of danger. Risk includes the likelihood of a hazard developing into an actual adverse effect causing loss, injury or some other form of danger. Kaplan and Garrick (1981: 13) use the following example to illustrate this difference:

> The ocean is a hazard. If one attempts to cross the ocean in a small row-boat, a great risk is incurred. If the crossing is made aboard the *Queen Elizabeth*, the risk is reduced — all else being equal.

The ocean-going vessel is a device used as a safeguard against the hazard. In general, risk may be diminished by increasing the safeguards, but may never, as a principle of mathematics and logic, be reduced to zero, unless the hazard itself is removed.

A hazard is an undesirable event that may occur, such as an injury to health. A risk can be either the probability of this event occurring to an individual in a population, *or* the number of times that event could occur in a population (Whipple 1987). When risk refers to the expected number of mishaps in a population, it is the product of the size of population at risk and probability that an individual will experience the hazard. To summarize, hazards are generally thought of as sources of danger, whereas risk is viewed as the possibility or probability of loss from a hazard.

The link between hazard and risk suggests that hazards can only be defined in terms of their impact on human society, and hence must be seen as the joint product of events occurring in the world, and existing human adjustments to those events. Hazards are potential disasters waiting for a breakdown in a system (either a natural system or a human system) to occur. As such, disaster will occur when there is a fluctuation in, or a malfunction of, the physical and social systems which are said to govern the laws of nature and society. If the status quo is preserved, no disaster can be triggered.

Technically, the word "risk" refers to situations in which a decision is made, the consequences of which depend on the outcome of future events having known probabilities. In fact, the probabilistic approach to assessing risk attempts to minimize uncertainty by generating mathematical models. Vlek and Stallen (1980: 237) list six definitions of risk which are common in the "probabilistic-oriented" literature:

1. Risk is the possibility of a loss.
2. Risk is the size of a possible loss.
3. Risk is a function of the product of probability and size of loss.
4. Risk is equal to the variance of the probability distribution of all possible consequences of a risky course of action.

5. Risk is the semi-variance of the distribution of all consequences.
6. Risk is the weighted linear combination of the variance and expected value of the distribution of all possible consequences.

These probabilistic conceptions of risk are intimately linked to the positivist conception of science, and as such, only represent a narrow, incomplete picture. Additionally, this approach fails to appreciate a central truism — namely, decision under risk implies that probabilities are known; decision under uncertainty implies that probabilities are unknown.

The uncertainty which shrouds discussion of environmental issues is itself of moral importance, since it is often cited both as a basis for minimizing the future development of a technology or use of resources, and also as a basis for discounting dire predictions of environmental decay on the grounds that they can't be proved. For example, the debate about whether climactic changes associated with the production of the so-called "greenhouse gases" is leading to a global warming trend or another "Ice Age," is based on how uncertainty is interpreted in science. Unfortunately, such uncertainty often leads to a loss of confidence in the central institutions (read: government), thus paralysing the state and sabotaging the public policy process (Thomas 1986).

Douglas and Wildavsky (1982), in their book *Risk and Culture*, suggest that risk is defined and perceived differently by people in different cultural and social contexts. The authors of this seminal text note that the current consideration of risk has three peculiar aspects. First, the disagreement about the problem of defining and addressing the issue of risk is deep-seated and widespread in the Western world. Secondly, different people worry about different risks — for example, pollution, employment, war, inflation and civil unrest. The third peculiarity is that substantial disagreement about the nature of particular risks prevents action designed to reduce loss and damage. According to Douglas and Wildavsky, risk is a product of how public policy, technology and public opinion blend together to reduce costs and minimize danger. This mix implicitly assumes that the primary function of government is to protect its citizens against certain kinds of risk. This argument is based on the rather obvious fact that risk (and benefit as well) is the end result of regulation (Otway and Ravetz 1984).

Policy designed to deal with environmental risk must account for the ambiguous nature of risk and the unsettling issue of uncertainty (Linnerooth 1984). An unavoidable consequence of this form of policy making is that what the public sees of the regulatory process is the formal decisions and, to some extent, depending on the national culture, the debates that led up to them. Corruption, "cover-ups" (for example, see Rosalie Bertell's book, *No Immediate Danger: Prognosis for a Radioactive Earth*, 1985) and the influence of powerful special interest groups, may influence this political process.

Canadian society contains several examples of social institutions and agencies which have evolved, at least in part, to control the risks from new technological developments. The Canadian Standards Association (CSA), building codes, labour unions and regulatory agencies like the Atomic Energy Control Board (AECB) are just a few examples of institutions and agencies which must grapple with the question: How safe is safe enough? It takes such forms as: Does this toaster have enough safeguards built into it to prevent house fires? Should schools with asbestos ceilings be closed? Is the work environment in this factory safe for workers? Should there be additional containment shells around nuclear power plants? These types of questions represent "acceptable-risk decision problems" which require a choice among alternative courses of action. Often the decision between two courses of action becomes muddled by disagreement between "expert" risk assessors (engineers, mathematicians, biologists, etc.) and members of the "lay public."

Risk Assessment: An Overview

It is commonly known that the "public" does not see risks in the same way as experts (Fischhoff et al. 1981). Studies or anecdotes showing the fallibility of lay judgements are frequently cited as evidence for reducing the role of lay people in the risk-assessment process (e.g., Gould et al. 1988). Some have gone so far as to suggest that public overreaction to technological risk shows "pathological elements of phobic thinking" (Sjoberg 1987). William Clark (1980) asserts that "Society's attitudes toward risks such as cancer and nuclear reactors are not readily distinguishable from its earlier fears of the evil eye" (p. 305). This argument often presumes that experts are immune to judgemental biases. The so-called risk experts lament that if only the public could understand risk assessment and its logic, the public-policy process would better reflect the needs of the society, thus ensuring "good" government. After all, as the argument goes, risk assessment is a tool used by policy makers to make informed decisions (Chambers 1989). Additionally, risk assessment is a necessary component for virtually any effort to interpret and apply scientific data, in the development of federal regulations, where health and environment are concerned. These differences are at the heart of political debates between rival interest groups when it comes to assessing and managing environmental risk (Douglas 1992).

Many regulations impose costs on society in exchange for reducing the impact of certain risks; for example, the number of cancer-related deaths linked to emissions from a solid-waste incineration plant can be reduced by spending money on clean, scrubber-fitted smokestacks (Fisher, Lauraine, and

Violette 1989). Due to the large number of risk-producing technologies, and the relatively high cost of reducing the risk for each of them, government agencies face difficult resource-allocation decisions. Often, environmental risks are placed in a specific frame of reference in order to address issues of likelihood, consequence and cost. A key factor in judging the acceptability of a given risk, is how it compares to other risks (Slovic 1986). For example, the risk associated with a specific technological disaster (such as a nuclear power plant accident) is often compared to an equivalent natural disaster (such as an earthquake). Within this comparison is the tacit assumption that a technology is socially acceptable if the risk of death associated with it does not exceed the risk of death from its equivalent natural disaster. However, a comparison between "risky" technologies suggests that as one becomes riskier, the other becomes safer. For example, as chemical plants become more dangerous, nuclear power plants appear to become relatively safer. This relational construct implies that as environmental risks become more common and extreme, greater tolerance for dangerous technologies will follow.

This kind of logic is used by risk assessors to evaluate a technology. What these types of assessments fail to account for, is the way in which risks possess cumulative properties which may, or may not, combine synergistically. It should be obvious that the sum of a large number of small risks could be significant. For example, it would hardly be comforting to learn that, although no single chemical in drinking water poses a cancer risk in excess of 10^{-7} (less than one in 10 million), that there were one million such chemicals in the same water (Sjoberg 1987).

Professional risk assessors rely heavily on risk compendiums (aggregate estimates which compare risks in quantitative terms) in order to help the lay public understand these kinds of numbers. A classic example of a risk compendium comes from Wilson (1984) who lists several activities which are equivalent in terms of the increased chance of death in any year by a factor of one in a million. For example, smoking 1.4 cigarettes, spending one hour in a coal mine, travelling six minutes by canoe, receiving one chest x-ray, consuming 40 tablespoons of peanut butter, and living one year near a typical nuclear power plant are all considered to be equivalent risks. The lay public is expected to understand these comparisons, and to support policy based upon these types of numbers. Perhaps the risk-assessment community fails to appreciate the fact that most lay people do not compare events strictly in terms of actuarial risks.

One of the most challenging areas in statutory interpretation is the problem of setting cut-off levels for risk regulation. A concept used to help sort out this difficult task is known affectionately by risk assessors as *de minimis* risk, which is a level of risk too trivial to warrant the expenditure of resources for assessing or controlling them (Whipple 1987). The Environmental

Protection Agency (EPA) in the United States has suggested a *de minimis* individual lifetime risk level range of 1×10^{-5} (one in 100,000 chances of dying from exposure to a specific risk) to 1×10^{-4} (one in 10,000 chances) for small populations, and a range from 1×10^{-7} (one in 10 million chances) to 1×10^{-6} (one in a million chances) for large populations (Travis et al. 1987).

These numbers suggest that exposure to specific environmental risks is treated according to the population being considered. However, the distinction between small and large populations is a questionable one. For example, a metropolitan area (normally considered a large population) can be divided into several so-called "small populations." As a result, the policies designed to protect the individual from harm, may end up being susceptible to the politico-economic pressures exerted by corporations and government-run corporations (for example, Ontario Hydro) which maintain a vested interest in reducing cost and increasing profit. Furthermore, the standard debate pertaining to job loss and creation always seems to be intimately involved in the issues of assessing and managing environmental risk. As well, risks from risky technologies are not borne equally by the different social classes; risk assessments ignore the social-class distribution of risk.

Risk and Economy

The juxtaposition of environmental risk with economy leads to several assumptions about the nature of risk: all risks can be reduced provided that sufficient resources are employed; resources available for risk reduction are limited; risks should be evenly distributed over the population — a type of "risk democracy"; zero risk is neither possible nor desirable; and finally, risk is a natural, permanent and necessary part of everyday existence. All of these statements sound inherently logical and true. However, it is the strong link between our understanding of risk and concerns about economic issues (a type of "rational," cost/benefit form of thinking) that makes it difficult to spot some of the more profoundly disturbing aspects of this conceptualization. With the increase in risk and associated public concern, risk assessment has grown up, giving advice and (usually) legitimating the decisions of the élites in the private and public sectors.

Policy makers, who consider the link between economic concerns and environmental risk, often assume that any restrictions on a risky activity should be based on the relative gains and losses of the activity, as compared with its alternatives. The classic comparison made within this trade-off notion deals with improving mortality and morbidity rates. These comparisons are collectively known as "value-for-life assessments" (see Fisher, Lauraine, and Violette 1989), or the "human-capital approach." In brief, value-for-life assessments are calculations of lost pre-tax income to a community or region from the premature deaths of individuals exposed to a specific risk. A life is worth $300,000 (one study concluded), less if you are over 60, and even less if you are otherwise enfeebled (Perrow 1984). These calculations are then used as a basis for determining how much money should be spent in offsetting the risk to the individual in order to minimize financial strain on a community or region. It should be obvious that this type of calculation lends itself particularly well to inequitable risk-reducing budgets. For example, it may be the case that communities with a high proportion of upper pre-tax income individuals would either be spared the exposure to certain risks (it is also possible that such a community has politically influential members), or that such a community would be provided with safeguards so as to reduce the risk. Essentially, risk may not be equitably distributed across all segments of society.

One method of reducing this apparent disparity, is to allocate funds to individuals and communities who are willing to embrace additional risk. Nuclear power plants, liquid natural gas plants, municipal incineration plants and waste disposal sites are examples of technologies and locations where communities and individuals (working at one of these locations) are given financial incentives to generate risk-embracing attitudes and behaviours. For example, the Nuclear Power Generating Station in Pickering, Ontario provides residents of Pickering with well-kept parkland, recreation centres, day-care facilities, and high paying jobs in the plant. Residents living near this power plant are motivated to accept the risk of the reactor by embracing these financial inducements. Additionally, lower cost housing makes living near a potentially dangerous site more attractive, or possible.

The distribution of environmental risk appears to be moulded by the relative ability of a community, or an individual, to financially buffer and resist these types of inducements. For example, workers in "dangerous" occupations are willing to accept more risk, if adequately compensated. This is called a "risk wage." Fire-fighters sent to quell the fires in the oil fields of Kuwait in 1991 earned $500,000 (U.S.) for their hazardous employment (*The Wall Street Journal*, 27 March 1991, p. A1). Elephant handlers at the Philadelphia Zoo receive an extra $1,000 compensation annually because elephants are said to pose a risk to handlers they do not like (*The New York Times*, 13 May 1991, p. D1). Fisher, Lauraine, and Violette (1989) suggest that a "willingness-to-pay" (WTP) estimate be included in risk assessments in order to determine how much extra money is required to attract a worker to a risky workplace environment. This is based on a well-known axiom that people will generally tolerate greater voluntary risks than those imposed on them by others (Slovic 1986); payment for accepting additional risk will probably ensure increased tolerance and acceptance. Unfortunately, the amount of money available for such inducements plateaux, and then decreases, as safeguards become more expensive to implement. That is, workers must, at this point, be willing to take a reduction in pay so as to maintain a safe work environment. A typical question consistent with this approach to risk might be: What should a 60-year-old worker in the asbestos industry be willing to pay, in terms of a reduction in income, to lessen the risk of lung cancer by some given factor — say, by half?

This wage-risk approach relies on two main assumptions. First, it assumes that workers are aware of differing risks across jobs. Workers who over- or underestimate the risks of a particular occupation will gravitate to, or be repelled from, certain industries in a way inconsistent with actual levels of risk. Secondly, this approach assumes that workers are able to move freely between jobs. However, the location of these risk-generating facilities in financially disadvantaged areas makes the prospect of leaving a well-paying job somewhat difficult for many workers. As a result, the cost effectiveness of these existing human-capital-risk assessments varies widely, from $200,000 per life saved for airplane cabin fire protection, to as much as $132-million per life saved for a 1979 U.S. regulation of DES (diethylstilbestrol) in cattle feed (Viscusi 1992). The technical literature is fond of pointing out that millions of dollars are spent on safety devices to save one nuclear power plant worker (e.g., emergency-core-cooling system). Yet there is a reluctance to spend $80,000 to save the driver of an automobile (e.g., seat belts) (Morgan and Henrion 1990). These wide-spread discrepancies not only reflect differences among agencies in their risk-cost balancing, but differences in the character of risk-reducing opportunities. Additionally, statutes and policies related to risk reduction are often formulated by different committees, at different times and under different political pressures. In essence, social conflicts over risks arise because some parties see risks where others see opportunities.

The relationship between risk and opportunity is clearly illustrated by the example of how female nuclear-powerplant workers have reacted to the nuclear-industry's policy restricting the type of work in which they can engage. It is common knowledge that exposure to certain types of ionizing radiation results in genetic damage to the unborn. Therefore, the nuclear industry restricts female employees of reproductive age from certain types of work that involve "higher than average" levels of radiation exposure. Every institution is now aware of its liability to prosecution for exposing its employees to risk. However, the increased risk associated with such jobs is financially rewarding to many of the male employees willing to accept an increased dose of radiation. Consequently, female employees falling within the confines of the "reproduction" restrictions, have less opportunity to earn money in some of the higher-paying positions. Ultimately, technologies and their associated risks influence the distribution of power and wealth in society.

The Political Processing of Risk

The link between risk and economy suggests that risks will become more frequent and dangerous as economic conditions continue to deteriorate. This section of the paper will examine several issues which demonstrate how the relationship between risk and economy is mediated by political processes. As such, issues of social inequality, the role of the mass media, "risk agendas" and related international issues, will be covered to illustrate this point.

One of the major policy issues which emerges from this formulation has to do with the social inequalities of risk exposure. A good orienting question to ask is: Will the already disadvantaged be forced to suffer more as financial considerations exceed concerns about risk and its possible consequences? Historically, the poor have often carried the blame for disasters and mishaps, and the brunt of the consequences. According to Paul Slack (1988) in the Great Plague, the majority of the English well-to-do saw the incidence of plague as a sign of God's judgement directed against the sins of the poor. The spread of HIV infection (AIDS), and the paucity of research conducted to cure this dreaded disease (Richards 1988), is a modern-day example of how "moral" posturing influences how a certain segment of the population is treated.

The political processing of risk in Canada also tends to discriminate against certain segments of society — namely, non-Christian, ethnic groups. For example, Health and Welfare Canada, following the lead of the Food and Drug Administration (FDA) in the U.S., is planning to ban the importation of the betel-nut — the nut and leaf from the areca palm, used by those of the Hindu faith in religious ceremonies — due to the increased chance of developing oral cancers in those who eat this product (CBC-TV News, 27 July 1993). Why is there so much concern about banning a product which is infrequently used by a small segment of the population, when something like tobacco, and its associated health risks, continues to be widely available? This movement in the study of risk towards issues like equity, represents a more general evolution of thought and reflection on the topic of risk. Denis Duclos (1989) examined how the political dimensions of risk are related to the development of knowledge bases. By contrasting the attitudes of the lay public, engineers, scientists, government and industry, Duclos discovered that members of the lay public define risk, not so much in terms of potential disaster and possible death, but according to the impact that technology will have on their lives in two ways: a concern about the oppression by authority empowered by such technology; and a fear that nothing can be hidden from the new electronic means of scrutiny. In other words, risk means more to people than just fatalities, and its meaning can change from one technology to the next (Otway and Ravetz 1984).

Traditionally, the mass media have been both thanked and blamed for many of the perceptions and beliefs held by members of the general public. In terms of the political processing of environmental risk, however, the mass media plays more of a role than that of merely providing information. With all due respect to Noam Chomsky, it is relatively obvious that the coverage, content and perspective taken by the national news agencies in both Canada and the United States mirror the often "hidden" intentions of the politically active, wealthy entrepreneurial class. This "agenda-setting" function of the news, engenders in the public, a set of attitudes which encourages the consumption of goods and services, reinforces the assumptions which connect economic concerns with the concept of environmental risk (for example, the hot debate between British Columbia loggers and environmentalists) and generates politically inactive, perhaps even comatose, members of society. A result of this linkage between the biased role of the mass media and the political processing of risk, is that environmental issues exist — in terms of public consciousness — mainly through the filters of the media (Sandman et al. 1987). Any issue inadequately covered, or simply ignored, will tend to have a parallel reduction in the perception of risk and consequence. In essence, the mass media tends to portray debates about environmental risk as being conducted between those who are heavily risk averse, and those who are risk takers.

The mass media, by their very nature, maintain a one-way flow of information. Television news and programming, newspapers, magazines and radio, silence opposition by not allowing a "level playing field." Additionally, many of the large corporations that produce electronic items (e.g., The Sony Corporation), work on defense contracts (e.g., General Electric), and provide satellite communications, own and/or control much of the mass media. The existence of interlocking boards of directors also demonstrates the degree to which vested interest goes into the information broadcasting arena. These extensions into many sectors of the economy and political arena are insidious and well disguised. Wilkins and Patterson (1987), in an examination of the news media's treatment of the Bhopal and Chernobyl disasters, proposed that the news media view risky events as political events. Union Carbide, the large multinational corporation responsible for the deaths of hundreds of people in Bhopal, India in 1984, was virtually exonerated. A small, relatively unsuccessful class-action lawsuit on behalf of the disaster victims and their families clearly demonstrates this. After all, it was just an accident! Calling something an "accident" tends to mitigate the burden of responsibility. On the other hand, the Chernobyl nuclear-powerplant disaster was more than just an accident; it represented, as presented by the mass media, a general failure of the socialist system (for example, see David Marples's book, *The Social Impact of the Chernobyl Disaster*, 1988).

Those involved in the public-policy debates on environmental risk use the term "risk communication" to refer to a process whereby technical and scientific information is transmitted from élites to the general public (Krimsky and Plough 1988). According to Covello et al. (1984), risk communication is a purposeful exchange of scientific information between concerned parties regarding health or environmental risks. Accordingly, information about the "true" nature of a particular risk is intended to reduce opposition and increase compliance. So-called "experts" are often used to lend some credibility and respectability to this information-exchange process. Unfortunately, very little "exchange" actually occurs in the risk assessment and risk management process. The one-way information-flow model ensures that the opinions of supportive "experts" overpowers the small, voiceless resistance of the "phobic" individual. Ostensibly, risk communication is intended to inform the public, stimulate "rational" debate and encourage public participation. However, the rhetoric of the risk assessor, and the low value placed on the opinions of "non-experts," tend to subvert these good intentions. It almost seems as if those involved in environmental risk policy see the public as a lost, pitiful mass of confused desires.

Isaac Levi (1980) examines how bias in the risk assessment process is intimately linked to the political processing of knowledge. Levi reviewed the statistical arguments used by the U.S. Nuclear Regulatory Commission (NRC) to support the continued use and development of nuclear power plants for generating energy. He suggested that the conclusions of the report, entitled "The Reactor Safety Study," were based on inadequate data. Furthermore, Levi regarded the construction of the argument used to interpret this data as one "incapable of rendering a negative verdict on the practice of evaluating risks of a serious accident in nuclear plants...." (Levi 1980: 441).

Almost all risk assessments are plagued by inadequate data (Shrader-Frechette 1985). The purpose of such data is to demonstrate cause-effect relationships between exposure to a hazard and adverse effects on humans. However, causation is difficult to determine because people are exposed simultaneously to multitude potential hazards, making it difficult, or impossible, to identify a single causative agent. The issue of uncertainty is used by experts and policy makers to discount dire predictions about exposure to environmental risks and associated accidents and catastrophes.

Charles Perrow (1984) examines how and why catastrophic accidents occur in advanced, technologically-oriented societies. Perrow uses the Three Mile Island nuclear-powerplant disaster, as well as several other examples like aircraft accidents, marine accidents and petrochemical-plant accidents, to demonstrate that such accidents are consequences of the way we launch industrial ventures. The environmental and health risks associated with these types of accidents are the product of a complex, "structural" relationship between the development and use of a particular type of technology, and the political forces which are present.

It should be no surprise that the appearance of so many catastrophic human-made processes has mobilized public concern and influenced how risk assessments are conducted. As societies confront new and potentially dangerous risks, the role of the public in risk assessment is bound to increase. Linnerooth (1984) noted that public policy agencies are increasingly relying on input from industry, government and public-interest groups in the development of standards and rules for the environment.

The development of standards and rules, which relate to environmental risk, not only vary within a sociopolitical region, but also within the international arena. Consider some of the environmental problems that give rise to political controversy: air pollution, toxic-waste disposal, nuclear-powerplant safety, depletion of non-renewable energy resources like oil and gasoline, global warming and the shrinking fresh-water supply. These environmental problems tend to involve risks that extend beyond the immediate country of their origin to other parts of the world. The risks from such "travelling" environmental hazards have direct effects on the "host" country and its people: an increased risk of developing respiratory diseases from air pollution; cancers and immune-system problems from radiation exposure (Lambert 1990); and famine associated with global warming and its effects on local food crops, water supplies and weather patterns. However, another type of environmental risk also "travels" from one country to another. The international market economy, supported by the political power wielded by multinational corporations and the military power from governments of economically advantaged nations, promotes a trade in environmental risk.

Risks that present a substantial threat to the health and well-being of those living in Western, industrialized nations, are often eliminated or reduced by participating in this international "risk-trade" marketplace. Typically, this market involves a one-way transfer of risk from the wealthy core countries to the less-wealthy periphery countries. Income differences, cultural differences and laxer standards in underdeveloped countries, allow for the international redistribution of risk to proliferate. This global politico-economic-systems argument (see Wallerstein 1980) contends that trade in risk is largely coterminous with the emergence of centre/periphery structures, in which highly developed polities maintain regional inequalities by avoiding many of the consequences associated with a chosen path of development. The perseverance of such regional inequalities is rooted in the nature of development itself (Gill and Law 1988). Environmental risks increase in magnitude and frequency due to the industrial activities of developed nations. As a result, a modification in core/periphery relations ensures that the standard model of international trade (developed nations trading manufactured goods on a mass scale with relatively underdeveloped zones that provide raw materials) includes a provision for participating in the "risk trade."

The influence of Western, industrialized nations on lesser developed parts of the world often involves economic, political, and sometimes military pressure. Robert Guimaraes (1991) examines how Brazil is suffering from an ecological crisis resulting from a development strategy, which sacrifices the environment for economic prosperity. Guimaraes suggests that this ecological crisis can be examined on two levels: local and international. Locally, the Brazilian government, like all governments, must grapple with the issue of balancing economic development with the formation and implementation of environmental policy. However, Guimaraes laments that the Brazilian government advocates the position that development should not be sacrificed in the name of a cleaner environment; and, furthermore, that the environmental degradation associated with such development should be taken care of by the developed countries of the world. This argument is based on the assertion that wide-scale clearing of the Amazonian rain forests is necessary in order to generate foreign currency, which can be used to service Brazil's substantial foreign debt. Unfortunately, the environmental consequences of such development practices encourage the poor of the world to destroy precisely those resources which are essential to their long-term subsistence and well-being. Additionally, environmental risks resulting from these practices tend to have an immediate impact on the local populations, and a delayed impact on other parts of the world. For example, locally, deforestation in the Amazonian River basin leads to soil erosion and reduced fertility of the land. The effects are most damaging to the poor who eke out a living from the land. Globally, the relationship between deforestation and the greenhouse effect is a slow and unpredictable process.

The relationship between the so-called developed nations and lesser developed nations tends to be of a coercive nature. That is, developed nations which depend heavily on natural resources for economic prosperity need not face economic disaster if they deplete their own resources, so long as world supplies remain available. This notion of scarcity is central to the international trade in environmental risk. Developed nations trade technology to lesser-developed nations, in exchange for raw materials, access to foreign markets, foreign labour pools which are willing to embrace more risk than labour pools in developed nations, and politically active groups that will tolerate greater environmental risks in order to parallel Western development patterns.

This transfer of technology, given a favourable environment, is assumed to reproduce a developed, Western society wherever it may be placed. However, lesser-developed nations are unable to cope with the fact that more technology is needed to meet the emergencies that technology has created in the first place (Grant 1986). Additionally, technological risks are related to issues of national security and sovereignty.

Ethan B. Kapstein (1992) persuasively argues that scarce and non-renewable natural resources, like petroleum, create economic interdependencies between nations which tend to threaten national security. For example, the Organization for Arab Petroleum Exporting Countries (OAPEC) placed an embargo on the sale of oil to the United States in October 1973 to protest the U.S. support of Israel during the Yom Kippur War. The gasoline shortages, quadrupled price of oil and the resulting "stagflation," made the American public and its corresponding political bodies acutely aware of the impact that dependence on foreign energy supplies has on its economic well-being. Kapstein notes that such threats involve a complex relationship between political action within a country, and international action between countries. These national security concerns often involve environmental risk. The world trade in non-renewable energy resources, like petroleum, is essential to the economic well-being of many industrialized countries such as Canada and the United States.

The 1990 conflict between Iraq, Kuwait and the United Nations is a good contemporary example of how national security concerns created specific environmental risks. The oil-well fires, which were a product of this international battle over access to petroleum, wreaked environmental havoc. A burning question remains: Who should be responsible for cleaning up this ecological nightmare?

In fact, the globalization of the world economy has accelerated the destruction of ecosystems in many parts of the world, due in part, to the belief that national security is dependent on the international trade in natural resources. However, another kind of environmental risk emerges from this national security paranoia: namely, risks from alternative technologies, like nuclear energy, which ostensibly reduce dependency on foreign petroleum sources.

Canada, as well as the United States, is heavily committed to nuclear power as an alternate source of energy to fossil fuels. Canadian Crown corporations, such as Atomic Energy of Canada Limited, sell nuclear reactors to countries like India, South Korea, Mexico, Egypt and Romania. However, controversy about the safety of nuclear-powerplants, disposal of nuclear waste and the link between nuclear energy production and nuclear weapons proliferation, has disenchanted both politicians and the public. The environmental risks of this alternative-energy source must also undergo political scrutiny and cost/benefit analysis.

In both Canada and the United States, nuclear power is being marketed as a "clean," locally available substitute energy source for imported petroleum. In essence, if local nuclear power production can provide all of a country's domestic energy requirements, the global trade in non-renewable energy sources, like petroleum, will be substantially reduced. Although this issue goes beyond the scope of this paper, it is apparent that environmental risks, associated with this shift from energy importation to energy production, need to be examined in more detail. For example, will increased nuclear-power production actually reduce air-pollution levels and decelerate the process of global warming? Will the increased risks associated with nuclear-power production and the mining of uranium, transportation of ore and waste products and the disposal of spent nuclear fuel rods be equivalent to the risks associated with the use of non-renewable fossil fuels? Alternative energy sources create new sets of interactions which expose individuals and ecosystems to new environmental risks.

To summarize, a sociological understanding of environmental risks illustrates the tension between two different ways of conceptualizing risk. On the one hand, risk as a "rational," "objective" calculation of probability and consequence can be used most effaciously by political decision makers to weigh alternative courses of action. On the other hand, if the assessment and management of risk represents constellations of values, aspirations and political orientations, environmental-policy decisions should account for perceptions of risk which consider the social acceptability of particular technologies. A truly democratic society should not expose its population to environmental and health risks which have poor public support. Environmental sociology, as a field of inquiry, should help establish what values need to be included in a nation's strategy for protecting the natural environment, human health and economy.

DISCUSSION QUESTIONS

1. The definition of risk seems to be a complex and inconsistent one. How do the various notions or perspectives on risk influence its definition?
2. Mehta argues in his paper that risk is a socially and politically constructed concept based on the interests of conflicting social groups. Provide examples of how such social conflict influences the assessment and management of risks. Discuss one such example in detail, bringing in concepts like cost-benefit analysis, value-for-life assessments, and sustainable development?
3. Since risk assessments are now a standard part of most environmental assessments in Canada, how do such assessments attempt to bring the public into the decision-making process?
4. Some have suggested that new technologies create social structures and forms of organization which exclude the lay public from the everyday

functioning of government. How does a macrosociological perspective on risk address these conflicts between the ideals of democracy and the need for expert advise on technological issues involving risk?

References

Beck, Ulrich. 1992. *Risk Society: Towards a New Modernity*. London: Sage Publications.

Bertell, R. 1985. *No Immediate Danger: Prognosis for a Radioactive Earth*. Summertown, Tenn.: The Book Publishing Company.

Chambers, D.B. 1989. *Risk Assessment as an Environmental Tool*. Paper presented to the American Mining Congress Mining Convention, 20 September, San Francisco, California.

Clark, W. 1980. Witches, Floods, and Wonder Drugs: Historical Perspectives on Risk Management. In *Societal Risk Assessment: How Safe is Safe Enough*?, ed. R.C. Schwing and W.A. Albers. New York: Plenum Press.

Covello, V.T., L.B. Lave, A. Moghissi and V.R.R. Uppuluri, eds. 1984. *Uncertainty in Risk Assessment, Risk Management, and Decision Making*. New York: Plenum Press.

Douglas, M. 1992. *Risk and Blame: Essays in Cultural Theory*. London: Routledge.

Douglas, M. and A. Wildavsky. 1982. *Risk and Culture: An Essay on the Selection of Technical and Environmental Dangers*. Berkeley: University of California Press.

Duclos, D. 1989. *La Peur et la Savoir: La Societe Face a la Science, la Technique et leurs Dangers*. Paris, France: Editions de la Decouverte.

Fischhoff, B., S. Lichtenstein, P. Slovic, S.L. Derby and R.L. Keeney. 1981. *Acceptable Risk*. New York: Cambridge University Press.

Fisher, A., G.C. Lauraine and D.M. Violette. 1989. The Value of Reducing Risks Of Death: A Note of New Evidence. *Journal of Policy Analysis and Management* 8: 88–100.

Gill, S. and D. Law. 1988. *The Global Political Economy: Perspectives, Problems and Policies*. Baltimore, Md.: The Johns Hopkins University Press.

Gould, L.G., G.T. Gardner, D.R. DeLuca, A.R. Tiemann, L.W. Doob and J.A. Stolwijk. 1988. *Perceptions of Technological Risks and Benefits*. New York: Russell Sage Foundation.

Grant, G. 1986. *Technology and Justice*. Toronto: Anansi Press.

Guimaraes, R.P. 1991. *The Ecopolitics of Development in the Third World: Politics and Environment in Brazil*. Boulder, Colo.: Lynne Reinner Publishers.

Jasanoff, Sheila. 1986. *Risk Management and Political Culture*. New York: Russell Sage Foundation.

Kaplan, S. and B.J. Garrick. 1981. On the Quantitative Definition of Risk. *Risk Analysis* 1(1): 11–27.

Kapstein, E.B. 1992. *The Political Economy of National Security: A Global Perspective*. Columbia: University of South Carolina Press.

Krimsky, S. and A. Plough. 1988. *Environmental Hazards: Communicating Risk as a Social Process*. Dover, Mass.: Auburn House Publishing Company.

Lambert, B. 1990. *How Safe is Safe? Radiation Controversies Explained*. London: Unwin Paperbacks.

Levi, I. 1980. *The Enterprise of Knowledge: An Essay on Knowledge, Credal Probability, and Chance*. Cambridge, Mass.: MIT Press.

Linnerooth, J. 1984. The Political Processing of Uncertainty. *Acta Psychologica* 56: 219–31.

Mandl, C. and J. Lathrop. 1983. Risk and Decision. In *Risk Analysis and Decision Processes*, ed. H.C. Kunreuther et al., Chap. 7. New York: Springer-Verlag.

Marples, D.R. 1988. *The Social Impact of the Chernobyl Disaster*. New York: St. Martin's Press.

Morgan, M.G. and M. Henrion. 1990. *Uncertainty: A Guide to Dealing with Uncertainty in Quantitative Risk and Policy Analysis*. New York: Cambridge University Press.

Otway, H. and J. Ravetz. 1984. Technology: Examining the linear model. *Futures* June: 1–9.

Perrow, C. 1984. *Normal Accidents: Living with High-Risk Technologies.* New York: Basic Books Inc.

Plough, A. and S. Krimsky. 1987. The Emergence of Risk Communication Studies: Social and Political Context. *Science Technology and Human Values* 12: 4–10.

Richards, D.A. 1988. Human rIghts, Public Health and the Idea of Moral Plague. *Social Research* 55(3): 492–528.

Sandman, P.M., D.B. Sachsman, M.R. Greenberg and M. Gochfield. 1987. *Environmental Risk and the Press: An Exploratory Assessment.* Oxford, U.K.: Transaction Books.

Shrader-Frechette, K.S. 1985. *Risk Analysis and Scientific Method.* Boston: D. Reidel Publishing Company.

Sjoberg, L., ed. 1987. *Risk and Society.* Boston: Allen & Unwin.

Slack, P. 1988. Response to Plague. *Social Research* 55(3): 433–53.

Slovic, P. 1986. Information and Educating the Public About Risk. *Risk Analysis* 6(4): 403–15.

Thomas, P.B. 1986. Uncertainty Arguments in Environmental Issues. *environmental ethics* 8: 59–76.

Thompson, P.B. 1986. Uncertainty Arguments in Environmental Issues. *Environmental Ethics* 8: 59–76.

Travis, C.C. et al. 1987. Potential Health Risks of Hazardous Waste Disposal. *Journal of Hazardous Material* 14(3): 309–20.

Viscusi, W.K. 1992. *Fatal Tradeoffs: Public and Private Responsibilities for Risk.* New York: Oxford University Press.

Vlek, C. and P.J. Stallen. 1980. Rational and Personal Aspects of Risk. *Acta Psychologica* 45: 273–300.

Wallerstein, I. 1980. *The Modern World System.* Vol. 2. New York: Academic Press.

Welch, Sharon D. 1990. *A Feminist Ethic of Risk.* Minneapolis, Minn.: Fortress Press.

Whipple, C. 1987. *De Minimis Risk.* New York: Plenum Press.

Wilkins, L. and P. Patterson. 1987. Risk Analysis and the Construction of News. *Journal of Communication* 37(3): 80–92.

Wilson, R. 1984. Commentary: Risks and Their Acceptability. *Science, Technology and Human Values* 9(2): 11–22.

ENVIRONMENTALISM AND THE POLITICS OF INVISIBILITY

Ted Schrecker

Westminster Institute for Ethics and Human Values, London, Ontario

More than is customary in academic writing, this article is a reflection on, and of, personal experience. Much has been said of late about the need for societies to make the transition to what is now called sustainable development. I have contributed to this discourse in various working capacities over the past 16 years, yet at the same time, I have gradually begun to feel, at least metaphorically, like a visitor from another planet. That planet is considerably poorer than the one inhabited by the people whose preoccupations structure the discourse of sustainable development. The denizens of the planet I call home share economic insecurities rather than overabundance. When we achieve affluence, it is precarious; when we confront material scarcities, they threaten to become permanent.

These inchoate reservations became more clearly defined when I read what one member of parliament had to say about wealth and poverty in the late 1980s. Discussing the supposedly declining political salience of class in Canada, this observer claimed that affluence had become universal, to the point that "obesity and not malnutrition is a medical problem. The

---

An earlier version of this chapter appeared in *Alternatives: Perspectives on Society, Technology and Environment* 20(2), 1994. The research assistance of Seanna-Lin Brodie-Keys in preparing the revisions is greatly appreciated.

undernourished are more likely to be anorexic teenagers, whose lives may be empty, but whose refrigerators are full."

The author of this quotation was sociologist Lynn McDonald (1987: 245). She served for several years as the environment critic for the federal New Democratic Party, which used to give the impression of taking poverty seriously. In the prerecession year of 1987, when this statement was published, slightly more than one million Canadians under the age of 18 (one out of every six Canadians in that age bracket) were living in a household whose income fell below Statistics Canada's poverty line (National Council of Welfare 1992: 8). For example, in 1993, a single parent living below that poverty line in Metropolitan Toronto or Vancouver, supported two children on a before-tax income from all sources of less than $26,624 (National Council of Welfare 1994: 75). Those who question the relevance of semi-official poverty lines as indicators of economic hardship, should try doing this for a while.

In terms of its obliviousness to economic reality, McDonald's utterance is not atypical of what environmentalists say — neither is the phenomenon of recent origin. Two decades ago, one American academic wrote that "the private worlds of most Americans are now complete with expansive homes and gardens, full garages, kitchens and closets" (Johnson 1973: 175). Quite apart from the class bias of this claim, one might expect, in the U.S. context, that its implicit racism would have been commented upon, but so far as I know, this has not happened.

Class, Environment and the Politics of Exclusion

How can one best describe and analyse what is going on in situations like this? Extensive and growing literature tries to identify the social characteristics (or, more accurately, the class position) of environmental activists. My approach here involves concentrating instead on the "content" of a particular discourse: what environmentalists or advocates of sustainable development actually say, and what the implications are. I will loosely follow a set of categories developed by Artemis March, as she identified three ways women have been rendered "invisible" in a corpus of androcentric–sociological theory. The first of these is exclusion: "Women are completely ignored or neglected." The second is pseudo-inclusion: "Data and interpretation appear to take account of women, but a dissection of the analysis reveals that the interpretation can really only have been drawn from males." Finally, there is alienation, in which the inclusion of women and their concerns in the analysis is nevertheless compromised by "androcentric theory, method and/or language [which] mediated between women's experience and its interpretation" (March 1982: 100).

I am concerned here with the politics of class, rather than those of gender. However, the use of feminist categories outside the context in which they were originally developed is appropriate, since it is feminists (almost exclusively) who have articulated the meaning of persistent and pervasive marginalization in contemporary society. March's sensitivity, and that of other feminist theorists, to the politics of invisibility, thus provides a valuable conceptual framework for exploring the subtle class biases of much environmental advocacy — the ways in which it reflects the perspective of an (over)privileged minority, while ignoring or devaluing the situation of the economically marginalized and the economically insecure. In today's open, rapidly changing and politicized economy, this means more and more of us.

A brief digression is necessary here. Some people find the concept of "class" problematic and imprecise, or at least anachronistic. They argue that it may have been applicable to the social arrangements characteristic of the early decades of the Industrial Revolution, but is no longer relevant. I disagree strongly. In a superb reply to such criticisms, Jon Elster has defined "class" in terms of what he calls endowment structures: "The endowments include tangible property, intangible skills and more subtle cultural traits." The behaviour is defined in such economic terms as working versus not working, selling versus buying labour power, lending versus borrowing capital, renting versus hiring land and giving versus receiving commands in the management of corporate property. A class is a group of people who, by virtue of what they possess, are compelled to engage in the same activities, if they want to make the best use of their endowments (Elster 1986: 147). In practice, "best use" is an abstraction that, for hundreds of millions of people worldwide, and for expanding numbers in North America, means finding food, clothing and shelter as best they can.

For purposes of understanding environmental policy and the associated discourses, the advantages of Elster's approach to defining "class" are twofold. First, rather than focusing on conceptions of economic and political structure that neglect human cognition and agency, it directs attention to the causal relations among individuals' economic options, the political choices available to them, and the range of life situations they consider normal or unexceptionable. Second, it serves as a valuable counterpoint to admittedly well-intentioned descriptions of environmental conflicts in dichotomous terms that pit corporations against the rest of us, or "developers" against "the public" (e.g., see Shrader-Frechette and McCoy 1993: 149–85). The real politics of such situations are substantially more complex.

Class, in Elster's individualist sense, matters, as well, in the way connections between economic and environmental issues are defined. Lynn McDonald's claim about poverty in Canada exemplifies the process by which the less-than-affluent are excluded from the universe of political discourse by way of simple refusal, on the part of the affluent, to admit that they exist. As another example, an award-winning 1990 series of *The Toronto Star* articles by journalist Paul McKay, spoke of the Canadian economy as "the ecological equivalent of 'shop till you drop,' " and warned that: "Canadians can voluntarily learn to live with less, or we can ride out a consumption junket that comes with its own lethally toxic hangover" (McKay 1990: 3, 5). Along the same lines, the cover of a recent issue of *Alternatives: Perspectives on Society, Technology and Environment*, devoted to waste reduction and social change, featured a cartoon of a young man in beach clothes and designer running shoes, with a portable cassette player clipped to the belt of his shorts. He is standing in front of a mountain of discarded consumer goods, including a personal computer, compact discs, unopened holiday gifts and other less-identifiable items. A cartoon rat is looking very interested in a discarded fast-food hamburger and fries, an image which incidentally trivializes the situation of people, for whom rats are an everyday health hazard that comes with their residential territory, or with their forced proximity to garbage dumps, that are negligently managed and inadequately policed. The cartoon man is saying:

> You should see what I picked up at the mall ... a new release of that video game and that shoe place finally got in the hightops with the orange stripes ... the junk for my hair was on sale ... good deal on those colour monitors for the computer in the spare room ... remember those shirts — 10 for $39.99 ... the release of the Falcons' new CD on Wednesday ... so I picked up a pair....

The issue's guest editor explains the theme as being "that insatiable consumerism is not viable on a limited planet" and describes the cover as "pointing to the types of lifestyle changes that will be necessary if we hope to reduce the waste we generate" (Sinclair 1993: 1).

But who are the insatiable "we?" "Which" Canadians must learn to live with less? How much less, and who can legitimately claim the moral and political authority to tell them so? The *Alternatives*' cover implies that the population, as a whole, has available to it a level of consumption that, on an impressionistic assessment, is probably available only to the richest segment of the Canadian population. This includes those who were members, as of 1990, of census families (average size: just over three people) with before-tax labour incomes of $90,000 and over, and unattached individuals with incomes of over $40,000 (Statistics Canada 1992). Another perspective on this issue involves looking at the distribution of discretionary income in Canada; a Statistics Canada study concluded that in 1986, fewer than three out of 10 Canadian households had any discretionary income at all (Owens 1991).[1] Half of the households with discretionary income were headed by a professional or manager, underscoring their atypicality in demographic terms, and fewer than five per cent of Canadian households controlled more than 35 per cent of the country's available discretionary income (Owens 1991: 4.4). Nevertheless, the myth of universal affluence is persistent: McDonald spoke approvingly of Charles Taylor's discussion of "the cult of production" and the "endless propensity to consume" (McDonald 1987: 243–44). Words and images like these define out of existence not just the economic situation of a marginalized and easily forgotten minority, but the economic reality faced everyday by the clear majority of Canadians. A computer in the spare room — really?

The limited number of examples I have cited suggest that, now as then, environmentalists presume that affluence and economic security are universal within their own societies, despite abundant empirical evidence of which they should be aware. This is not a new observation. Twenty years ago, Hans Magnus Enzenberger pointed out in a classic, "Critique of Political Ecology," that promoters of the then-popular "Spaceship Earth" metaphor "deny once and for all that little difference between first class and steerage, between the bridge and the engine room" (Enzenberger 1974: 15). Do environmentalists fail to make this distinction because of their own privileged class positions? Consider the views of Rudolf Bahro, a long-time activist in the West German Green Party, on the class position activists in the European Green movement. "If you define the movement sociologically," he says, "those people are most correct who represent us as some combination of the new middle class and certain sections of young people who have not yet been definitively classified." He goes on to say: "But the movements are not an organ for the particular interests of these classes and, on the contrary, this stratum which is being formed is going into battle ... for common interests which actually affect the whole of society" (Bahro 1986: 82).

Bahro seemingly has no qualms about the claim of an economically privileged and culturally distinct segment of the population to speak for the "whole of society." An equally revealing claim is made by Robyn Eckersley, who acknowledges the privileged class position of many active environmentalists, yet defends it with reference to the "autonomy" associated with their economic security, and to the unique critical capacities supposedly enhanced by higher education. "[I]t is the new class, more than any other class, that is ... likely to be more aware of the scale and depth of the social and ecological problems facing the modern world and more able and motivated to become politically engaged in overcoming them" (Eckersley 1987: 222). For Eckersley, perhaps even more than for Bahro, the privileged-class position of a particular set of speakers is a source of cognitive or moral authority, rather than a cause for scepticism about that authority.

Jobs and Environment, Revisited

The arrogance of Bahro and Eckersley is not uncommon, although their candour is atypical. More frequent and subtle is the example of a recent newspaper article in which two university professors protested proposals to expand logging in the old-growth forests of Northern Ontario's Temagami region.. However good an idea it may be to preserve old-growth forests, refusal to permit expanded logging there, and in a number of regions throughout Canada where similar conflicts have arisen, will undoubtedly result in localized job losses and economic dislocations. Perhaps some arguments for protecting wilderness areas would justify imposing such losses, either with or without compensation. However, the authors of the article are never, in their lifetimes, likely personally to face the economic hardship they implicitly accept on behalf of the Temagami region's loggers, sawmill workers and small-business people.

Another example comes from *The Toronto Star* series, in which McKay generally took a thoughtful and balanced view of the inescapable trade-offs among growth, employment and environmental protection that are associated with transitions to sustainable development. Indeed, McKay quite rightly criticized environmentalists involved in disputes about British Columbia's proposed Windy Craggy copper-mining project, for being insensitive to their own relatively privileged situation. Nevertheless, he concluded that Canada should: "Establish and endorse one simple law: zero discharge of persistent, toxic and cancer-causing pollutants from any industry by the end of the decade. No exceptions. If not achieved by the year 2000, the industry must by law cease to operate" (McKay 1990: 17).

This prescription is beguiling in its simplicity, and may well be approachable in engineering terms with respect to the routine operations of a variety of manufacturing processes (U.S. Congressional Office of Technology Assessment 1986: 77–109). However, even were it possible as a matter of engineering to operate a metal mine in accordance with this principle, as a matter of economics no mining operation could ever be expected to meet this criterion and remain viable in the international economy. Either the criterion would have to be defined in a manner so elastic as to render it meaningless, or its adoption would mean shutting down most of the mining towns in Canada. What about income support and retraining for the tens of thousands of Canadians whose economic futures would be shattered as the result of shut-downs? What about compensation for savings invested in houses that become worthless? There are policies which could address these issues; they are administratively, if not necessarily politically, feasible (see e.g., Wykle et al. 1992). Simplistic prescriptions for zero discharge obscure the complex and perhaps insurmountable political barriers to providing safety nets for those who would lose the most, were such policies actually implemented.

The limitations of brave rhetoric about the compatibility of employment and environmental protection have by now become glaringly obvious. Environmentalists have, quite correctly, identified and decried the corporate use of "job blackmail" as a strategy for resisting regulation (Kazis and Grossman 1982). The problem is that within the existing structure of property rights, which has proved remarkably resistant to change, a conflict between jobs and the environment is real when the owners of the plant or the logging rights say it is. Although job blackmail is often used as a bargaining tool in conflicts that are fundamentally about the profitability of particular industrial undertakings, the most strident advocates of bluff-calling strategies seldom have much directly at stake in economic terms. However intense their commitments to environmental protection, Toronto environmentalists are unlikely to lose their homes, jobs and social support networks (indeed, they are unlikely to lose a penny), when a transnational forest-products firm responds to aggressive pollution-control policies by closing a mill in Northern Ontario.

It is therefore not surprising that people directly vulnerable to such economic losses are often sceptical about claims that, although jobs-environment conflicts may be real at the plant or industry level, they are unreal at the economy-wide level. Hugh Mackenzie, the Canadian research director of the United Steelworkers of America, has written a devastating critique of these and related arguments, in which he points out that whatever their numerical accuracy, which is usually hard to verify, the question of whether wage and benefit levels will be comparable is hardly ever addressed. A recent news article points out that a job in tourism pays about one-third as much as an average worker will earn logging the forests of Clayoquot Sound in British Columbia, and that the multiplier effects associated with tourism employment are predictably less substantial (Matas 1993).

Even more importantly, explicit attention is seldom paid to questions of whether the workers whose jobs are lost will be able to get one of the new jobs in a more environmentally benign sector of the economy, or of who will pay such transition costs as moving expenses and the loss of property values (Mackenzie 1987). The legitimacy of imposing such costs in the name of protecting "our" environment is assumed, without further inquiry into the distribution of those costs. Although concerns for employment impacts are acknowledged as legitimate in the policy discourse, they are dealt with in a way that ignores or devalues the political and economic constraints faced by the people whose jobs, savings and livelihoods are at risk. Class defines the space between those doing the talking and those being talked about.

The Social Construction of Environmental Problems

How should we account for the persistent inattention to class in Canadian discussions of environmental protection and sustainable development? This part of my analysis is more exploratory and speculative than the preceding one; it is also more important. A useful starting point is Annette Baier's aphorism that: "Morality is the culturally acquired art of selecting which harms to notice and worry about" (Baier 1986: 49). I would emphasize cultural acquisition. However much the promoters of particular solutions to social problems might like to believe this, social problems do not possess innate characteristics that compel us to pay attention to them and to consider them as significant. This is a point missed by Eckersley, among others, in her argument that critiques of environmentalism, like Enzenberger's, are misplaced. Instead, Eckersley invokes the "actual, multifaceted social and ecological crises facing the world today," and claims that "the Green movement ought to be accepted as a response to ... objective crises" (Eckersley 1987: 223). Ecological politics, in her view, can apparently rely on the existence of an Archimedean point from which crises, social or ecological, can be "objectively" defined.

This position is unexceptionable in one respect, since presumably the effects of multiple stressors on the interactions of species within ecosystems, or of carbon dioxide emissions and deforestation on global climate change, are, if unpredictable, at least independent of the particular observer. In another respect, however, this uncritical invocation of objectivity is what Donna Haraway has perceptively called a "god-trick," and one which has the potential to conceal the exercise of power in a variety of ways (Haraway 1988: 582). Contemporary feminist epistemology emphasizes what Haraway calls situated (and necessarily partial) knowledges. Scientific knowledge, from this perspective, cannot be separated entirely from the social situation of the knower, described in class, gender and other terms. Haraway warns her readers about automatically attaching authority to certain epistemological positions, and suggests that they guard against the dangers of accepting the view of reality available from "the brilliant space platforms of the powerful" (Haraway 1988: 583).

The practical relevance of Haraway's perspective can be illustrated with the familiar example of the enhanced satellite images, showing changes in tropical vegetation that supposedly result from global warming. Such images represent a God's-eye view, in the most literal terms. The background policy discourse against which these images are displayed, involves the destructive consequences of global warming, that will ensue in the absence of radical changes in patterns of human activity, to reduce greenhouse gas emissions. Indeed, the changes in the way people live, that are associated with global warming, will no doubt be substantial. However, as counterpoint to this image, consider the front-page story in the *New York Times* of January 21, 1994, which reported the plight of that city's homeless people as they sought meagre heat sources on the streets in a winter of record cold (Purdy 1994). If asked, might they view global warming as a benefit rather than a blight? Does the God's-eye view merely reflect the insulation of those who take it from any long-term contact with economic unpleasantness?

According to Enzenberger, "[t]he ecological movement has only come into being since the districts which the bourgeoisie inhabit and their living conditions," rather than just those of the working class, "have been exposed to those environmental burdens that industrialization brings with it" (Enzenberger 1974: 10). This dynamic is very similar to that identified by Susan Moller Okin, when she comments that much of the political theory which forms a part of the conventional curriculum "is about men with wives at home," simply because they wrote it (Okin 1989a: 10; see also Okin 1989b). The perspective of those men was, and to a great extent still is, accepted as neutral and normative. The remarks of Eckersley and Bahro suggest that the same sort of thing is going on with respect to the environmentalists' perspective on issues of class, even when they might be expected to know better in terms of an academic understanding of the contemporary class structure.

I have tried to illustrate, using a necessarily limited number of examples, that the discourse and institutional practice associated with sustainable development are best understood by way of an epistemology that emphasizes the situation of knowers and doers. I have also suggested that the policy agenda of environmental protection and sustainable development, in its present form, reproduces and legitimizes the perspective of the denizens of the space platforms, rather than that of people below who look up at their dark undersides, and maintain the machinery that keeps them operating. This point is now rather familiar when made about the divergent priorities of rich and poor countries (Soto 1992); it is less familiar with respect to the politics of environmentalism within nations themselves.

Deliberately reorganizing one's assumptions about the world is never easy. The task is especially difficult when it requires confronting myths that are as powerful and convenient as that of universal affluence, and have become so integrated into the iconography of popular culture and imagination. The interaction of material and symbolic factors in constructing that iconography is an unexplored theme in the work of Ronald Inglehart. He argues that the rapid spread of so-called postmaterialist values can be attributed to the childhood political socialization of people growing up in postwar industrial societies, which "are a remarkable exception to the prevailing historical pattern: the bulk of their population does *not* live under conditions of hunger and economic insecurity" (Inglehart 1981: 881; see also Eckersley 1987: 216–17). More recently and specifically, Inglehart has argued that the intergenerational reproduction of postmaterialist values is not uniform. Rather, it reflects the class structure of contemporary societies: "Postmaterialists tend to be those who start life with all the advantages; to a considerable extent, that is why they are postmaterialists" (Inglehart 1990: 171). In defining what income and consumption patterns are considered normal or routine in a particular society, surely "those who start life with all the advantages" will have the upper hand, and it is with reference to their backgrounds that the normative material dimensions of everyday life (Lynn McDonald's full refrigerators) are defined for all of us.

This is just one set of issues, among many, raised in this article that suggest directions for more extensive research. Dorothy Smith and others have identified the importance of hierarchies of credibility: who is to be believed, about what and why (Smith 1990; Code 1988). Why is the image of universal affluence so seldom challenged? Is it because those in the best experiential positions to do so (indeed, in positions far better than mine for this purpose) are invisible, or not considered credible, for a variety of reasons? More examination is needed of the often-fluid hierarchies of credibility that have come to characterize the sustainable-development-policy discourse. More systematic and extensive research is also required on what proponents of environmental protection or sustainable development actually say, or fail to say, about class and economic inequality. My examples were chosen on a highly unsystematic basis. There was no sampling strategy, no attempt to check for "representativeness." More systematic research might show that I am all wrong about the politics of invisibility, by indicating the atypical nature of those examples. However, I do not think so.

Conclusion: Invisibility in Discourse and in Institutions

So far, I have dealt with the realm of discourse: what people say. The concept of a discourse involves the overall set of images and concepts people use to think about public policy. Its political significance is immense; what people say cannot be isolated from how they think, what they do and how they think about what they do (Edelman 1964, 1988; Walker 1986). The inadequacy of existing opportunities for public participation was a key element of environmentalist critiques of existing political institutions (Sax 1970; Gibson 1975; Castrilli and Lax 1981), long before the more general emergence of "radical democracy" as a political objective (Mouffe 1993; Aronowitz 1994). It is therefore worth looking at the way in which the connections between class and invisibility shape and limit the persuasiveness of this critique. Can institutions for public participation be designed that would be conducive to a genuinely inclusive politics? Can they avoid favouring certain kinds of participants, and in the process, selectively reinforce particular, (over)privileged views of the way the world works? Perhaps more importantly, will people who promote expanded opportunities for public participation even be concerned about selective or biased inclusiveness?

The first two questions have not been satisfactorily answered in more than 20 years of discussion. In 1969, Sherry Arnstein classified microlevel political institutions in terms of an eight-rung "ladder of citizen participation." At the bottom of the ladder are manipulation and therapy; at the top are delegated power (that is, delegated by government agencies to citizens) and citizen control (Arnstein 1969). Arnstein's model does not provide a basis for examining the ways institutions select among potential participants, making effective participation (as distinct from the tokenism she rightly condemns) relatively easy for some, and impossible for others. This is not a criticism of Arnstein, who is consistently sensitive to issues of race and class (although not, interestingly, gender), and herself points out that "the typology does not include an analysis of the most significant roadblocks to achieving genuine levels of participation" (Arnstein 1969: 217). However, if environmentalists hold distorted beliefs about class and about the universality of affluence, they are likely to be relatively unconcerned about the ways in which time, money, vocabulary and networks of political contacts serve as prerequisites for effective political participation.

Against this background, a cynical view is that many proposals for expanded public participation in environmental policy are best understood as instances of pseudo-inclusion (March's word) or co-optation (Arnstein's). Superficial solicitude is shown for the concerns of citizens, the public, the community, the grass roots or some similarly decontextualized category. All these terms obscure the relevance of such issues as class and gender, and thereby ignore the manifold practical barriers to participation that confront many individuals. An alternative view is that environmentalists may be cognizant of such barriers, yet are willing to live with pseudo-inclusion (in March's words) at the institutional level, on the basis that it is (a) better than nothing at all, and/or (b) the best that can be hoped for.

At least one observer with a lengthy background in another contemporary social movement seems to believe that institutional barriers to a genuine politics of inclusion are insuperable. Monique Bégin's political career spans more than two decades: from having been research director for Canada's Royal Commission on the Status of Women in the late 1960s, she moved on to hold various cabinet posts in the national Liberal government between 1976 and 1984. She has described the expanding range of interest groups organized around women's issues as "defending the common good of a wide majority of women, *necessarily* middle class, while constantly bringing up front the case of lone mothers, of poor women, of older pensioners, of immigrants, of women who are disabled and those in situations of weakness" (Bégin 1988: 345–46, emphasis added). Unlike Bahro, Bégin at least concedes the potential existence of class conflict about ends and priorities. Like him, she seems to claim that nothing can be done, at least for the moment, about exclusions based on certain kinds of powerlessness. And like Bahro (and many others), Bégin seems quite unaware of the epistemological and political precariousness of claims to "speak for" the poor, the marginalized or the otherwise disempowered.

It is against the background of examples like this one that we should consider the claim that new social movements and their organizing principles "challenge, each in its own way, the privileging of class as the fundamental social cleavage" (Bleyer 1992: 103). The claim should be challenged on at least two grounds. First, even if one concedes that class is not the most fundamental social cleavage, there is a fine line between refusing to privilege class in this respect and implicitly privileging the perspective of an advantaged minority by ignoring class altogether. Thus, Claus Offe commented a decade ago that "[new] middle class politics" arguably exemplified by environmentalism, "in contrast to most working class politics ... is typically a politics *of* a class but not *on behalf of* a class" (Offe 1985: 833, emphasis in original). It is therefore interesting that a recent collection of essays on "evolving ecological socialism" contains remarkably few specific mentions of class in discussions of such complex issues as trade-offs between economic development and environmental protection in a resource-dependent economy (Vorst et al. 1993).

Second, even when discussions of eco-socialism do pay explicit attention to class as a conceptual lens through which to view individuals' economic opportunity structures, they tend to ignore the objective dimensions of class, which are at the core of Elster's definition, in favour of a focus on the "subject positions" that constitute class identity (Adkin 1992: 152). Subject identities do matter, of course, in the context of individual agency. However, the objective constraints that define class position matter, too: "a Rockefeller cannot change his class status by taking a subordinate management job" (Elster 1986: 147), and the privileged position of those to whom Bahro and others are willing to assign the task of speaking for the whole of society, is a matter of far more than subjective identity.

A great deal of basic rethinking is necessary if social movements, which claim to be progressive, are not to deserve identification as cynical defenders of privilege. In the meantime, two warnings must be kept in mind in the context of environmental protection or sustainable development. First: Don't be fooled by the rhetoric of common interests and the image of saving an endangered planet. On environmental questions that affect growth, economic distribution and economic opportunity, we are most certainly *not* all in this together. Second: A special sensitivity to issues of class is urgently required on the part of all those who take environmental issues seriously, unless they are willing to accept with equanimity the deepening gulf between rich and poor in terms of both material well-being and political power. Caricatures that perpetuate the misleading myth of universal affluence are profoundly destructive of that sensitivity.

Endnote

1. A household was considered to have discretionary income if its after-tax income was at least 30 per cent above the economy-wide average expenditure for goods and services of comparable households, based on "a U.S. Department of Labor survey which had concluded that, to allow for a relatively comfortable standard of living, a household's budget had to be about 30% higher than the average consumption expenditure for similar households" (Owens 1991: 4.2). The survey's findings are therefore vulnerable to the criticism that they ignore the country's overall level of affluence. Nevertheless, substantial discretionary income, as defined in the StatsCan survey, is almost certainly necessary for the routine non-credit purchase of the items depicted on the *Alternatives*' cover.

DISCUSSION QUESTIONS

1. Why is Schrecker critical of environmentalists' attitudes toward income and wealth? Do you agree or disagree? Why?
2. Why is the concept of a discourse important in understanding the world-view of environmentalists in understanding public policy?
3. What is the concept of "situated knowledge?" Why is it important?
4. How could existing social and political institutions be changed to make decision making about environmental issues more inclusive?

References

Adkin, Laurie. 1992. Counter-hegemony and Environmental Politics in Canada. In *Organizing Dissent: Contemporary Social Movements in Theory and Practice*, ed. W. Carroll, 135–56. Toronto: Garamond.

Arnstein, Sherry. 1969. A Ladder of Citizen Participation. *American Institute of Planners Journal* 35(July): 216–24.

Aronowitz, Stanley. 1994. The Situation of the Left in the United States [with responses by Amarpal Dhaliwal, Barbara Ehrenreich, Barbara Epstein, Dick Flacks, Michael Omi and Howard Winant, and Eli Zaretsky]. *Socialist Review* 23(3): 5–150.

Bahro, Rudolf. 1986. *Building the Green Movement.* Translated by Mary Tyler. London: Heretic Books/GMP.

Baier, Annette. 1986. Poisoning the Wells. In *Values at Risk*, ed. D. MacLean, 49–74. Totowa, N.J.: Rowman & Allanheld.

Bégin, Monique. 1988. Debates and Silences — Reflections of a Politician. *Daedalus* 117(Fall): 335–62.

Bleyer, Peter. 1992. Coalitions of Social Movements as Agencies for Social Change: The Action Canada Network. In *Organizing Dissent: Contemporary Social Movements in Theory and Practice*, ed. W. Carroll, 102–17. Toronto: Garamond.

Castrilli, J.F. and Clifford Lax. 1981. Environmental Regulation-Making in Canada: Towards a More Open Process. In *Environmental Rights in Canada*, ed. J. Swaigen, 334–405. Toronto: Butterworths.

Code, Lorraine. 1988. Credibility: A Double Standard. In *Feminist Perspectives*, ed. L. Code et al., 64–88. Toronto: University of Toronto Press.

Eckersley, Robyn. 1987. Green Politics and the New Class: Selfishness or Virtue. *Political Studies* 37: 205–23.

Edelman, Murray. 1964. *The Symbolic Uses of Politics.* Urbana: University of Illinois Press.

———. 1988. *Constructing the Political Spectacle.* Chicago: University of Chicago Press.

Elster, Jon. 1986. Three Challenges to Class. In *Analytical Marxism*, ed. J. Roemer, 141–61. Cambridge: Cambridge University Press.

Enzenberger, Hans Magnus. 1974. A Critique of Political Ecology. *New Left Review* 84: 3–31.

Gibson, Robert. 1975. The Value of Participation. In *Environmental Management and Public Participation*, ed. P.S. Elder, 7–39. Toronto: Canadian Environmental Law Research Foundation.

Haraway, Donna. 1988. Situated Knowledges: The Science Question in Feminism and the Privilege of Partial Perspective. *Feminist Studies* 14: 575–99.

Inglehart, Ronald. 1981. Post-Materialism in an Era of Insecurity. *American Political Science Review* 75: 880–900.

———. 1990. *Culture Shift in Advanced Industrial Society.* Princeton: Princeton University Press.

Johnson, Warren. 1973. The Guaranteed Income as an Environmental Measure. In *Toward a Steady-State Economy*, ed. H. Daly, 174–89. San Francisco: W.H. Freeman.

Kazis, Richard and Richard Grossman. 1982. *Fear at Work: Job Blackmail, Labor and the Environment.* New York: Pilgrim Press.

Mackenzie, Hugh. 1987. A Labour Perspective: Finding Common Ground. *Alternatives: Perspectives on Society, Technology and Environment* 14(August/September): 13–17.

March, Artemis. 1982. Female Invisibility in Androcentric Sociological Theory. *Insurgent Sociologist* 11: 99–107.

Matas, Robert. 1993. Clayoquot: The Sound and the Fury. *The Globe and Mail*, 22 May: A6.

McDonald, Lynn. 1987. *The Party that Changed Canada.* Toronto: MacMillan.

McKay, Paul. 1990. *Plundering the Future.* Toronto: Atkinson Foundation. [Originally published as a series of articles in *The Toronto Star* during November 1990.]

Mouffe, Chantal. 1993. Radical Democracy: Modern or Postmodern? In *The Return of the Political*, 9–22. London: Verso.

National Council of Welfare. 1992. *Poverty Profile, 1980–1990.* Ottawa: Supply and Services Canada.

———. 1994. *Poverty Profile 1992.* Ottawa: Supply and Services Canada.

Offe, Claus. 1985. New Social Movements: Challenging the Boundaries of Institutional Politics. *Social Research* 52: 817–68.

Okin, Susan Moller. 1989a. *Gender, the Public and the Private.* Legal Theory Workshop Series WS 1989–90, no. 4. Toronto: Faculty of Law, University of Toronto.

———. 1989b. *Justice, Gender and the Family.* New York: Basic Books.

Owens, Donna J. 1991. Tracking Down Discretionary Income. *Canadian Economic Observer*, March: 4.1–4.14.

Purdy, Matthew. 1994. In Sea of Wealth, Homeless Defy the Cold. *The New York Times*, 21 January: A1, A12.

Sax, Joseph L. 1970. *Defending the Environment: A Strategy for Citizen Action.* New York: Random House.

Shrader-Frechette, K.S. and E.D. McCoy. 1993. *Method in Ecology: Strategies for Conservation.* Cambridge: Cambridge University Press.

Sinclair, J. 1993. Waste and the Final Frontier. *Alternatives: Perspectives on Society, Technology and Environment* 19(January/February): 1.

Smith, Dorothy. 1990. *The Conceptual Practices of Power*. Toronto: University of Toronto Press.

Soto, A. 1992. The Global Environment: A Southern Perspective. *International Journal* 47: 679–705.

Statistics Canada. 1992. *Family Incomes: Census Families 1990*. Ottawa: Statistics Canada.

U.S. Congressional Office of Technology Assessment. 1986. *Serious Reduction of Hazardous Waste,* OTA-ITE-317. Washington, D.C.: U.S. Government Printing Office.

Vorst, Jesse, Ross Dobson and Ronald Fletcher, eds. 1993. *Green on Red: Evolving Ecological Socialism*. Winnipeg: Fernwood.

Walker, Georgia. 1986. Reforming Medicare: The Limited Framework of Political Discourse in Equity and Economy. *Social Science and Medicine* 23: 1237–50.

Wykle, Lucinda, Ward Morehouse and David Dembo. 1992. *Worker Empowerment in a Changing Economy: Jobs, Military Production and the Environment*. New York: Apex Press.

SUSTAINABLE DEVELOPMENT: A SOCIOLOGIST'S VIEW OF THE DEFINITION, ORIGINS AND IMPLICATIONS OF THE CONCEPT

Jean-Guy Vaillancourt

Université de Montréal

In this paper, I shall try to do three things. First, I shall briefly examine the concept of sustainable development, as originally presented in the Brundtland Report.[1] Then I shall look into the origins and the historical antecedents of this rich and vague concept. In the third and final part of my presentation, I shall attempt to show that sustainable development implies more than a simple reconciliation of economics and ecology, and of socio-economic development and environmental protection. Because of the limitations imposed by this type of presentation, I shall of necessity have to be brief in each of the three parts of this paper.

In the conclusion, I shall briefly touch on the question of the operationalization and the applications of the concept of sustainable development, but that is another, even more important, issue that would require another, more lengthy, presentation.

Paper presented at the International Studies Association's annual meeting, March 1994, in Washington D.C.

The Concept of Sustainable Development

Since the publication of the Brundtland Report in 1987, the expression "sustainable development" has caught on and propagated itself among environmental and development NGOs, and more rapidly still among government bureaucrats and vote-hungry politicians, as well as in public-relations departments of major corporations. The concept of sustainable development has even become the rallying cry of nearly all those in the world who are interested in simultaneously pursuing both economic development on the one hand, and environmental protection and resource conservation on the other. Proponents of environmental protection and defenders of economic development have always entertained rather difficult relationships with each other, but sustainable development tries to reunite them in a durable fashion.

According to the Brundtland Report: "Sustainable development is development that meets the needs of the present without compromising the ability of future generations to meet their own needs."[2] It is a process of economic development based on renewable resources which respects the basic ecological processes, biological diversity and the natural support systems for human, animal and vegetative life. But it also implies, as we shall explain later, a certain degree of equity between rich and poor nations, and between generations. It even implies the necessity of disarmament as a prerequisite to liberate the funds needed for its implementation, as we can see in Chapter 11 of the Brundtland Report.

It is quite clear that the Brundtland Report did not intend to be an exercise of legitimation of current economic operations of major industrial corporations bent on extracting rapidly non-renewable resources in order to make short-term mega-profits, nor a call to have them make a few cosmetic changes in their activities. The Brundtland Report is rather critical of such enterprises, and it calls for profound changes in their *modus operandi*. On the other hand, it does not legitimate either the bucolic vision of certain utopian deep ecologists, who want to preserve all of nature in its pristine state, and who dream of an impossible return to an idealized neo-paleolithic mode of life, to a paradise lost, to a reconstructed Garden of Eden, where humans would live in complete harmony with an untouchable and quasi-deified nature. If the concept of sustainable development excludes these two extreme positions, namely the ideology of non-stop economic growth on the one hand, and extra-deep ecology on the other, it is still vague and encompassing enough to include the positions of most conservationists, preservationists, environmentalists, and social and radical ecologists who constitute the various tendencies of today's Green movement.

Sociologically, the concept has been playing many functions for many categories of economic, social and political actors in the environment and development fields. The sociologists Ann P. Hawkins and Frederick H. Buttel argue, for example, that sustainable development must be understood in terms of four roles that it can play: (1) as an environmental movement ideology; (2) as a development fad and symbolic pivot of conflict over development policy; (3) as a guidepost for designing and evaluating development policies and programmes; and (4) as a theoretical category.[3]

My own view is that it is mainly a vague and complex theoretical category, which also serves partly as an ideology, and partly as a loosely operationable concept, for various groups of interested actors.

Hawkins and Buttel have put forward the view that Third-World progressives have used the concept of sustainable development as a round-about way to circumvent the blockage of their more clear-cut social justice appeals and demands concerning prevailing development policies, and as a means to mobilize the support of powerful allies (particularly environmental groups) in the industrialized countries.[4] At the same time, sustainable development, like other previous development fads, is also superficially embraced and institutionalized by organizations like the World Bank and the International Monetary Fund, the same way that previous fads like "basic needs," "appropriate technology," "participation in development" and "women in development" were bureaucratically used earlier by these same organizations.

The concept of sustainable development seems to me to be like a blanket, partly scientific and partly ideological, that everyone is pulling at; it is a rallying cry for those who are interested in both development activities and environmental protection, in the harmonization of economics with ecology. It is a compromise formula which emerged slowly between the early 1970s and the late 1980s between environmental activists and proponents of development (especially, but not exclusively in poorer countries). Brundtland and the members of her commission found a way to reconcile environmentalists and developmentalists from rich and poor countries on the one hand, with forward-looking government bureaucrats and enlightened liberal business people on the other.

The Brundtland Commission was instituted by the United Nations General Assembly in 1983. It comprised six Westerners, three East-Europeans, and 12 persons from Southern countries,[5] all fairly independent from their respective governments. This composition probably accounts for some of the vagueness of the definitions given in the report, and also for the wide acceptance of the report by many categories of people all over the world.

Origins of the Concept

In order to better understand the concept of sustainable development and its phenomenal success, it is necessary to examine its origin and the similar concepts which preceded it. Throughout history, from the early Greek and Roman philosophers of nature, to the Medieval Benedictine monks, to political economists like Malthus and Ricardo, to the early conservationists in Europe and America, there have always existed people who have promoted a wise and moderate use of nature's resources. For example, in 1915, the Canadian Commission of Conservation which had been created in 1909, encouraged the respect of natural cycles, and suggested that each generation had the right to profit from the interest on nature's capital, but that this capital had to be maintained intact for future generations to enjoy and to use in a similar fashion.

In 1923 in Paris, the major topic at the International Congress for the Protection of Nature, was the relationship between the protection of nature and the use of its resources. Paul Sarasin, the Swiss conservationist and key organizer of this meeting, in his paper on "The International Protection of Wildlife," pleaded for conservation, but also for a rational use of wild animals. In 1948, during a conference organized by UNESCO and the French government at Fontainebleau, the International Union for the Conservation of Nature and Its Resources was founded on the premise that both nature and its resources should be protected for the benefit of existing and future generations.

In 1951, that organization published a remarkable document entitled *The State of the Protection of Nature in the World in 1950*, which comprised 70 reports on so many countries. This work, which was updated in 1954, can be considered one of the "precursors" of the Brundtland Report. This international organization also co-operated with UNESCO and with other international organizations like the WHO, FAO and the ICSU (International Council of Scientific Unions) to hold, in Paris in 1968, an Intergovernmental Conference of Experts on the Scientific Bases for the Rational Use and the Conservation of the Resources of the Biosphere. This conference's report, published in 1970, says that: "conservation methods ... must aim at maintaining ... the conditions that permit the best qualities of man to flourish."[6]

The first report to the Club of Rome in 1968 launched a heated debate with its concept of zero growth. This concept antagonized those who wanted development to continue, in the Third World particularly, but also in industrialized countries. In preparation for the Stockholm Conference of 1972, a seminar was held in Founex, France, in June 1971, on the topic of development and environment, in order to find a middle way between growthism and the new ideology of zero growth. Maurice Strong, who was the convenor of this preliminary meeting, wanted to avoid a kind of developmentalism which did not care about the environment, and an environmentalism which neglected the need for development, especially in Third World countries. In fact, in Founex, some of the participants to the seminar proposed to engage in a type of development which would respect natural resources.[7]

That same year, an International Institute for the Environment and for Development was created and placed under the presidency of Lady Barbara Ward Jackson, who then co-authored the book *Only One Earth* with René Dubos.[8] This book, written by a humanist economist and a biologist-ecologist, served as a sort of manifesto for the 1972 U.N. Conference on the Human Environment in Stockholm. The idea of "eco-development" was put forward at that time as a way to reconcile pro-environment Western activists on the one hand, and pro-development Third-World radicals and Western industrialists on the other. In the 109 points proposed in the Plan of Action of the Stockholm Conference, a type of development respectful of the environment and of the management of natural resources, was presented as a priority. The need to preserve the capacity of the Earth to produce vital renewable resources was asserted. The conference asked the U.N. to create various institutional mechanisms in the environmental area.

The result was the United Nations Environmental Programme (UNEP), which was then instituted as a means to complement the United Nations Development Programme (UNDP). Its offices were located in Nairobi, Kenya, as if to show that the protection of the environment and Third-World development were intimately linked. UNEP has been financed ever since through voluntary governmental contributions deposited in its environmental fund. At the first meeting of UNEP, its executive director Maurice Strong insisted on the need for eco-development, a type of development which takes into account environmental constraints in the long term. Strong was able to reassure both Third-World countries and Western industrialists that environmental protection did not mean zero growth, but could be compatible with development as long as it preserved natural resources for the use of future generations.

In 1973, Ignacy Sachs, a Polish social scientist working in Paris, founded the International Center for Research on the Environment and Development, and started publishing a journal called *Cahiers de l'écodéveloppement (Ecodovelopment Notebooks)* a year later. In 1977, it started putting out another periodical, a newsletter called *Nouvelles de l'écodéveloppement (Ecodevelopment News)*. In 1980, Sachs summarized his ideas in a book entitled *Stratégie de l'écodéveloppement (Ecodevelopment Strategy)*.[9] Sachs talks of harmonizing ecology and economics, of submitting economic decisions to the urgent necessity of conservation. By his work during the 1970s to promote ecodevelopment, especially in the Third World, Ignacy Sachs, along with Maurice Strong, became one of the pioneers in the effort to harmonize environment and development. However, it was not their concept of ecodevelopment which caught on, especially in the English-speaking world, but rather the concept of a sustainable society. During the 1970s, there was a lot of talk about a new international economic order, about "a different kind" of development, and especially about ecodevelopment. These were all different ways to promote the idea that socio-economic development and environmental protection had to be harmonized. In the end, it was the idea of sustainability which gained the most popularity.

Sustainability and Its Implications

The concept of a sustainable society emerged gradually in the 1970s, for example, in the work of Barry Commoner on the need to use renewable resources and low-impact technology, and of Herman Daly on steady-state economics,[10] of Edward Goldsmith, Sicco Mansholt and Kimon Valaskakis on the conserver society, in the 1976 Manifesto of the Ecology Party of Great Britain, and in the work of Amory Lovins on soft-energy paths.[11]

It also appears prominently in the work of Lester Brown, particularly in his book *Building a Sustainable Society*.[12] For Brown, a sustainable society entails a stable population, the conservation and wise use of the land and of renewable resources, especially in the field of energy, the protection of fisheries and forests, and of biodiversity and the web of life.

The earliest mention of the expression "sustainable development" that I can find, is in the World Conservation Strategy launched in 1980 by the International Union for the Conservation of Nature and Its Resources, the World Wildlife Fund and the United Nations Environmental Programme (UNEP), with the backing of UNESCO and FAO:

> This strategy anticipates that sustainable development must take into account social and ecological factors as well as economic factors, the biotic and non-biotic resource base, and also the advantages and inconveniences, both short term and long term, of alternative solutions.[13]

After the failure of the 1982 U.N. Conference on the Environment in Nairobi, the U.N. General Assembly decided, the following year, to create a World Commission on Environment and Development, the famous Brundtland Commission. Its 1987 report entitled *Our Common Future*, popularized the term sustainable development and paved the way for the U.N. Conference on Environment and Development, and the Global Forum, which were held in Rio de Janeiro in June 1992.[14]

The concept of sustainable development, as it appeared in the Brundtland Report, was not only an effort to reconcile economics and ecology, or even socio-economic development and environmental protection and conservation. It also implied sociopolitical dimensions, particularly social equity, as many people, especially social scientists, have indicated.

Although the Brundtland Report does not exclude the necessity of some form of economic growth, especially in poorer countries, it views development as a complex process that goes much beyond simple economic growth:

> Development involves a progressive transformation of economy and society.... Even the narrow notion of physical sustainability implies a concern for social equity between generations, a concern that must logically be extended to equity within generations.[15]

The Brundtland Report also stresses the idea, in its 11th chapter, that sustainable development cannot be achieved without disarmament. Although most environmental sociologists still see sustainable development as "an attempt to reconcile and synthesize concerns about environment and socio-economic development,"[16] many sociologists have also insisted strongly on the idea that socio-economic development is in no way equivalent to economic growth. The human and social dimensions inherent in the expression socio-economic development, and even in the word development itself, have pushed some sociologists, like Michael Redclift, to present it as a concept which goes well beyond economics and which must "take into account the inter-connectedness and embeddedness of human society in natural ecology."[17]

Peter Jacobs, Julia Gardner, and David Munro even suggest that we should use the expression "sustainable and equitable development" rather than just "sustainable development."[18] But the third dimension beyond environment and development that is increasingly being mentioned, even goes beyond the mere idea of equity. It is a wider sociopolitical dimension which includes the idea of equity along with other socio-political aspects like disarmament, human rights and democratic participation.

Already, in 1981, Robert Riddell had described eco-development, as being characterized by three elements: economic equity, social harmony and an environment in equilibrium. Barry Sadler, in various essays, has represented the three dimensions of sustainable development as three intersecting circles, or as three triangles imbedded into a larger equilateral triangle[19] (see Figures 1 and 2). And according to David Pearce et al., sustainable development comprises not two but three elements: (1) the value of the environment; (2) the sustainability of economic development; and (3) inter- and intragenerational equity.[20] Among the forms of intragenerational equity which the Brundtland Report mentions, are equity between North and South — i.e., between rich and poor countries; but it goes without saying that this is not the only type of equity which is implied in the concept of sustainable development put forward by the Brundtland Report and by its most authoritative commentators in the social and natural sciences of the environment, and in various organizations devoted to promoting sustainability. The Brundtland Report states: "Such equity would be aided by political systems that secure effective citizen participation in decision making and by greater democracy in international decision making."[21] I have constructed the following triangle to show that sustainable development's third sociopolitical dimension

FIGURE 1 Sustainable Development

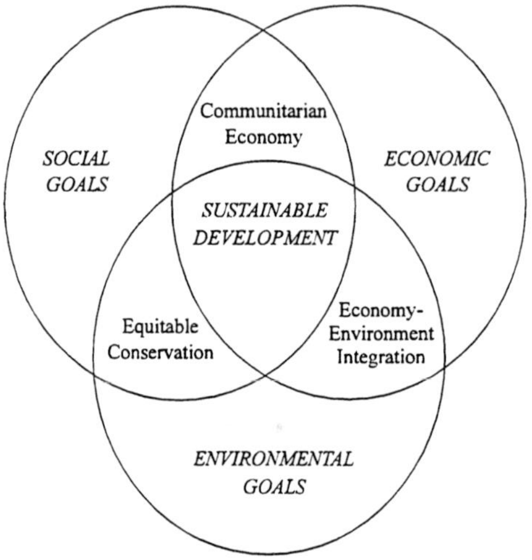

Source: Peter Jacobs and Barry Sadler. *Développement durable et évaluation environnementale: perspective de planification d'un avenir commun*. Ottawa: CCREE, p. 10.

FIGURE 2 Decision Making in Matters of Sustainable Development

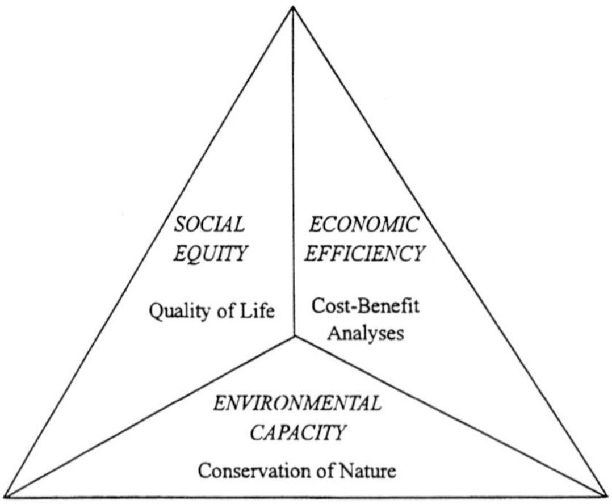

Source: Adapted from Peter Jacobs and Barry Sadler, *Développement durable et évaluation environnementale: perspective de planification d'un avenir commun*. Ottawa: CCREE, p. 18.

FIGURE 3a Sustainable Development: A Ternary Approach

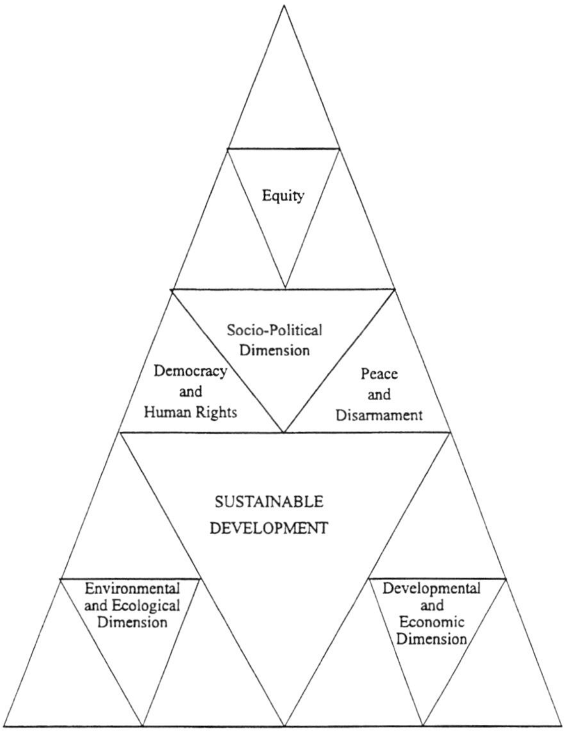

itself comprises three major dimensions, namely democracy and human rights, disarmament and peace, and equity. Equity, in turn can be of three types: intergenerational (i.e., between young and old, and between present and future generations), international (i.e., between North and South and between rich and poor countries), and intragenerational and intranational (i.e., social equity between men and women, between social classes and categories — rich and poor, powerful and powerless etc. — and between ethnic and religious groups). (See Figure 3.)

Conclusion

The exercise in conceptual clarification that I have attempted in the three above sections of this paper, should not be considered an end in itself. The most important thing about sustainable development is not how it is

FIGURE 3b Sustainable Development: A Ternary Approach (*cont'd*)

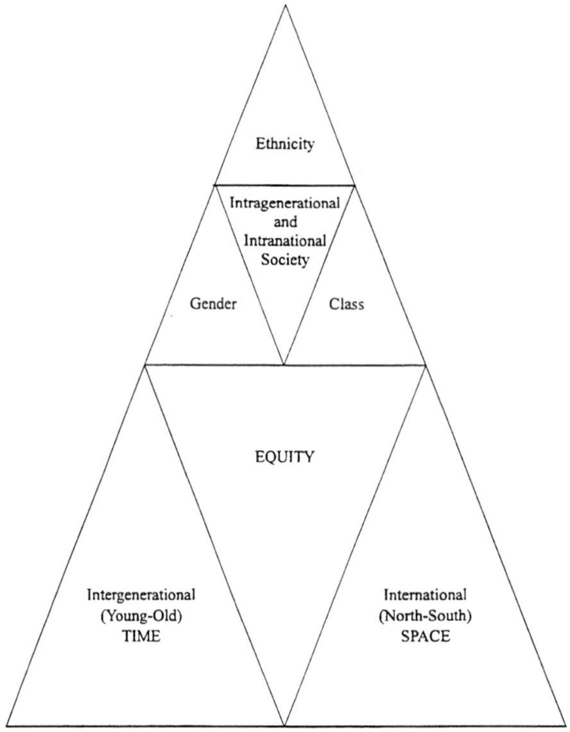

conceived, but how it should be applied, at the international, regional and local levels, and in different parts of society and of the environment. In order to go from theory to practice, the concept of sustainable development must first be defined, dimensionalized and operationalized, and then it must be applied in a concrete fashion to various sectors of human activity. It is only when the concept has been clearly presented that the success of this operationalization and of these applications can be measured accurately with appropriate indicators.

Even before the Brundtland Report was published, many countries, provinces, cities, political organizations and economic enterprises were devising plans to put sustainable development into effect. This is an ongoing process which was given a new impetus with the Rio summit's Agenda 21 and the Global Forum's alternative treaties. It would take another lengthy paper to describe some of these experiences, starting with the Commission for Sustainable Development, created in the aftermath of Rio by the U.N. General Assembly. In the case of Canada, we would have to mention the National Round Table on the Environment and the Economy, and the Green Plan, as well as dozens of other federal, provincial and municipal institutions, programmes, publications and projects.

A local experience, like the Sustainable Development Laboratory in the Saguenay-Lac St-Jean region in northeastern Quebec, illustrates very well how sustainable development is not only a concept that needs fine-tuning, but a process that is slowly but surely being implemented at all levels. The only question that remains to be asked is the following: In light of the mounting evidence that environmental problems are increasing at all levels, are all the efforts being made to achieve sustainable development worthwhile, or is all of this too little, too late?

DISCUSSION QUESTIONS

1. What is the Brundtland Report's definition of "sustainable development?" Do you agree or disagree with it? Explain.
2. What function does the concept of sustainable development play for Hawkins and Buttel?
3. Name the concepts similar to that of sustainable development used prior to 1987. How are they different from the modern use of the term?
4. Is sustainable development merely a reconciliation of economics and ecology, or is there an important third dimension that must be addressed? Discuss.
5. What are the main types of equity which must be taken into account in order to implement sustainable development?

Endnotes

1. World Commission on Environment and Development, *Our Common Future* (Oxford: Oxford University Press, 1987).
2. Ibid., p. 43.
3. Ann P. Hawkins and Frederick H. Buttel, "Sustainable Development," in *Concise Encyclopedia of Participation and Co-Management*, ed. G. Széll (New York: Walter De Gruyter, 1992), p. 831.
4. Ibid., pp. 835–36.
5. WCED, op. cit., pp. 352–56.
6. Conférence internationale d'experts sur les bases scientifiques de l'utilisation rationnelle et de la conservation des ressources de la biosphére (Paris: UNESCO, 1970), p. 307.
7. *Development and Environment* [Founex, 4–12 June, 1971] (The Hague: Mouton, 1972).
8. Barbara Ward and René Dubos, *Only One Earth* (New York: W.W. Norton and Co., 1972).
9. Ignacy Sachs, *Stratégie de l'écodéveloppement* (Paris: Editions ouvrières, 1980).

10. Barry Commoner, *The Closing Circle* (New York: Knoft, 1971); Herman E. Daily, ed. *Steady-State Economics* (San Francisco: Freeman, 1973).
11. See Ecology Party, *Manifesto for a Sustainable Society* (Sheffield, England: September 25–26, 1976), and Amory Lovins, *Soft Energy Paths: Toward a Durable Peace* (New York: Ballinger Publishing Co, 1977).
12. Lester Brown, *Building a Sustainable Society* (New York: W.W. Norton and Co., 1981).
13. Union internationale pour la conservation de la nature et de ses ressources, *Stratégie mondiale de la conservation: la conservation des ressources vivantes au service du développement durable* (Gland, Suisse: U.I.C.N., 1980).
14. Jean-Guy Vaillancourt, "Earth Summits of 1992 in Rio," *Society and Natural Resources*, vol. 6 (1993), pp. 81–88.
15. WCED, op. cit., 43.
16. Frederick H. Buttel, "Sustainable Development" in *Concise Encyclopedia of Participation and Co-Management*, ed. G. Szëll (New York: Walter de Gruyter, 1992), p. 831.
17. M. Redclift, *Sustainable Development: Exploring the Contradictions* (New York: Methuen, 1987).
18. Peter Jacobs, Julia Gardner and David A. Munro, "Sustainable and Equitable Development: An Emerging Paradigm" in *Conservation with Equity: Strategies for Sustainable Development*, ed. P. Jacobs and D. Munro (Cambridge: Cambridge University Press, 1988).
19. Peter Jacobs et Barry Sadler. n.d. *Développement durable et évaluation environnementale: perspective de planification d'un avenir commun.* Ottawa: Conseil Canadien de recherche sur l'évaluation environnementale.
20. David Pearce, Anil Markanda and Edward B. Barbier, *Blueprint for a Green Economy* (London, Earthscan Publications, 1989).
21. WCED, op. cit., p. 8.

THE SOCIOLOGY OF SUSTAINABILITY: FEMINIST ECO/NOMIC APPROACHES TO SURVIVAL

Melody Hessing

Simon Fraser University

Approaching the year 2000, the sustainability of life on earth is increasingly under threat. Efforts to prevent human starvation, decrease the rate of species extinction and reduce poverty, require increased state intervention. Imagine, if you will, the following television scenario:

> Captain Brundtland-Strong is at the helm of the Starship Sustainability, plotting a course for planetary survival. The payload is largely conceptual: distributive justice, stabilized human population growth, the preservation of biodiversity. The youthful spandexed crew responsible for the implementation of this mission — six men and a woman — strut in and out of the control room, preoccupied with new projections, correcting the models, exuding a high-tech efficiency.

Yet the special effects pale against the mundane and global disasters of everyday life. Skyrocketing populations, accelerating environmental degradation, starvation, economic polarization and civil wars result in environmental degradation and social conflict. Meanwhile, delays in government funding, a decline in international aid, public apathy, and budgetary shortfalls glue the Starship to the launch pad, its planned trajectory under continuing revision.

Although the conditions for supporting future life on earth are increasingly under threat, there is little attention paid to these issues in most sociology textbooks. As the above scenario illustrates, demographic expansion and resource depletion continue to escalate dramatically. Efforts to implement sustainability remain limited, floundering in a sea of market-determinism, budgetary inertia, incremental institutional adjustments and political impasses. The polarization of wealth, decline of the resource base and extinction of species continue to accelerate even as appropriate responses are puzzled over. The concept of sustainability — ensuring the survival of humans and the natural environment — remains, like other ecological and environmental issues, marginal to the discipline of sociology. The notion of sustaining human and other life remains largely consigned to either the status of science fiction (an esoteric problem for future generations absent even in Star Trek episodes) or to an anomalous academic pursuit directed to the aggrandizement of institutional toe-holds. The lack of sociological attention to sustainability minimizes concern about environmental degradation and social change, and diminishes sociological contributions to the dialogue on sustainability.

There are a number of possible explanations for the lack of attention by Canadian sociologists to sustainability, and more generally, ecological sociology. The abundance of the Canadian land and resource base, as well as the concentration and relatively small size of human populations, have reduced the perception and degree of environmental problems from popular concern. While a Canadian political economy tradition has contributed to the recognition of natural environments as resources to be used for extractive purposes, little attention has been paid to their cultural and aesthetic significance. A focus on socio-economic dependence and the manoeuvrings of labour and the economic élite within a staples economy, have further deflected academic concern from physical/material infrastructure. A negotiative regulatory climate, integrating corporate and state interests, has obscured the extent of industrial damage to natural environments. The characterization of the environmental movement as white and middle class has contributed to its trivialization and neglect by those on the "Left." Emphasis on areas such as modernization, urbanization, human ecology, industrialization and technological change reveal an anthropocentrism which directs attention to built, rather than natural environments. While the sociological focus on the organization and activities of human beings has complemented the preoccupation of mainstream science with "natural" phenomena, it has done so at the price of ignoring the ecological foundations of social life.

The absence of a gendered analysis has further discouraged sociological interest in the issue of sustainability, while it has reduced its popular appeal and restricted potential implementation. Consider the fictional sustainability "enterprise" described above. In this scenario, gender may not appear to be directly consequential to the lives or activities of actors, even though women are underrepresented, and children are invisible. There is little attention to gender inequalities in the representations of men and women in their work, or in the performance of domestic work. Women's roles have miraculously disappeared in the interest of efficiency, taut bodies, tight scripts and a neutered and scientific postmodernism.

Like Star Trek, models of sustainability are typically based on mainstream macroeconomic approaches which ignore the perspectives and contributions of women. Although women have been disproportionately and adversely affected by social and economic processes, there has been little analysis of the implications of gender for economic development and environmental sustainability. While problems of social inequality, such as the lack of distributive justice, have been identified as necessary for the achievement of sustainability, these issues have primarily been addressed as issues of class and race and especially as north-south issues. An extensive literature on Women in Development (WID) has dealt with the economic development of women in the south. The WID argument is "that current development policies fail to recognize gender relations despite the fundamental roles women play in maintaining the household in informal, rural and even market economies and in their management of natural resources" (Harcourt 1994: 3). The more recent Women and Environment and alternatives to Development (WED) literature is critical of the entire traditional process of development, arguing that the exclusion of both women and nature from most development paradigms precludes a genuine sustainability (Shiva 1989; Harcourt 1994). Yet even this work remains peripheral to the mainstream dialogue on sustainability and neglects the variation of women's situations and ecological contexts in Western cultures.

Our understanding of women's impact on natural environments, and the ways in which women's relations to ecosystems inform their lives in all cultures remains, at present, incomplete. The connections between women and the environment, in terms of their common economic marginalization and exploitation, remain largely unexamined, in spite of research indicating that "the environment is experienced very differently by women and by men ... (which) will lead (them) to different priorities as well as different responsibilities for environmental concerns" (Brown and Broom 1994: 5). Attempts to plan for sustainability (with the exception of recent WED analyses) assume the validity of a universal approach to social issues and deal with women's relations to the environment only tangentially. Yet, only when the analysis of society and environment begins with an understanding of women and the environment of which they are a part, will the problem of how to sustain life be adequately addressed. As a means of contributing to this analysis, this paper will explore a feminist ecological economics as the basis for sustainability. It provides a way of integrating sociological perspectives of gender with ecological issues. I will first describe the gender bias of dominant economic models of sustainability, prior to discussing an alternative feminist approach.

MODELS OF SUSTAINABLE DEVELOPMENT

The concept of sustainable development challenges us to explore the implications of contemporary socio-economic patterns for long-term human and ecological survival. There has been extensive research into the social and environmental consequences of contemporary demographic and economic forces by researchers such as Anne and Paul Ehrlich (1991) and many others. The 1987 report by the World Commission on Environment and Development, chaired by Gro Harlem Brundtland, served as a global catalyst in recognizing the potential crisis of current socio-economic, demographic and environmental trends. The concept of sustainable development identified by the Brundtland Report "meets the needs of the present without compromising the ability of future generations." The report, entitled *Our Common Future*, is ground breaking, not only in its formal recognition of impending social and ecological crisis, but in its integration of social struggles for fairness and equality with those of environmental protection. The concept of sustainability focuses attention on the interrelationship between humans and the environment, considers the consequences of contemporary demographic and political trends for future generations, and recognizes the significance of social and economic forces for the continuation of not only human life, but the viability of all ecosystems.

Liberal Versions of Sustainability

Mainstream concepts of sustainability are based on a neoclassical model which emphasizes individual choice and market regulation of goods and services. The harnessing of development to an environmentally enlightened market, promises to provide new opportunities, thereby fostering greater opportunity for the "trickling down" of wealth through increased employment and investment opportunities. New products and technologies, propelled by a market demand reflecting an increasingly Green *Zeitgeist*, are viewed as compatible with the protection of environmental quality. The powers of consumer choice, as well as appropriate regulation, drive the production and consumption of environmentally friendly products: unbleached paper, recycled running shoes, organic produce. Governments subsidize industry initiatives, and impose higher standards for environmental protection, while political parties adopt environmental slates. Institutional change provides greater political opportunities for women and encourages the representation of women's interests on the political agenda. Within this climate of optimism, the slowness of current systems to adapt to change is explained by institutional lag, budgetary restraint and lack of political will. It is assumed that continuing political lobbying, shifting market forces, electoral processes, and networking through conferences and strategic initiatives, will continue to forge new policies to bring about greater environmental protection.

Yet, for many, this version of sustainability is inadequate. The neoclassical model is based on private ownership of capital and accumulation of profit. It assumes the inevitability of a socio-economic hierarchy, and the assessment of value by, and in, the market. For many observers, the sustainability of ecological systems, and the redistribution of wealth, are antagonistic to a model which assumes growth and the continuing privatization of profit. Neo-classical economics avoids the long-term consequences of environmental degradation — the decline and diminishing vitality of the resource base, the extinction of species, and the social consequences of excess pollution, such as increased costs to human health. The liberal economy is seen to contribute to the benefits and privilege of some interests at the expense of others: humans to natural systems, rich to poor, white to visible minority, north to south, men to women.

Many women have profited by the implementation of the liberal model — institutional change, affirmative action programmes and training programmes have contributed to the improvement of women's access to, and participation in, the labour force. Yet the continuing segregation of women's work, their increasing numbers in poverty and in deskilled labour, and the increased stress reflected in their combined domestic and waged labour, reflect a continuing economic marginalization. Feminist critics observe their "disappearance" in mainstream models — the invisibility of their working lives, and the implications of their lack of economic power. Mainstream market approaches fail to adequately recognize women's subordination and segregation in their paid work, as well as the value of their domestic work (Ferber and Nelson 1993). "Feminists raise questions not because economics is too objective but because it is not objective enough. Too many assumptions and methodological ideals have been exempted from scrutiny because ... economists have perceived them as universal and impartial" (Ferber and Nelson 1993: vii). The neo-classical model remains, as Martha Macdonald (1994: 159) notes:

> overwhelmingly concentrated on the public sphere, on the buying and selling of goods in the market, and on the production of these goods for sale.... This focus on market prices means that much of what women do is ignored in the analysis ... (reflecting) the real invisibility of women's labour in the market economy. Rather than revealing the true relation between the unpaid work of women and the more visible, paid market transactions, economics replicates and reinforces the separation of the market/non-market spheres in the real world.

While gender inequality may be understood by this model as a matter of socialization practices and institutional resistance to change, it is also viewed as a product of luck, individual opportunity (or lack of same) and circumstances. Liberal models also treat gender differences as largely surmountable with incentives (education, affirmative action) which can provide a more competitive playing-field between men and women. For women in the south, educational programmes, birth control and the reallocation of agricultural resources are among the measures directed towards sustainability. Yet, in this model, "there is no systematic, structural basis for inequalities or discrimination or the sexual division of labour" (Macdonald 1994: 160). Mainstream models of sustainability, derived from the same assumptions as liberal economics, thus understate the significance of gender inequality, while they underestimate structural resistance to its elimination.

Radical Alternatives

Neoclassical models assume that the primary issues on the sustainability agenda — environmental degradation and socio-economic inequality — will be addressed and corrected by market forces and government remedies. Government programmes and a broad variety of incentives, however, continue to ignore the systemic discrimination faced by women, and fail to address root sources of inequality. In response, alternative models such as radical and ecological economics specifically address these concerns. The radical perspective is derived from a Marxist position which emphasizes the class structure and economic imperatives of an industrial capitalist society. It exchanges the liberal belief in individual choice and opportunity, for a critique of class hierarchy and inequality. From a radical perspective, the vision and strategies of sustainability are impaired by their location within an economic system which is antithetical to both environmental sustainability and to distributive justice. Under capitalism, the accumulation of surplus wealth is necessary to stimulate and reward investment, supporting continuing growth. Profits are appropriated by the wealthy, who perpetuate their own position, resulting in an increasing gap between rich and poor.

In terms of the natural environment, capitalism also profits through its utilization of resources as a form of "natural capital." Natural resources are exploited for productivity and profit. Market economies are directed to short-term, private benefit rather than to long-term collective well-being (Ophuls 1992). In this light, for instance, inadequate silvicultural practices are not an aberration, but reflect the maximization of short-term profits. The side- and long-term effects of private enterprise, known as externalities, are excluded from corporate accounting systems. The costs of mitigating environmental externalities, such as pollution, the deterioration of air, water and soil, and the extinction of species, are externalized, or passed on to both humans and other species, while the benefits of production accrue more narrowly to the owners of capital, labour and consumers. As well, social externalities are treated as separate from the economic system. The stress of family breakdown, divorce, alcoholism, suicide and unemployment are disassociated from plant operations and profit margins, and absorbed by individuals and the state.

Yet the failure of radical economics to incorporate both women and environment in the economic equation has also prevented their successful implementation. Class analysis precedes and overpowers that of gender in most Marxist accounts. While theoretical attention to the benefits of reproductive labour to capital includes women within the economic paradigm; "there has been an emphasis on class relations ... over gender relations and a tendency to pursue most economic analysis without reference to women" (Macdonald 1994: 161). In terms of environment, the materialism of radical approaches implies a recognition of the value of natural environments to, and as a source of, capital. But this relation is undeveloped in comparison to its focus on class differences, reflecting an underlying anthropocentrism unable to perceive the complexities of ecological relations.

While radical perspectives recognize the structural limitations of neoclassical approaches, the vision and implementation of alternative models also remains inadequate. Their abstraction obscures problems which may arise with implementation. The recent dissolution of state-centred regimes, and their legacy of environmental destruction and continuing gender inequality, reflect problems of bureaucratization, state control and gender bias in mass societies ignored by many critical observers. Women continue to be underrepresented in political office, segregated and subordinate in the paid workplace, and responsible for the majority of household work. Production in so-called socialist states has been characterized by an Orwellian industrialism, malignant to both humans and the natural environment, with environmental degradation (illustrated by Chernobyl and urban industrial pollution) continuing unabated by concerns for the future productivity of natural and social systems. The natural environment is perceived in most centralized states through resourcist and utilitarian lens, not unlike those of Western economies. The implementation of an alternative critical model has been unsuccessful, especially in its continuing devaluation of both women and the natural environment.

Ecological Economics

An ecological economics represents another alternative to traditional liberal paradigms. Bill Rees, for example, argues that while economics and ecology share the same semantic roots and substantive focus, neoclassical economics "produces analytic models based on ... assumptions about resources, people, firms, and technology that bear little relationships to their counterparts in the real world (1992: 2). While neoclassical economics understands the economy as separate from the "environment," an ecological perspective views human economics as integrated with, and dependent on, the ecosphere.

Ecological economics range from the incorporation of environmental factors into mainstream equations, to a critique of mainstream market approaches to natural environments. Recent critical ecological approaches begin with the natural environment as the source of economic and other social systems, and explore the embeddedness of contemporary economies and societies in natural ecosystems (Daly and Cobb 1989; Paul Hawkens 1993). The natural environment is understood as the reproductive, as well as the productive base of all ecosystems, and environmental degradation — habitat loss, species extinction, pollution — threatens ecological survival. The continuing viability of human society and biodiversity requires a greater recognition and protection of the natural environment, and includes the costs of replenishing and maintaining biodiversity and the quality of ecosystems.

This more critical version of ecological economics understands environmental systems as preceding, supporting and influencing market economies. The natural environment is the source of ecological capital, used as a biological basis for nourishment and regeneration. While ecological capital generates profits on the economic market, its regenerative and other ecological functions in its natural state, are also acknowledged. The recognition of reproductive value of ecological systems is significant as well, for the valuing of human childbearing and child-rearing work. Thus ecological economics creates a more extensive and inclusive economic paradigm, by including the environments in which we live, and the ecosystems of which we are a part, as the context and source of everyday activity.

Yet ecological economics remain inadequate for a number of reasons. Whether or not ecological economics represents a genuine departure from traditional economics, or an ecological "revisionism," is subject to some debate (Harcourt 1994: 14). The resistance of existing market systems to the fundamental shift which could be required by a radical ecological position remains unacknowledged by many observers. Visionaries like Hazel Henderson, while alerting us to the crisis and pointing the way to sustainable economics, is optimistic in her prognosis for shifts to ecological sustainability. "We will more carefully model the problem of production in its larger social and ecological dimensions.... People can lead the world's economies back to sanity and to the basic realities in which human wealth and well-being have always been rooted ..." (1981: 148). Daly and Cobb (1989) and Hawkens (1993) recognize that a major shift is required in contemporary economies to adapt to sustainability. Hawkens states, "At some time in the relatively near future we will achieve a 'balance' between what we are consuming and the capacity of the earth's ecosystems to provide those needs, although under existing models of production and consumption, it is likely to be far different and cause far more suffering than we are presently willing to admit ... a restorative economy means thinking big and long into the future" (1993: 212).

While new ecological models could include non-consumptive and regenerative elements of natural systems, the underlying hegemony of a market system favouring existing economic interests continues to translate the natural environment into human terms, rather than vice versa. We have learned to value trees as timber, and are beginning to appreciate the potential for tourist dollars in a wilderness park, but we are slower in costing out regenerative and ecological functions. This is not because new models are technically difficult, but because the present model serves existing, and powerful economic interests. Much environmental research is directed to applied (e.g., market) use, and has ignored species and issues which remain economically invisible. As Hawkens notes, even conceptualizing ecological sustainability is difficult because its "language is unfamiliar ... because most Americans are more often taught to identify types of cars than types of birds; we can identify one thousand corporate logos but less than ten native plants" (1993: 214).

Yet, the most significant shortcoming for the purposes of this paper is that many ecological approaches continue to underestimate the significance of gender in conceptualizing the impact of human activities on natural environments. The significance of household activities in economic terms (as consumer and productive units) remains largely unexplored. Ecological economics generally bypasses the issue of distributive justice, especially with respect to women, because it has not yet adequately differentiated human impacts on the natural environment by gender (or by other demographic criteria). Ecological economics also continues to devalue — both economically and socially — the work and status of women. Furthermore, while it can be argued that an ecological economics begins with the valuing and nurturing of life, these functions, especially in human social systems remain largely undeveloped. There is little recognition of the fundamental gender inequalities which continue to be reflected in contemporary economic systems, of the interest of patriarchy and capitalism in maintaining them, nor is there an indication of how these would be erased or mitigated by a new economic system.

FEMINIST APPROACHES TO SUSTAINABILITY

The above radical and ecological approaches to sustainability would transform the dominant neoclassical paradigm through recognition of the ecological and class-based foundations of both existing and alternative economic systems. Yet, like mainstream liberal models, neither of these alternatives has been successful in its emphasis on women's relations with ecological systems as they are significant to sustainability. In contrast, feminist analyses explore the mutual connections, subordination, values and contributions of women and natural environments as fundamental to the development of a vision of sustainability. This interconnection between women and environments is reflected in women's socio-economic status, their work (both paid and unpaid), and their representation in decision making.

Women and Redistributive Justice

The Brundtland Report understands the need for an emphasis on greater distributive justice as a precondition, and objective, of sustainability. In addition to north/south differences, the growing socio-economic polarization within industrial nations, limits our ability to engage in more sustainable forms of social organization and everyday living patterns. The linkages between environmental degradation, poverty and powerlessness affect all women, while variations of culture, race and class accentuate differences among and within countries. Women's experience and perceptions of environments reflect differences of class, race and age. In Canada, women in aboriginal groups are more likely, than other women, to live in poverty; they experience the results of resource degradation directly as the loss of both a culture and a way of making a living. Rural women, especially farmers, experience natural environments as residence, work site and source of revenue, while the use of pesticides, physical isolation and poverty, have varying, but direct, effects on these women's lives. For urban women, built environments provide social amenities and supports, while deteriorating air and water quality, population density, and crime levels exert a negative influence on the quality of their lives. We need to know not only much more about how physical environments affect women's lives, but how the context and activities of their lives impact natural ecosystems.

There are links between socio-economic status and environmental quality. In Canada, one-third of women under 65, almost one-half of senior women, and 60 per cent of female single parents, live in conditions of poverty (Khosla 1993) in which their own living and work environments are substandard and inadequate. The feminization of poverty has meant that women, especially older women and single parents, are disproportionately represented among the poor. But it is not just unemployed, senior and single mothers who are poor. The employed poor, receiving inadequate wages, represent a growing group of women, and there is an increasing contrast between their salaries and those of men.

While there is evidence that the gap between women's and men's earnings is decreasing, married women and part-time wage earners continue to earn significantly less than their male peers. Women continue to be employed in low-paid and short-term work, and are at the low end of a spectrum of wages. A National Action Committee report points out that when we take into account all full *and part-time* wage earners, "women's average wages are just $18,050, compared to $29,328 for men. All working women, on average, earn 61.5 per cent what men earn" (Khosla 1993: 8). Although men's jobs in the industrial sector have been negatively impacted by technological innovation and the offshore movement of capital, women are also being displaced from jobs in clerical and service work due to fiscal restraint and other aspects of economic restructuring.

What are the implications of wealth and poverty for women's sustainability? Wealth provides increased ability to exercise environmentally friendly options, reflecting greater choice as well as the potential to exert greater market influence. Those who are wealthy can more directly impact environmental decision making in state or corporate forums in positions of employment and influence. Wealth, which is associated with greater educational opportunity, also brings with it lower birth rates and the ability to exercise increased choice over the size and spacing of families. Wealth provides individuals with insulation against environmental degradation — increased mobility, residential distance from toxic dumps, organic food. Nonetheless, as the north-south argument indicates, the greater consumption patterns by the wealthy exercise a strong environmental impact which arguably, surpass Green lifestyle choices. Recycling one's waste and choosing not to water one's lawn do not adequately counterbalance the cost of resources associated with the purchase of three televisions, two cars and a summer home.

For those who are poor, relations to the natural environment are direct and often damaging. Higher birth rates, lack of educational opportunity and inadequate employment all contribute to the feminization of poverty. The poor in all countries are more likely to live in areas which are environmentally polluted. Poor women are more likely than men to live in conditions of poverty, which means that they and their children are more likely to experience this degradation firsthand. Furthermore, poor women are often forced to further degrade the environment as they scrape together the essentials of everyday survival. While "environmentally friendly" options beckon at the supermarket, the welfare cheque cannot cover the amenities of recycled paper products or organic produce. For women in poverty, interaction with the environment is experienced directly — as diminished sources of fuel for warmth and cooking, inadequate food for dinner or exposure to toxic wastes.

The connection between the subordination of women and that of the planet has been identified by many ecofeminist writers. As Marilyn Waring notes, "It is no accident that our much-loved and much-pillaged and raped planet is often called Mother Earth" (1988: 11). A feminist economic model accordingly understands that the work performed by women and that of ecosystems is undervalued by a system which does not just overlook this work, but which profits from it. Accordingly, the class- and gender-based interests of the mainstream economic framework, benefit a male élite as well as a larger economic system which flourishes at the expense of its key supportive ingredients — women and natural environments. When the source of social inequalities and ecological degradation is recognized as structural, not just incremental or contingent, the potential resistance of existing interests to change is better understood.

Women and Labour Force Participation

Women's relationship to the natural environment is forged through their work, both in the household and in the paid labour force. Women's participation in the paid labour force has tripled in the past five decades, and they constitute 45 per cent of today's labour market (Armstrong and Armstrong 1994: 15). Yet their paid employment continues to be segregated by gender. Pat and Hugh Armstrong observe that:

> most women continue to do women's work at women's wages. Two-thirds of women in the labour force are in jobs in which more than 70 per cent of the workers are women. They are concentrated in just thirty-five of ... 200 occupations ..., where the income is low and the opportunities for promotion are limited (1994: 10).
>
> ... Women are still overwhelmingly slotted into industries and occupations characterized by low pay, low recognized skill requirements, low productivity and low prospects for advancement (1994: 15).

The kinds of occupations in which women are employed are significant for their relations to the environment. Women continue to be strongly overrepresented in clerical, educational, health and service work, and underrepresented in work in the primary sector. This means not only that women receive less pay and recognition for the work they do, but that they experience natural environments less directly than many of their male counterparts, and are more likely to work within a built environment — in a shopping mall, a lab, a school, an office building. The lack of power and control in their jobs may put them at environmental risk, when they work in conditions injurious to their health. In Canadian universities, women continue to be underrepresented in "environmental" disciplines such as ecology, biology, economics and geography which prepare them for future employment. In occupational terms, women are underemployed in the primary sector, comprising only 15.9 per cent of employees in mining and 14.9 per cent of forestry workers (Armstrong and Armstrong 1994: 25). While the Canadian economy has moved into a postindustrial phase, its staple legacy of resource extraction is still important, and it is men who hold the majority of jobs in mining, logging and fishing. In British Columbia in 1988, women comprised only

9.1 per cent of fishery (and trapping) workers and 5.6 per cent of forestry workers (Alfred 1989), jobs with the most immediate contact with natural ecosystems.

The work sectors typically identified by sustainability programmes reflect economic and government sectors with an explicit environmental focus, such as agriculture, fisheries, forestry, mining, energy, manufacturing, transport and tourism. These are primarily male-dominated sectors. Women's employment in them is treated as marginal, as Martha Macdonald notes with regard to debates in Nova Scotia about fishery policy:

> The actors in these discussions always seem to be the companies, the fishermen and the government. Where are fish-plant workers, the majority of whom are women? The fishing industry runs on the unpaid and low-paid work of women: as wives of fishermen, helping to outfit the boat, sell the fish and cook for the men; as fish-plant workers doing the most tedious, low-paid jobs; and as workers in the service sector earning enough to enable their husbands to keep fishing for low returns (1994: 167).

While women are underrepresented in the primary sector, this does not mean that women's work does not have an environmental impact. Nor does it suggest that the physical environment does not enter into the work that women do, or that women's work and responsibility remove them from any contact or familiarity with the environment. We have not yet identified the links between women's work (as well as men's) and the environments on which this work relies. Yet, we can assume that most women do not have the same direct experience in working in, and with, natural resources as many of their male peers. This exclusion from resource-extractive labour, especially its technically driven aspects, has adversely affected work experience, skills and salaries. In an era of rapid economic transition, with increased reliance on capital-intensive labour, planning and regulation, women need to be equally positioned at all levels and phases of resource work. We need to identify the barriers to this gender restructuring, and we need to be conscious of the consequences of this for our continuing relationship with the natural systems on which we rely.

Many women work in the tertiary sector (clerical, information, support) to facilitate their partners' employment in resource industries. While women's "outside" or "town" work may originally have been a supplement to farm and fishery income, to ensure a basic provision of family needs, this work has become increasingly essential to households. As resource-based employment has suffered from shortages, especially in the fishery, women's paid employment has become the foundation of family support. While women's relation to the natural environment is thus less occupationally direct, it nonetheless often complements and supports this work.

Seventy-one per cent of women work in five occupational areas: teaching, nursing and related health occupations, clerical, sales and service (Khosla 1993: 11). Much of the work that women do is in one way or another an interface between contemporary lifestyles and natural environments — resource product sales, clerical support work in resource sectors, environmental education and service work. Women's employment is often a component in the vertical operation of resource industries — it supplies the paperwork, administers licenses, ensures requisitions have been made, liaises with regulatory officials — work that precedes and envelopes the actual scaling, cutting and export of timber and minerals. "These work sectors, namely health, education, social welfare and administrative and retail services represent the human resource capital, the social support mechanisms and the regenerative processes essential to environmental management" (Brown and Broom 1992: 11).

Women's work in the paid and unpaid workforce ensures that clients, patients, and family members are housed, fed, and attended to. In health care roles — as doctors, midwives, nurses, and other health practitioners — women increasingly mediate the externalities — the social and physical effects of pollution and increasing damage to the environment caused by efficient and cheap industrial production processes. Finally, while the ruggedness and danger of men's work in resource industries — mining, timber falling, fisheries — is recognized by the public, women's work environments are also significant to their health. Excessive contact with video display terminals, "sick building" syndrome, eye strain, chronic fatigue syndrome and other problems are ongoing problems which remain underresearched and little understood by the medical community.

Women and Household Work

There are several ways in which women's household roles are significant for ecological sustainability. Patricia Mingie has observed that:

> the Greek root of both "economics" and "ecology" is *oikos,* meaning "house." Perhaps we should view the study of economics as that of "keeping house." That would make each of us earthkeepers, as much as we are each other's keepers. Economics would become imbued with this sense of interconnectedness and shared responsibility (1994: 5).

Women's work in the household and the labour force involves the conversion of natural resources into products and processes which are utilized by humans. In household work, women convert natural resources — agricultural products, fish, timber — into products which can be used or consumed by family members. Women's work not only brings natural environments into the household and converts them for household use, it also insulates households from the environment, by providing its members with shelter and clothing, protecting them from "the elements."

Women continue to be responsible for the majority of domestic work, in spite of efforts to redistribute the burden of household labour. Gender has implications for economic behaviour, although the patriarchal benefits of women's continuing responsibility for household labour, and their subordination in the labour force are invisible in mainstream models. Rather, these models treat households as individual units in which utility is maximized through male altruism. This model overlooks women's structured disadvantage, as well as the impact of household relations, and especially gender, on natural systems.

The domestic work performed by women and the regeneration of ecosystems parallel a dual economic invisibility and lack of power. Lee-Smith and Trujillo, for instance, understand subsistence as a "useful paradigm for planetary ecological management ... (by which) women manage family or household needs for sustenance on an everyday basis ... but also (which) encompasses the management of the local ecosystem" (1992: 78). The maintenance of biodiversity is necessary for the preservation of life, while the complex nature of ecosystemic functions — pollination, erosion, germination, the production of carbon dioxide, precipitation — remains invisible and undervalued.

Women give birth; they perform the majority of child-care work; they socialize and educate the next generation; they provide physical maintenance and emotional support for household members. Women's mundane and everyday household maintenance and support are essential to its sustainability; this work is the foundation for any model of sustainability, but it remains invisible because it is unpaid and has no economic value. As Waring notes, "Every time I see a mother with an infant, I know I am seeing a woman at work. I know that work is not leisure and it is not sleep and it may well be enjoyable. I know that money payment is not necessary for work to be done ... (yet in terms of mainstream economic models) those mothers I see with their infants are not working" (1988: 26).

Household roles are undeveloped in models of sustainability, which also curtails our understanding of women's connections to natural environments. Shiva (1990), for instance, explains how the invisibility of domestic processes in mainstream economic models obscures the deterioration of social and natural wealth, in that the introduction of market production in subsistence economies destroys both natural and household systems. The translation of the value of household labour into mainstream economic terms, reveals a major contribution to the GNP which is otherwise ignored. Yet, work in the home remains unpaid and undervalued, while it also restricts women's mobility in the labour force. In terms of reproduction, women's roles in population growth and family planning offer the most direct means of achieving sustainable futures through the stabilizing of the birth rate. Furthermore, women are environmental educators, socializing and educating family members in environmental processes.

As household consumers, women comprise a primary market for environmental products. Consistent with the liberal emphasis on market-based strategies, is an emphasis on individual, household-based environmental sustainability. Women's responsibility for the majority of household work, calls for them to implement recycling techniques, construct compost bins and do the laundry with phosphate-free detergents. Women's household caretaking roles are expanded by the additional responsibility of planetary caretaking. Yet, while women's environmental spending is consistent with market approaches to sustainable development, it remains inadequate as a strategy for achieving sustainability, because it:

> reduces people to consumers, a reductionism that follows the trend in resource economics to reduce natural systems ... to market commodities whose value is based on human consumption.... Second ... consumption in the West steals natural resources from Third World countries, depends on the labor of their children, and contributes excessively to global climate change ... (Hynes 1991: 5).

Household economics, while still ignored in mainstream models of sustainability, are significant to our relations with natural environments. The home, as the residence and site of consumption, comprises activities which include the conversion of, and the protection, insulation and separation from, natural environments. Households impact ecological systems, not only in terms of consumption, but in terms of the buildings we live in, energy sources we utilize, the waste products we create. Helga Moss (1994: 241) describes how her everyday consumption patterns reveal a global social and ecological dependence:

> on a very complex structure, a web of seemingly infinite concrete relations to the varying ecosystems and working people of the world ... I am caught in a serious ethical dilemma. If I want to live in harmony with nature, its plants and living beings, including women and children, and men, all over the planet, I and my children would have to starve, freeze and become outcasts in my own society.... My subsistence work — my household — is totally dependent upon consuming from this global household even if I reduce our consumption to the things most essential for survival....

Our understanding of households thus needs to incorporate not only the implications of gender, race and class relations, but the conversion and distribution of natural environments on a global scale, into terms of human use and consumption as central and ecologically significant processes.

Women and Decision Making

Women have been excluded and underrepresented in formal decision-making bodies and processes, and their lack of access to public and corporate power has limited their ability to influence environmental change. Women's underrepresentation has also shaped a political agenda of sustainability which neglects women's economic and social contributions. Women represent over 15 per cent of members of Canadian Parliament and legislatures, and they comprise, in some governments, such as British Columbia's NDP regime, more than 25 per cent of MLAs, and seven out of 19 ministers (Stainsby 1994: C11). Yet, at federal and provincial levels of government, women continue to exercise less power and control than their male counterparts in portfolios such as economics and finance, which impact directly on the environment. Few Canadian cabinet ministers have been women, and half of them have held more traditionally women's portfolios, such as health care, social service, tourism, or culture, only rarely and fleetingly attain top leadership positions.

The global picture is even more extreme. The appointment of Gro Harlem Brundtland to chair the World Commission on Environment and Development is an exception to the continuing pattern of women's underrepresentation in the political arena, especially in environmental politics. In international terms, Western countries far outrank their Third World counterparts in terms of the representation of women at the ministerial level, as illustrated by Norway and Finland, where women comprise more than 25 per cent of elected officials (Stainsby 1994: C11). Ten per cent of the world's parliamentarians, those elected to positions of political power, are women (Landsberg 1991: 1). Only 8.9 per cent of ministerial positions in Western European countries were filled by women, while the percentage is only 2.5 in African states, and only 1.6 in Asian and Pacific countries (O'Neil 1991: 47). In terms of corporate decision making, only 2–3 per cent of chief executive officers in Canada and the United States are women (Anderson 1991: 219). Thus, in spite of the rhetoric of "a more representative political system," women continue to be almost invisible in the contemporary political structure.

Although women make extensive contributions to both domestic and public economies, these remain politically underrepresented, while their lack of equal access to corridors of power reinforces their marginality. Governments undertake research, compile statistics and implement policies which reinforce a male view of the world by underestimating and ignoring women's social and economic contributions (Waring 1988). Underrepresented as legislators and bureaucrats, women are less likely to be in positions to direct policy instruments to environmental protection. Decisions which directly affect the environment — toxic-waste siting, environmental assessments, pesticide application permits, annual allowable cuts of timber — have been traditionally made by men. The pervasive underrepresentation of women throughout the environmental decision-making process — in law reform, standard setting, lobbying, administration, negotiation, monitoring — is indicative of larger structural and procedural biases. The technical nature of environmental decisions, women's lack of scientific training in "environmental" fields, and active, if often covert, discrimination against women, restricts their participation in the decision-making process.

Nonetheless, women have been active and successful in environmental politics, especially at the NGO level. Canadian women such as Vicky Husband, Colleen McCrory, Tzeporah Berman and Elizabeth May have lobbied successfully on issues of wilderness protection, pressuring politicians and corporations, and educating the media to the concerns of their representative organizations. On an international level, the World Women's Congress for a Healthy Planet (held in 1991 in preparation for Rio) and a range of other conferences in preparation for the Beijing Conference in 1995, have been highly effective in promoting the increased representation of women and women's needs within the international political agenda. As an alternative to mainstream politics, women's membership and leadership of the Green Party have contributed to an internal organization, policies, leadership, and strategies which consciously address feminist principles of non-hierarchical organization and decision making. The increased representation of women in positions of environmental law and administration in government agencies and other organizations, is another step in the feminization of environmental politics. Women's actions are necessary to get women and their needs, roles and perspectives on the political agenda, and to reflect the diversity and complexity of women's relations to natural environments.

Feminist Visions / Feminist Strategies

The preceding discussion explores some of the intersections between the socio-economic situation of women and that of the natural environment. I have discussed some of the ways in which women's work impacts natural environments, as well as the ways in which natural systems influence the work that women do. The gender and other socio-economic divisions (such as class and race) between household and waged labour, parallel the split between contemporary lifestyles and natural environments. Linkages between the subordination of women in both household and paid work, the invisibility of household work, and the devaluation of natural systems, have been ignored by both mainstream and alternative economic paradigms of sustainability. The continuing socio-economic inequality for women, not only in wages and occupational opportunity, but in terms of their ongoing responsibilities for the domestic sphere, has consequences for the natural environment. While women are the gatekeepers for household-based consumption, their ability to shift patterns of socialization and consumption are limited by their lack of resources (time, energy) and empowerment.

If we are to move towards a more appropriate model of sustainability, we must take into account the value of women's work, and ecosystemic contributions to economics, but we must also expand the paradigm beyond its currently narrow economic framework. Furthermore, we need to examine cultural, racial and class-based assumptions about work, value and wealth. There are a number of feminist critiques and alternatives to neoclassical economics, many of which recognize the necessity of integrating ecological and feminist contributions to sustainability (Mingie 1994). Marilyn Waring's critique of neoclassical economics, *If Women Counted*, describes the exclusion of women, their work, and their value from mainstream economics and from international systems of accounting which measure productivity, value and worth. She refers to two feminist critiques of mainstream economics, and the alternative approaches which they have put forward (Waring 1988: 299–302). Hazel Henderson understands neoclassical economic models as characterized by "linear concepts of the monetized sector, measured by GNP, and market 'supply' and 'demand' equilibrated by price" (1981: 29). The current system treats as "exogenous" to its model both the nonmonetized social system (laws, customs, culture, nurture, environmental interactions, non-monetized jobs and roles) and the unpaid work (voluntarism, hidden social costs and infrastructure costs) which comprise equally significant contributions and costs to the economy.

Waring introduces Henderson's model of the economy as a layered cake, which includes household, voluntary and other non-monetary work, as well as environmental costs, as integral to, and comprising, the foundation of the economy (1988: 300). Henderson's model includes four layers, beginning from the top and including: (1) the private sector, which includes jobs in factories, buying and selling and savings; (2) the public sector, including federal, state and local government spending; (3) an unpaid work sector, including child care, housework, gardening and volunteer work; and (4) nature, which is the bottom layer, and absorbs pollution, recycles wastes and reproduces. As a descriptive model, this model makes visible the economic contributions of women and nature that are missing in the mainstream model. Its incorporation of natural "ingredients" is especially significant as a means for factoring the contributions of natural environments into economic equations. However, because its layering provides a somewhat stratified approach, and because the private sector, in its proximity to the icing on the cake, appears to retain the most economic status, it remains problematic as a prescriptive model.

The second model presented by Waring is adapted from Hilkka Pietilä's presentation at the Nairobi Conference for women in 1985 (1988: 301). This model includes three concentric circles, beginning with A, the largest, at the core. The second circle, B, is significantly narrower, while C, narrower yet, is the most external ring. A, the central circle, represents the "free economy," the non-monetary, invisible work of supporting and caring for families. B, the "protected sector," "consists of production and work for the home market as well as public services ... including food production, construction of houses ... schools, health" which are primarily, but not exclusively, state-provided services (Waring 1988: 300). The third tier includes large-scale production for export, sometimes referred to as the open economy, but referred to by Pietilä as the "fettered economy," meaning fettered or connected to the world market through pricing, demand and international competition. While we tend to think of this third category as the basis and majority of all economic activities, Pietilä reduces it to the least significant sector, in terms of both expenditures of actual time and money. In this model, the everyday work of maintaining and supporting the lives of household members, community and society is central, and the key to all remaining activity. These services require and perform the bulk of economic activity, rather than being marginal or invisible.

These models are a preliminary means of portraying a feminist ecological economics, which makes visible the contributions of both women and nature (as well as their costs), and places them at the centre of economic activity. Yet these models remain problematic, and require continuing adjustments in a number of ways. While "internalizing" women's economic contributions, the portrayal of ecological functions is subordinate, rather than the essential ingredient which makes human society possible. Secondly, and more significantly, to internalize the contributions of women and environment is to risk an economic reductionism which will perpetuate existing inequalities. That is, unless the entire basis of the model is shifted to include values extending beyond current market mechanisms, such as those of intrinsic and aesthetic worth, the contributions of both women and ecological systems will be reduced in scope and significance. Furthermore, the difficulty of a transition from this neoclassical framework is not taken into account, and thus both models neglect the hurdles which may impede their own implementation. Finally, the dynamics of relationships need to be considered, to explore and diminish current inequalities, to recognize the ways in which women also are involved in the exploitation of ecosystems and of others. These would include the full range of women's relations to natural ecosystems as well as the diversity and inequities among women — north to south, and divisions of class and race within societies.

A feminist ecological model takes as a central objective the protection of life through the preservation of existing ecosystems and the more equitable distribution of wealth and justice. The mainstream market orientation to commodity production and capital accumulation for private profit rather than for social needs, would thus be replaced by one emphasizing the preservation of life and more equitable distribution of wealth and well-being. Social and environmental costs are included within the economic paradigm, while the paradigm itself is expanded to include the (re)productive, support and integrative work of both women and ecological systems (ensuring better child care as well as wildlife-habitat protection). Women and natural environments are viewed in terms of the full range of their relations to the sustaining of life.

TOWARDS A FEMINIST ECO/NOMICS

In this paper I have discussed some of the foundations for a feminist analysis of the economics of sustainability. I have discussed how the connections of women to the environment have been ignored by both mainstream and alternative economics. Market approaches have failed to predict, explain and resolve the escalation of contemporary social and environmental problems — especially the impoverishment of women, the decline of our resource base, the extinction of species, and the negative effects of excess pollution. Yet radical economic models continue to neglect both the role of women in the economy, and the environmental consequences of "business as usual." The emergence of a new ecological economics reconstructs our concept of the natural environment as both an economic arena and process. These and other alterations to the neoclassical paradigm attempt to redirect the forces of contemporary capitalism to sustainable objectives through an overhaul in economic concepts and practices. These models, nonetheless, underestimate the significance of women's labour and women's perspectives in their analyses.

Both mainstream and alternative models of sustainability have marginalized women and nature, and thus perpetuate their extra-economic characterization as devalued, primitive and secondary. Nelson has argued that in economics "the abstract, general, detached, emotionless, 'masculine' approach taken to represent scientific thinking is radically removed from, and clearly viewed as superior to, the concrete, particular, embodied, passionate, 'feminine' reality of material life" (1993: 25). But the marginalization of women is more than ideological. Rather, it reinforces existing systems of power and control which, in turn, perpetuate the subordination of women and of natural systems.

A feminist model thus does more than add women and nature to existing equations. It asks that we re-examine the relations among all actors and systems in the process. It warns against the simplistic and universal reduction of differences of race, class, geography and species to rational, instrumental "actors" which is characteristic of mainstream models. A feminist approach asks us to critically examine the relations of women to natural environments, to better understand their diversity and complexity.

The perpetuation of life itself, an essential preoccupation of all natural ecosystems, and the primary responsibility socially assigned to women, is devalued by the present economic system, while it is the foundation of sustainability. In this context, the concept of "value-added" (referring to the addition of processing and finishing activity to resource extraction) reflects all the stages and processes necessary to the sustaining of life. It requires us to change the dominant paradigm to a value-added approach. This will include spiritual, aesthetic, moral, intrinsic and other values conducive to the more equitable survival of life on earth.

The current faltering economic climate presents new challenges for sustainability — increased global dependency, a continuing polarization of wealth (both internally within nation states, and on an international basis), increased unemployment, resource scarcity, state fiscal crises and continuing recession. In Canada, as elsewhere, these trends have been accelerated by technical innovation and increased automation, southern and offshore production, continuing and shifting economic dependencies, the GST and Free Trade deals. A feminist environmental analysis would include other issues in its economic programmes which might include some of the following: reviews of the taxation of child-support payments; assessment of habitat loss for endangered species; reduction of poverty for older women and sole-support mothers; increased financial and emotional stress on dual-income families; increased youth-employment options; implementation of air-quality regulations; and the creation of a child-friendly society. This scenario urges us to adopt a feminist model of sustainability which begins with women and natural systems at the core; which redefines productivity and attends more to process; which decreases the hegemony of market competition and indiscriminate commoditization, and which increases the values of life support, caring, co-operation and well-being.

Imagine, if you will, the new Feminist Starship. Captain Brundtland-Strong is still at the helm, and to her policy and technical skills is added the wisdom of age. A new crew, made up of women, men and children amble into the control room, dressed in recycled civilian clothes, looking nothing short of ordinary. They are of all races, ages, body sizes and appearance. They pull down the computer "menu," choose by consensus and five of them go off to the galley to prepare dinner. The remaining crew is busied with social and technical maintenance duties — programing the console, dealing with diaper rash, reviewing homework, planning transportation to the next basketball game, writing research proposals, providing technical assistance in nutrition and quantum physics. The fall schedule features the maintenance of peace in Beijing, acute air crises in Manchester and Los Angeles, wage negotiations for child-care workers, new bacterial strains of interstellar flu, the addition of two Asian raptors to the endangered species list and the escalation in mortality rates in the Antarctic penguin population. The plot continues to pinball from crisis to crisis, but survives budget cuts through innovation, maintains high ratings because of its resonance with viewers' lives and maintains a glimmer of optimism sparked by success.

DISCUSSION QUESTIONS

1. Give three explanations, either from the paper or from your own observations, for the traditional lack of concern about ecological issues in Canadian sociology. Why do you think this is changing?
2. In what ways does a neoclassical approach to sustainability ignore women's contributions to the economy? Give two examples.
3. Identify the primary contributions of radical and ecological alternatives to the neoclassical approach to sustainability. What is their common shortcoming?
4. Name four ways in which women's labour has consequences for environmental quality, and give an example of each.
5. Should a feminist ecological economics internalize women's and ecosystemic contributions, or change the value system on which present economic models are based, or both? Why? Give an example to support your argument.

References

Alfred, Diane. 1989. *Women in the B.C. Labour Market.* Vancouver: Economic Services Branch, Employment and Immigration Canada.

Anderson, Doris. 1991. *The Unfinished Revolution: The Status of Women in Twelve Countries.* Toronto: Doubleday Canada.

Armstrong, Pat and Hugh Armstrong. 1994. *The Double Ghetto.* Toronto: McClelland and Stewart Inc.

Brown, Valerie A. and Dorothy H Broom. 1992. *Engendering Sustainable Development: Position Paper.* Barton, Australia: National Women's Consultative Council.

Daly, Herman E. and John B. Cobb, Jr. 1989. *For the Common Good: Redirecting the Economy Toward Community, the Environment, and a Sustainable Future.* Boston: Beacon Press.

Ehrlich, Anne H. and Paul R. Ehrlich. 1991. *Healing the Planet.* Reading, Mass.: Addison-Wesley.

Ferber, Marianne A. and Julie A. Nelson, eds. 1993. *Beyond Economic Man: Feminist Theory and Economics.* Chicago: The University of Chicago Press.

Harcourt, Wendy, ed. 1994. *Feminist Perspectives on Sustainable Development.* London: Zed Books.

Hawkens, Paul. 1993. *The Ecology of Commerce: A Declaration of Sustainability.* New York: Harper Business.

Henderson, Hazel. 1981. *The Politics of the Solar Age: Alternatives to Economics*. New York: Anchor Books.

Hynes, H. Patricia. 1991. The Race to Save the Planet: Will Women Lose? *Women's Studies International Forum* 14: 5.

Khosla, Punam. 1993. *Review of the Situation of Women in Canada*. Toronto: National Action Committee on the Status of Women.

Landsberg, Michele. 1991. Overview. *Official Report: World Women's Congress for a Healthy Planet*. New York: Women's Environment & Development Organization.

Lee-Smith, Diana and Catalina Trujillo. 1992. The Struggle to Legitimize Subsistence: Women and Sustainable Development. *Environment and Urbanization* 4: 1.

Macdonald, Martha. 1994. Becoming Visible: Women and the Economy. In *Limited Edition*, ed. Geraldine Finn. Halifax: Fernwood.

Meadows, Donella, Dennis Meadows and Jorgen Randers. 1992. *Beyond the Limits: Confronting Global Collapse, Envisioning a Sustainable Future*. Post Mills, Vt.: Chelsea Green Publishing Company.

Mingie, Patricia. 1994. An Open Letter to Women in Economics: Feminist and Ecological Critiques. Paper presented at the Conference on Women and Sustainable Development. May, Vancouver.

Moss, Helga. 1994. Consumption and Fertility. In *Feminist Perspectives of Sustainable Development*, ed. Wendy Harcourt. London: Zed Books.

Nelson, Julie. 1993. The Study of Choice or the Study of Provisioning? Gender and the Definition of Economics. In *Beyond Economic Man: Feminist Theory and Economics*, ed. Marianne A. Ferber and Julie A. Nelson. Chicago: University of Chicago Press.

O'Neil, Maureen. 1991. Men and Women Together? *Journal of the Society for International Development* vol. 1.

Ophuls, William. 1992. *Ecology and the Politics of Scarcity Revisited: The Unraveling of the American Dream*. New York: Freeman and Company.

Rees, William. 1992. Ecological Footprints and Appropriated Carrying Capacity: What Urban Economics Leaves Out. Paper presented to Globe, '92, March 16–20, Vancouver, B.C.

Shiva, Vandana. 1989. *Staying Alive: Women, Ecology and Development*, London: Zed Books.

———. 1990. Development as a New Product of Western Patriarchy. In *Reweaving the World: The Emergence of Ecofeminism*, eds. Irene Diamond and Gloria Feman Orenstein, 194–95. San Francisco: Sierra Club.

Stainsby, Mia. 1994. Women in Politics: Dancing with Wolves. *The Vancouver Sun*, 19 November, C10–11.

Waring, Marilyn. 1988. *If Women Counted: A New Feminist Economics*. San Francisco: Harper & Row.

THE DESIGNED ENVIRONMENT IN ENVIRONMENTAL SOCIOLOGY

Charles Gordon

Carleton University

Preamble

Some years ago, I submitted a paper to the environmental sociology section of the Canadian Sociology and Anthropology Association for presentation at the annual conference. The convenors of that session decided that their concern was with the natural environment and environmental movements, and forwarded the submission to the session on urban sociology. That process illustrates part of the issue that I wish to address — that is, the place of the built environment within sociological discourse. At the moment, it falls across the margins of a number of sociological subdisciplines: environmental sociology and urban sociology, as mentioned, as well as the sociology of the professions and culture studies. (There is a certain irony in the last-mentioned area, for as Gillian Rose [1988] points out, the discourse of "postmodernism" has its origins in architectural theory.) But the place of the built environment in urban studies, as described by Anthony King (1984), seems applicable across the board:

> Thus ... relatively little attention has been paid to the built environment as a physical and spatial reality; the notion of "space" frequently utilized in this literature is rarely adequately spelled out ... *and the actual tangible buildings and bounded spaces are somehow* **etherialized** *or, at best, transformed into abstract "scarce resources," consumption items or similar* [emphasis added]. That etherialization obscures the built environment as a human product, and as a factor in the process of social organization.

Ten years later, Magali Larson makes a similar point in locating the study of architecture within cultural studies (1994), describing an emphasis on the processes whereby cultural objects generally, and buildings particularly, are produced and consumed, while ignoring the fact of the forms themselves. She also points out that even within the world of cultural objects, architecture has a distinct character:

> We cannot regard architectural objects only as sculptures: they are buildings first, meant for use, not contemplation only.

The relation between form and function is a much-discussed matter in architecture, to the point of becoming a cliché. But it is that relation that locates the study of the designed environment within the bounds of environmental sociology. Ultimately, the nature of our species is based upon the ways in which we manage our relationship with the environment: the designed and built environment is a fundamental medium of that management.

Indeed, fundamental importance may have contributed to the relative inattention accorded to the built environment. Mennel, Murcott and van Otterloo (1992) point out that the sociology of food has been largely ignored precisely because it is such a basic substrate to all activity, and is therefore taken for granted. Shelter is equally basic in providing a setting, and indeed a condition for many human activities. As such, it may have received less attention than is due. As well, the interaction between these activities and their physical/spatial setting, may have been so downplayed in favour of a view which sees the physical as sufficiently determined by the social, that it needs less study (Saunders and Williams 1988).

Some Basic Premises

Peter Dickens (1992) lists the following "principle themes" as basic to an adequate understanding of the relation to the environment:

1. People are a species which, like all other living organisms, have needs like any other species and needs specific to human beings.
2. Nature is integral to people's bodies. There is therefore no logical way in which nature can be treated as separate from people.
3. Nature is socially constructed. It is always modified by people. And, in the process of changing nature, people change themselves.
4. Under capitalism, nature is privately owned and exploited. People therefore find themselves becoming alienated and estranged from the nature on which they work. They also become alienated from their own species.
5. The institutions and processes created through social relations and social processes can, however, come to have an apparent life or power of their

> own. They become fetishized or reified as things in their own right. As such, they become means by which people gain a sense of being in the world. They enable people to organize their lives and experience.

One of the basic needs is for protection of the body from external harm, and here, it is in that light that the built environment will be understood. It is the implications of that need, and the ways in which it is met, that form the concern. Indeed, following Dickens, those needs are part of the integration of people and nature, even as they form barriers between people and "nature," as it is often conceptualized.

A premise that we can derive from these principles is that design is a basic human process. That is because it is the way in which people relate to things; and the way in which they do that is crucial. As Sut Jhally (1987) puts it:

> The relationship between people and their things should not be considered a superficial or optional feature of life. It is in fact a definitional component of human existence. All societies are based upon the use of nature by humans. Humans as a species are only able to survive by the "appropriation" of the material elements that surround us.

Amos Rapoport (1984) defines "design" as follows:

> In fact, "design," as a process, needs to be seen much more broadly than is common: *design is any purposeful change to the physical environment* (emphasis in the original).

Such a definition resembles a Weberian definition of social behaviour as meaningful behaviour; indeed, one might argue that "design" so defined is a category of social action. But design is also a process of providing meaning:

> The etymology of *design* goes back to the Latin *de* + *signare* and means making something, distinguishing it by a sign, giving it significance, designating its relation to other things, owners, users or gods. Based on this original meaning, one could say: design is making sense (of things) (Krippendorf 1989).

Given these definitions of design, what then do we make of design as an activity? Firstly, it is an activity that is a constant part of everyday life. Indeed, it can be said that everyone is a designer. Jhally's statement applies not only to some notion of a simple society; to get through a day (and night), one is constantly designing. Examples abound, from the way in which personal space is marked in informal settings, to the personalization of work spaces (as well as to the "vernacular" planning of work by the worker); and from the need to control small objects in a limited environment by seniors, to the planning of safe routes (and safe clothing) for the trip home by women and children who have to traverse public spaces.

Second, like many other human endeavours, design has become a specialized professional activity. As such, it can be compared to other professions — not only architecture, which is the most obvious, but to the health professions, law and theology. As such, it involves a special language and body of knowledge. As with other professions, that knowledge monopoly is a matter of contest in society, as well as within the profession. But as it emerges as a professional activity, a specialist design becomes an exercise of power in relation to a number of kinds of people, clients and users in particular.

Finally, it must be remembered that design involves the design of production processes, as well as the design of products. Particularly, the roles of the architect and the industrial designer are implicated in the relations of people to their labour; i.e., it is part of the workings of the social class in terms of relations of production, as well as in terms of consumption. This is true historically. Adrian Forty (1986) provides a brilliant description of Wedgewood, demanding that his designers provide a production process tuned to the dumbest possible worker. Arguably, a version of industrial design was involved in the implementation of scientific management. That third sense refers to the embodiment and maintenance of social relations through the medium of the designed artefact, and the design of that embodiment.

Bruno Latour (1988) provides an intriguing discussion of this process that is on the one hand abstract and general, and on the other hand, a description of a most mundane artefact — an automatic door closer. He discusses the creation of a particular artefact as a metaphor for the labour process: if one has a wall, one must have a door; but if the wall (and the room) are to work, the door must be opened and closed. People are unreliable for this purpose; they don't obey norms, or they go on strike. And they are expensive. So the automatic door closer is developed. Latour refers to the process as "delegation to a non-human actor." Indeed, that is what design is. And the design process, which is essentially the means whereby that delegation is carried out, reflects a particular view of the human actor who is being controlled, as determined by the organizational, occupational and social location of the designer.

The exercise of power that comes with the specialization of design is not always as direct as that which Latour suggests. In addition, there is an indirect but significant power:

> ... monopolized discourse exerts effective influence silently and invisibly. Its effects are measured in the non-physical constraint of accepted definitions, of internalized moral and epistemological norms. It is in one sense

> impersonal, for it makes the most general knowledge claims; yet it is deeply personal, in that the individual who internalizes the general and special discourses of his or her culture experiences them as natural expressions or extensions of his or her own will and reason (Larson 1984: 36).

And as Giddens (1990) points out, daily life is possible because of the trust in expert knowledge, as embodied in such mundane designed objects as a run of stairs.

It is possible to view that power of monopoly knowledge and design in diverse lights. Fry (1989) sees it as problematic:

> The form of contemporary modernity as consumer society (de facto and desired) is readable as a social text, that is, as the designed world and the semantic structure of its forms, functions and mediated appearances. The designer in such a context has to be named and interrogated not as creative subject but as deployed labourer, as expression of the corporation and commodity.

Dilnot (1982), on the other hand, has a more positive view (one which sounds more like the traditional altruistic norm of the profession):

> ... design thinking and communication has its ends in the material transformation of the world, or better, has its ends in the desired transformation of human social relations achieved through (in design's case) transformation of the material surroundings.

But, altruistic or not, the transformation of human relations through the transformation of the relation to the natural environment (which is what the material surroundings are about) seems to be significant indeed. The understanding of those transformations is what a sociology of the designed environment seeks to elucidate. It is thus a significant part of an environmental sociology, and indeed of sociology at large.

The Built Environment as a Social Institution

Dolores Hayden (1976) argues that building has been a fundamental activity for many groups. Peter Wilson, in *The Domestication of the Human Species* (1988), goes much further. He argues that the "settling down" of the human species, and the creation of a relatively permanent built form, was a fundamental step in the development of social structure, first, because the new complexity of the buildings allowed physically for privacy and the manipulation and management of attention:

> Now attention may not be unique to human beings, but architectural interference with it is, so by living in a built environment, people create for themselves, quite unknowingly, problems of the senses vital and essential

> to the conduct of all other activities instrumental in survival subsistence and well-being.

And second, because:

> ... the anchoring of a person that comes with domestication results in the identification of a person with location and location with person, a merging of person and place which differs from the idea of focus that specifies the relation between hunter/gatherer and location. The merger of person and place and their reciprocal identity is the essence of property, of ownership and the right to exclude others.

Wilson also points out that buildings form a perfect medium for the expression of power: "The most readily available means whereby great quantities of labour may be used to convert to prestige, a means that is the great invention of domesticated society, is architecture in general and the house specifically."

James Marston Fitch (1972) notes that humans have developed two "ameliorating devices" to manage the interchange between the human organism and the surrounding physical environment — clothing and buildings:

> ... both modulate the play of raw environmental stresses on the body's surfaces and orifices; simultaneously, they create a third, meso-environment whose characteristics can be tailored to meet human requirements.

The principle difference between the two, he continues, is that buildings shelter social processes as well as the individual. I have discussed elsewhere (Gordon 1980) the effects of the parallels between such a notion of building and a Marxian definition of labour. Basically, I suggest that the inability to control one's shelter, and thus the inability to manage a relationship with the environment, generates another version of alienation. The difference between building and labour is in the nature of the interchange with the environment and the intervention therein. In the case of buildings, that interchange can be discussed in two ways: (1) the management of the environmental stresses (and the loss of energy) as describe by Fitch (1972); and (2) the protection of the social process, and the individual's participation therein.

To elaborate briefly: To see a process in institutional terms, we must understand it as a recurrent set of sanctioned social practices, which are of major significance in that they go towards meeting a perceived social need. The built environment can be defined institutionally in relation to meeting a need for "shelter" (e.g., protection and safety). We can discuss the protective "function" of buildings in terms of three dimensions:

1. What is being protected?
   - individual
   - organization
   - artefacts

2. What is the hazard?
   - "natural"
   - "other species"
   - other humans:
     - from within the same group
     - from outside the same group
3. What are the means of protection?
   - physical
   - mixed physical and symbolic
   - symbolic

The concern is to understand the translation of such a fundamental notion into the variation of social forms involved — that is, to understand the design of that protection as a human product, and its effects. As Forty (1986) points out, design casts basic ideas about who and what we are, into "permanent and tangible forms." Thus, the design of buildings, and its embedding in other social processes, represents a fundamental form of social reproduction. If buildings can be described in Fitch's terms, then the notions of safety and protection, and their enactment as buildings, are fundamental forms of social reproduction, and should be understood as such. In the choices taken as to what to protect, and what to define as hazard, we reproduce a structuring of the society as basic as that which we describe in terms of labour. For example, in something as mundane as building codes, we choose what we will regard as a hazard (snow, fire or disease), who or what will be protected (the structure of the building, the running of a business, or the health of children) and the means of protection (a roof, a particular material or a warning sign).

The Social Construction of Construction

It has been pointed out that the relation between humans and the environment is created and recreated as a process of social construction. The nature of the social construction that concerns us here involves the way in which the medium (buildings) involved is created, and the particular form that medium takes in a complex society. In particular, we are concerned with the way in which the development of specialist knowledge and organization determines the relation to the environment. As Markus (1993) notes, the Industrial Revolution changed the relationship between patrons and architects, and diluted the identity of values that was assumed to exist between them:

> However, the upheavals of the Industrial Revolution shook this secure relationship. A host of new and unexplored building types were needed by

> the state, municipal authority and entrepreneur. Both sponsors and architects came from a wider social background — identity of class interest could no longer be assumed. Moreover, users of buildings began to develop their own voices. It was now necessary to take steps to safeguard sponsors' interests and, at the same time, to camouflage the link between buildings, and power and bonds, by means of an unwritten pact which was based on the myth.

The organization of architectural production occurs in a variety of settings. If framed in the usual terms of organization and occupation, the import of that variety may be lost. The recent image of the architect as the lone hero, à la Ayn Rand, loses sight of the fact that the making of shelter has always taken part within a complex division of labour. But the more recent move of the architect into the framework of large organizations (whether professional firms or not) may mask the complex variety of work settings within which shelter is produced. A more general notion of production organization is required, and to that end, I would like to borrow the concept of "production regime" of Michael Burawoy (1985):

> ... The process of production is seen to have two political moments. First, the organization of work has political and ideological *effects* — that is, as men and women transform raw materials into useful things, they also reproduce social relations, as well as an experience of those relations. Second, alongside the organization of work — that is, the *labour process* — there are distinctive political and ideological *apparatuses of production* which regulate production relations. The notion of *production regime* ... embraces both these dimensions of production politics.

That notion of production regime covers a number of employment relations that may exist in the world of architecture and design, from employment by the state, through varieties of organizational forms, non-organizational forms of employment (contract and consulting, for example), to various kinds of relations to the client directly. To the extent that these regimes constrain the relation of the person to the product of his or her efforts, including its commodification, they may all be seen as potentially alienated or alienating. As well, the organizational field within which this production takes place, is of significance as a source of that constraint, along with its ideological context.

Indeed, it is possible to typify professions by the production regime within which they work. The typology provided by Brante (1990) is based upon a continuum of such regimes, based upon the extent to which they are located in market settings or bureaucratic organizations:

Free Professions	*Capital Professions*	*Political Professions*	*State Professions*	*Academic Professions*
Self-employed physicians, lawyers, accountants, engineers, psychologists	Privately employed engineers, accountants, physicians, lawyers	Government political élites, higher civil servants etc.	Publicly employed physicians, teachers, social workers etc.	Natural and social scientists, scholars of the arts

Each of these settings represent a different production regime, varying by the kind of ownership, the way in which intention is imposed and the kinds of internal relations. In Brante's examples, different occupations show up in a number of the settings. Indeed, while he does not include architects among his examples, it seems that they might fit into all of these forms. That said, this does not seem to be a typology of professions, as much as a typology of the production regimes within which professionals work.

Larson (1993) cites (from Weld Coxe) another organizational typology based upon the nature of the kind of service provided to the client:

> *Strong-idea firms* which deliver singular expertise or innovation on unique projects....
>
> *Strong-service firms* which are organized to deliver experience and reliability, especially on complex assignments....
>
> *Strong-delivery firms* which are organized to provide highly efficient service on similar or more routine assignments....

Larson argues as well that these considerations of the conditions of production result in differences in the form of the outcome:

Organizational Form	*Individual's Position*	*Orientation to Practice*	*Typical Product*
entrepreneurial loose structure	principal "guru"	strong idea, innovative	unique solutions
corporate partnership; structured project teams	design partner or director	strong service, diversified, high competence	high-repeat projects "personalized" for each client

She also comments on the noted architect Robert Venturi: "... we must not omit the fact that Venturi, with a teaching job at an élite university, a small office and little work, enjoyed a total though unwanted freedom from corporate dictates." Perhaps one might argue with this characterization; but the effects of the increasing move of architecture into academe are a matter of particular concern. This is true in terms of the "technical rationality" of which Schon writes. As well, it requires inquiry into the "political economy of academe" as described by James Mayo (1991).

Along with the structural considerations mentioned above, the issue of ideology needs discussion, as well. The study of organizational culture is developing rapidly within organizational theory; it would seem singularly appropriate to the study of production regimes so involved with the creation of meaning as well as artefacts. Jeffcutt (1993) provides a dual classification of the studies of organizational culture which is applicable:

1. culture as a corporate (managerial) possession (a.k.a. "Corporate Culture")
2. culture as a communal (collective) expression (a.k.a. "Culture in Work")

Jeffcutt sees these differences (as well as a variety of narrative forms employed in the expression of the culture) as being related to the politics of representation in organizations. This seems particularly appropriate to the study of architectural regimes.

The inter- and intraorganizational settings within which is shelter, create a particular kind of problem for the occupational definition of architecture as a profession (Larson 1993):

> ... insofar as architects consider themselves architects, they are implicitly participants in a common discursive field. In other words, when architects talk or think about architecture, they have primarily other architects in mind or across the table (even if only to curse what the "worthless rascals" are doing to the art, the profession or the business). The common discursive field holds autonomy and heteronomy together in a permanent and constitutive contradiction....
>
> The autonomous pursuit of architecture and the heteronomous conditions of its making insert a permanent contradiction into the heart of the profession's practice and even of its discourse.

As well, Larson lists another "peculiarity" of architecture, as profession and as art:

> ... it contributes to culture not through discourse and codified practices alone but also, and crucially, through *artifacts* that are useful and can be beautiful.

These contradictions, between heteronomy and autonomy, between the provision of a service and the relation to the production of an artefact, and indeed, between use and beauty, shape the occupational nature of architecture. They therefore shape the nature of doing architecture as a form of work.

If we take on a view of the designer as a worker (by extension of Fry's notion), what indeed is his or her product, and what determines (alienates) the relation to it. Particularly, what control does the designer have over the nature of his or her product. Forty (1986) describes Josiah Wedgwood telling his designers to design products that could be made by the dumbest possible workers; one can discuss at length the resistance (or lack thereof) of such workers. Should we discuss as well the resistance (or lack thereof) of the designers? I was present when a designer was asked by a design student about the source of his aesthetic. He replied that a manufacturer asked him to fill six feet of shelf space with products that were made of injection-moulded plastic and sold for $3.98; that was his aesthetic. I have been present in a similar discussion when the subject was the way in which fire safety regulations, as embodied in the design of a building, would determine the way in which a health facility would operate. It should be mentioned that there have been attempts to design both products and buildings in ways that took into account the labour process that was being created thereby, as in the efforts of William Morris and of the Bauhaus. The success of these attempts is a matter of debate.

Rosenbrock's (1985) description of the ignorance of the human aspects of the production process in design engineering is likely more typical. The extent to which the safety of the worker, as well as of the user, becomes a major matter of concern, is located within the development of the professional subculture. The relative emphasis on architectural style is likely to respond to the labour process within the occupation:

> This recurrent cycle of formalism and reform has been driven by tensions inherent in the disciplinary structure of professional design, tensions that reflect the nature of the profession's efforts to contain an awkwardly broad and culturally diffuse jurisdiction within a certain kind of social structure: a professional labour market (Brain 1989).

The designer's exercise of control over the product of his or her labours is constrained by processes analogous to those in other work situations. Becker, in "Art as Collective Action" (1974) has described the situation in terms of the division of labour, and the effects thereof, on artistic production. It seems clear in his account that the final form that is produced reflects the structure of that division, as well as the simple fact of its existence. In the case of the design professions, that division of labour is reflected in technical specializations within the profession, and to related professions in the design process (e.g., architect to structural or civil engineer [Cuff 1989; Ferguson 1992]), as well as in the construction/production process. It is also manifest in a number of "employment relations" which serve to structure the production of design, although they vary widely in organizational terms. There is a model, or perhaps one should say a cliché, of the architect as "hero and genius" (Saint 1983) based largely on the fictional protagonist of the Ayn Rand novel, *The Fountainhead.* That image is perhaps stronger outside the bounds of the profession. Prak (1984), however, devotes a book to the consequences of the view of the "artistic" stance of the architect, and Marquart and de Montlibert (1970) discuss the consequences of the "charismatic" self-image of the architect. The problems that these authors list are the result of the fact that the "artistic" and "charismatic" designers are trying to enact those images in an organizational setting. The lone practitioner, as in the novel, is much like the lawyer in single practice; the "star" designer is often at the head of some sort of organization, perhaps organized in an "atelier" fashion. Prak (1984) provides a useful perspective on design organizations:

> The architect heading his own small office is of necessity a jack-of-all-trades, contacting clients, doing design and often working drawings, writing specifications and certificates of payment, supervising jobs. In larger offices, there is a division of labour which can in principle be handled in two ways. Either the office is organized as a federation of small offices, each job having its own draftsmen, job captain and designer, or it is organised on the principle of the assembly line: the job goes from the design department to the drafting room for contract drawings etc. The "federal" type usually has a joint clerical staff and some other centralised functions, such as spec writing, job supervision or public relations. The large and very large offices are all run on the assembly line method, as far as I can tell.

Along with the design firms, there is a range of alternative employment possibilities: in-house corporate design, participation in development and construction-management firms, and employment by the state. As well, there are varieties of consultancy and contract work. From the standpoint of the designer, these arrangements determine the control over their product. In addition, there are a number of "players" in the creation of a building or product, ranging from other professionals to regulators, not to mention a considerable division of labour within what we label "the client." As Cuff (1989) points out, each of these "players" is "typically a team of actors representing a still larger organization."

The division of labour in offices has come to have its own logic and to be its own imperative:

> What was at first a separation of responsibility and division of labour resulting more from the growing needs of growing offices than from deliberate intention, soon became a basic principle of organization; and it was applied not only to large offices but even, in some cases, to small ones as well (Boyle 1977).

The effect of this was the development of a more finely honed division of architectural work and the "separation of the decision making from the work itself [which] led in turn to specialization on the part of the workers and their separation from one another" (Boyle 1977). And, Boyle continues, it also led to a career concerned almost totally with such a special task, rather than in the full range of professional activities.

The irony is that these organizational realities are not reflected in the aspirations of the designers. Cuff (1989) describes a continuum of "ideal types" of design: one pole is vernacular design, described in terms of a nostalgia for a slow evolution of design reinforcing:

> ... the world view, available technology, economy and social patterns of the very people whose hands shape the physical environment in which they dwell.

At the other pole is a type called "the organizational management of design," placed thoroughly in the organizational context that has been described above. In between, there is a third position, which is based upon the skills of the sole individual architect:

> ... aspersions cast upon a managerial approach are coupled with disdain for vernacular building because it produces naive results.

The point that Cuff makes is that none of these positions accurately reflect architectural practice. If she is correct, than the question must be raised as to where these conceptions come from. Indeed, some attribution must be made to popular conceptions of architecture. But I would suggest that even as professional education creates a subculture that may be at odds with popular conceptions, so too does it create the kinds of "ideal types" that Cuff outlines. As well, it provides (or fails to provide) knowledge in terms of environmental and safety issues, knowledge of technical options, knowledge of the user(s), and knowledge of form in an aesthetic sense. As such, it provides a ranking of the significance of such issues, in practice, and sets the terms whereby these issues are negotiated in the making of a particular building.

The Designed Environment as A Carrier of Knowledge

Anthony Giddens, in discussing the role of expert knowledge in modern society, has provided an interesting summary of the way in which faith in that knowledge is acted out:

> Most laypersons consult "professionals" — lawyers, architects, doctors, and so forth — only in a periodic or irregular fashion. But the systems in which the knowledge of experts is integrated influence many aspects of what we do in a *continuous* way. Simply by sitting in my house, I am involved in

> an expert system, or a series of such systems, in which I place my reliance. I have no particular fear in going upstairs in the dwelling, even though I know that in principle the structure might collapse. I know very little about the codes of knowledge used by the architect and the builder in the design and construction of the home, but I nonetheless have "faith" in what they have done. My "faith" is not so much in them, although I have to trust their competence, as in the authenticity of the expert knowledge that they apply — something which I cannot check exhaustively myself (Giddens 1990: 27–28).

When Giddens acts on his reliance, he does it by going up the stairs — that is, by using these particular designed objects. He does it without thinking and without looking, in the faith that the stairs will not fall in. On those stairs, and on many others (if I may be allowed to presume for him), he expresses assurance that the configuration of the stairs will be of a certain sort. That configuration allows him to take the physical process of climbing or descending more or less for granted. Thus, those stairs embody a diverse body of knowledge, even before we begin to discuss them in symbolic or aesthetic terms. One need not prejudge the quality of that knowledge; some of it is based upon science, some on professional practice and construction craft, some on political process, and some on hunches. However derived, that knowledge is summed up in those stairs as common artefacts, and we can't get through the day without walking on that knowledge, so to speak. As such, those stairs embody what Giddens terms "regimes": "regularized modes of behaviour relevant to the continuance or cultivation of bodily traits" (1991). Giddens expresses his faith in that knowledge, and in professional practices. He may forget, but we should not, that the knowledge must come to him as a designed artefact.

Early in the history of stairs, they shift from being a reshaping of the local terrain into designed objects:

> The shape of the stair begins to respond more to the deliberate decisions of the designer and the exigencies of the created structure to be served than to the natural contours of the land (Templer 1992a: 5).

We must then ask what the factors are that enter into those design decisions: what does the designer think of when the stairs are created, in the mind's eye and in the making. Those factors represent functional concerns:

> We must consider the epidemiology of accident; the biomechanics of human gait, reaction time and responses; work physiology and the energy demands of walking on stairs; etiology (the causes of accidents; kinematics; the influence of mass and force that occur during a fall); perceptual and cognitive aspects of behaviour on stairs; traffic engineering and pedestrian flow and the capacity of stairs to cope with crowds; materials engineering; and the

> many facets of human factors engineering that relate to illumination, field and ground identification and accident reconstruction (Templer 1992b).

In addition, there are aspirations and symbolic intentions that are built in as well:

> The stair has always been used to represent human spiritual aspirations and cosmography; to demonstrate secular power and authority, prestige and status; for aesthetic, architectural and spatial manipulation; to make adjoining floors seem close, and the ascent a gentle transition; or to accentuate the separateness of spaces, with the staircase acting as a bridge (Templer 1992a: 7).

If all of this knowledge, functional and cultural, had to be decided upon anew with the designing of every stair, the designer's work would be impossible. Thus, both the most banal and the most creative solutions embody a large number of taken-for-granted notions. They may be built into the culture at large, into the professional subculture of the designer, and into the regulations to which a design solution must adhere. All of these means of knowledge accumulation are summed up in the stairs which Giddens trusts, and the nature of that knowledge may represent the continuity and force of historical process, as much as it does the advance of scientific process. Templer (1992b) notes that the early anthropometric studies of Francis Blondel, done in the court of France at the end of the 17th century, remain with us today:

> Blondel's empirical anthropometric tenets have taken on the force of law in the 20th century — without adaptation to allow for changes in mensuration and human dimensions.

Surely, it must be an easy matter to change those laws. But equally surely, it is not. Partly, it is a matter of the clash of political will. The changes, when they have been recommended, have been resisted because of additional costs in commercial construction practice. In addition, because we have been working, walking and playing on such stairs for so long, we now take for granted the conformation of those stairs, because of the durability of stairs as artefacts. A considerable process of social learning would be required if a major change would be done. Such learning processes are possible (as with the shift in driving patterns in Sweden); but they are not taken on very often.

The complexity of such a change is staggering, precisely because of all those who would have to learn things anew. It must be remembered that the designer, the manufacturer and the user are but three of the figures in a complex division of labour that creates a designed object. The particular disciplines and subcultures related to each of those roles would have to change as well. For example, the very particular subculture of regulatory administration would have to change. The attempts to realize the unification of regulations in the European community provide a number of examples.

I have spelled out these notions in some detail in respect to stairs; they are applicable in a wide variety of ways. Built form, as it is part of a regime, provides us with both tacit and explicit conceptualizations of the environment, and our relations to it. Our quantification of time and space, for example, comes through the embodiment of such notions as numbers, measurement, mass and volume, as we experience them in the built environment. Those notions help determine our actions in the environment and on the environment. Moreover, they are part of our self-definition and socialization as persons.

The built environment embodies the institutional processes of cultural reproduction (Brain 1994), as they relate to problems like those of relations to the natural environment:

> They key point is that the social construction of a technological artifact is not simply a matter of selecting new solutions to existing problems, or new solutions to new problems, but a reconstruction of the relationship between problems and solutions such that a particular artifact, with recognized characteristics, is capable of representing itself as a technical solution to a problem which it is empowered to define.

Thus, to a large extent, the creation of the built environment defines the problem of the relation to the environment, as it is declared to be the solution.

If we are concerned with artefacts as the embodiment of knowledge, then we must be equally concerned with the derivation of that knowledge. The citation of the efforts of Blondel mentioned before, raises questions as to how the changes in mensuration were achieved, and how the changes in human dimensions are to be understood. More simply, we face the question that arises constantly in playground argument: "How do you know?"

The understanding of that question has been greatly advanced by the work of Thomas Kuhn (1962). In particular, he raises issues of the "normal" processes by which problems are generated and are seen to be solved. He raises questions of scientific revolutions, but in so doing, he also describes the social and political process of knowledge evolution. That process is described in terms of a "paradigm" — the ways in which problems are identified, knowledge is sought in aid of their solution, solutions are accepted, and new problems are generated. It is possible to speak of designers' paradigms, and given the complexity of the design and manufacturing processes, one must speak as well of the different kinds and sources of knowledge that enter into that paradigm.

Conclusion

The process of cultural reproduction through the medium of the built environment is a dynamic one, reflecting the fact that even large-scale social processes (such as institutions) are continually reenacted (Douglas 1986):

> This is how the names get changed and how the people and things are rejigged to fit the new categories. First the people are tempted out of their niches by new possibilities of exercising or evading control. Then they make new kinds of institutions, and the institutions make new labels, and the labels make new kinds of people. The next step in understanding how we understand ourselves would be to classify kinds of institutions and the kinds of classifications they typically use.

The making and use of the built environment "tempts people out of their niches" as they seek new ways to manage their basic and crucial relation to the natural environment. In its turn, the institution of shelter makes new people, as it provides labels for them and the spaces that they inhabit, and as it provides the means for their management of the environment. As Peter Dickens put it in the principles that were cited earlier: "In the process of changing nature, people change themselves."

DISCUSSION QUESTIONS

1. How is the built environment implicated in the basic human-environment relation?
2. How does shelter resemble labour in the human-environment relation? Can shelter be alienating in the same way that labour is?
3. How does the built environment function as a "non-human actor" in social systems?
4. Give a sociological reading of the "form-versus-function" debate.

References

Becker, Howard. 1974. Art as Collective Action. *American Sociological Review* 39: 6.

Boyle, Bernard. 1977. Architectural Practice in America, 1865–1965: Ideal and Reality. In *The Architect*, ed. Spiro Kostof, 309–44. New York: Oxford University Press.

Brain, David. 1989. Structure and Strategy in the Production of Culture: Implications of a Post-Structuralist Perspective for the Sociology of Culture. *Comparative Social Research* 11: 31–74.

———. 1994. Cultural Production as "Society in the Making": Architecture as an Exemplar of the Social Construction of Cultural Artifacts. In *The Sociology of Culture*, ed. Diana Crane, 191–220. Oxford, UK: Blackwell.

Brante, Thomas. 1990. Professional Types as a Strategy of Analysis. In *Professions in Theory and History: Rethinking the Study of the Professions*, ed. Michael Burrage and Rolf Torrstendahl, 75–93. Newbury Park, Calif.: Sage

Burawoy, Michael. 1985. *The Politics of Production*. London: Verso.

Cuff, Dana. 1989. The Social Art of Design at the Office and the Academy. *The Journal of Architectural and Planning Research* 6(Autumn): 3.

Dickens, Peter. 1992. *Society and Nature: Towards a Green Social Theory*. Philadelphia: Temple University Press.

Dilnot, Clive. 1982. Design as a Socially Significant Activity: An Introduction, *Design Studies* 3(3): 134–46.

Douglas, Mary. 1986. *How Institutions Think. Syracuse*. NY: Syracuse University Press.

Ferguson, Eugene. 1992. *Engineering and the Mind's Eye*. Cambridge, Mass.: MIT Press.

Fitch, James Marston. 1972. *American Building: The Environmental Forces That Shape It*. 2d ed. Boston: Houghton Mifflin.

Forty, Adrian. 1986. *Objects of Desire: Design and Society 1750–1980*. London: Thames and Hudson.

Fry, Tony. 1989. A Geography of Power: Design History and Marginality. *Design Issues*, 6(Fall): 1.

Giddens, Anthony. 1990. *The Consequences of Modernity*. Stanford, Calif.: Stanford University Press.

———. 1991. *Modernity and Self-Identity*. Cambridge, UK: Polity Press.

Gordon, Charles. 1980. Housing as Labour. In *Structured Inequality in Canada*, ed. John Harp and John Hofley, 459–73. Toronto: Prentice-Hall.

Hayden, Dolores. 1976. *Seven American Utopias*. Cambridge, Mass.: MIT Press.

Jeffcutt, Paul. 1993. From Interpretation to Representation. In *Postmodernism and Organizations*, ed. John Hassard and Martin Parker, 25–48. Newbury Park, Calif.: Sage.

Jhally, Sut. 1987. *The Codes of Advertising*. London: Routledge.

King, Anthony. 1984. The Social Production of Building Form: Theory and Research. *Society and Space* 2: 429–46.

Krippendorf, Klaus. 1989. On the Essential Contexts of Artifacts or On the Proposition That "Design is Making Sense (of Things)." *Design Issues*, 5(Spring): 2.

Kuhn, Thomas. 1962. *The Structure of Scientific Revolutions*, 2d. ed. Chicago: University of Chicago Press.

Larson, Magali Sarfatti. 1984. The Production of Expertise and the Constitution of Expert Power. In *The Authority of Experts*, ed. Thomas L. Haskell, 28–80. Bloomington: Indiana University Press.

———. 1993. *Behind the Postmodern Facade: Architectural Change in Late Twentieth-Century America*. Berkeley, Calif.: University of California Press.

———. 1994. How Should We Look at Buildings? A Preliminary Reflection on the Sociology of Architectural Objects. *Newsletter of the Sociology of Culture Section of the American Sociological Association* 8(3–4): 1–4.

Latour, Bruno. 1988. Mixing Humans and Nonhumans Together: The Sociology of the Automatic Door Closer. *Social Problems* 35(3): 298–310.

Markus, Thomas. 1993. Buildings as Social Objects. In *Companion to Contemporary Architectural Thought*, ed. Ben Farmer and Hentie Louw, 15–20. London: Routledge.

Marquart, Francois and Christian de Montlibert. 1970. Division du Travail et Concurrence en Architecture. *Revue Francaise de Sociologie* XI: 368–89.

Mayo, James. 1991. Dilemmas of Architectural Education in the Academic Political Economy. In *Voices in Architectural Education*, ed. Thomas A. Dutton, 65–82. New York: Bergin and Garvey.

Mennell, Stephen, Anne Murcott and Anneke H. van Otterloo. 1992. *The Sociology of Food: Eating, Diet and Culture*. London: Sage.

Prak, Niels. 1984. *Architects: the Noted and the Ignored*. Chichester, Eng.: John Wiley and Sons.

Rand, Ayn. 1942. *The Fountainhead*. New York: New American Library.

Rapoport, Amos. 1984. Culture and the Urban Order. In *The City in Cultural Context*, ed. John Agnew, John Mercer and David Sopher, 50–75. London: Allen and Unwin.

Rose, Gillian. 1988. Architecture to Philosophy — The Post Modern Complicity. *Theory, Culture and Society* 5: 357–71.

Rosenbrock, Howard. 1985. Engineers and the Work that People Do. In *The Experience of Work*, ed. Craig Littler, 161–71. Aldershot, UK: Gower.

Saint, Andrew. 1983. *The Image of the Architect*. New Haven: Yale.

Saunders, Peter and Peter Williams. 1988. The Constitution of the Home: Towards a Research Agenda. *Housing Studies* 3(2): 81–93.

Templer, John. 1992a. *The Staircase: History and Theories*. Cambridge, Mass.: MIT Press.

———. 1992b. *The Staircase: Studies of Hazards, Falls and Safer Design*. Cambridge, Mass.: MIT Press.

Wilson, Peter J. 1988. *The Domestication of the Species*. New Haven, Conn.: Yale University Press.

THE CASE-STUDY APPROACH

SOCIAL AND POLITICAL FACTORS OF ENVIRONMENTAL PROTECTION IN CANADA

Louis Guay

Université Laval

Introduction

A new map of the world is being remodelled by economic, political, cultural and environmental factors. Political remodelling has, and is still taking place, for instance, in Eastern European countries. On a larger scale, we assist, to an unprecedented degree, in economic circulation of goods, people and services. A new economic order is in the making and cultural factors are also at work. The world that slowly emerges will probably be very different culturally from the world of traditional cultural entities, such as nations, religious or territorial communities.

There is one element that seems, to me, to be reshaping the cultural and political map of the world: that is the awareness of large, even global, environmental problems. The protection of the environment is a planetary challenge. All nations, leaders and social groups do not agree on the proper course of action to best protect the environment, but there is a sense of urgency. This sense has been growing for some time, increasing its pace in the last 30 years. At the heart of it lies the awareness that some transformations,

The first version of this paper was presented at the Geography and Planning Research Seminar, London School of Economics and Political Science, 21 January 1993.

in particular the Industrial Revolution and the economic and political system that supports it, have been exacting a great price on the integrity of the environment.

It is not my intention to analyse the problem at the planetary level. I shall limit myself to the relation between some aspects of environmental degradation and the forms of environmental protection in Canada. I shall emphasize cultural and social factors, for the knowledge and the perception of some degradation, triggers a whole set of cultural and social reactions.

For nearly 30 years, ecological awareness in developed countries has been a new factor in politics, social movements and social values. Concerns over the state of the environment began with events — literary, scientific and historical — that sent a series of warnings to people in the developed countries. It became rapidly clear that environmental problems were not limited to the most-developed parts of the world. The whole planet was under stress — a stress that arises from the huge exploitation of natural resources, that comes from the poor management of various wastes, and that is linked, in some regions, to a rapidly growing population living in a local environment that may not sustain it indefinitely. There has thus been, as some geographers (borrowing from Karl Polanyi) have called it, a "great transformation" since the rise of our industrial and urban civilization more than 200 years ago.[1]

True, in the past, local environmental degradation led to serious social and political problems. Some have advanced the theory that the collapse of the Mayan civilization may have been due to an increase in population size, and its need for resources in a tropical and fragile environment that could not respond to human demands.[2] But environmental problems of the past, though they may have been severe, were mostly localized, although some may have affected very large regions. Now, some environmental problems are no longer local, but global, even though the extent to which they may endanger our very place on earth, is highly disputable.

Canada is a vast country, rich in resources and, partly because of its difficult climate, has a small population. The abundance of its resources — at least until now — and the vastness of its territory, have bred some behaviours and attitudes that, in the long run, were detrimental to protection of the environment. Canada has discovered, at the same time as other industrial countries, that its environment is no longer a land of plenty — at least for some resources — and that ecological degradation is taking place.

I would like, in the following, to report how awareness of this, triggered a series of responses in the political body, as well as, more broadly, inside Canadian society, and to tackle some of these environmental problems. I shall concentrate my analysis, not on the vast array of environmental problems themselves, but on how and why people and their political representatives responded and intervened in order to limit environmental degradation, as well as what institutions, rules and organizations were designed to modify people's behaviours and attitudes towards the environment. By institutional rules, I mean principles and procedures that become widely accepted inside a group, or a society, some of which may be enacted by parliaments. It is interesting to note that Canada, as many industrial countries, now possesses broad environmental legislations that attempt to control human intervention in the environment.

Here, a note of caution is in order. It is not possible in this short paper to review thoroughly all the environmental policies and legislations that have developed over the years. Though most problems are new, an impressive array of policies and legislation have come about. The federal structure of the Canadian political system is another reason why a full review is impossible. The federal, as well as the provincial, governments intervene in the environmental domain. When the Canadian Constitution was first drafted in 1867, environmental problems were unknown, or considered unimportant. Therefore, the environment was not included in the Constitution, and no particular tier of government inherited the whole area of environmental protection. However, inside their own constitutional responsibilities, the federal and provincial governments have taken initiatives regarding particular, serious environmental problems. For instance, the provincial governments, having jurisdiction over natural resources, designed policies to protect these resources. Thus, forestry and forest exploitation come under provincial jurisdiction. When forest exploitation led to environmental degradation, provincial governments were pressed to pass legislation to ensure that conservation practices accompanied resource extraction. We shall consider this aspect later on, for, as it is well known, the boreal forest in Canada is highly exploited, and its exploitation has not always been environmentally sound.

Moreover, if it is not possible to review all policies and legislation, it is also not possible to describe, even to name, all environmental problems. My analysis will thus be based on some selective problems — forest and water in particular.

Finally, a thorough sociological analysis of people's attitudes and behaviours is beyond the scope of this paper. Debates and values shape public behaviours. Not being able to record behavioural changes over a long period of time, I shall concentrate on some public debates that may have had an impact on public attitude towards the environment. I shall also try to present a picture of Canadian values towards the environment, stressing how environmental protection values may clash with, or complement, those that we may have towards economic growth.

Historical Context

The Canadian economy has been, since its first steps, highly based on the exploitation and extraction of natural resources. During colonial times, the Canadian economy was determined first by the French, then the British, empires. The commerce of staples — basic products — was one of the main factors in Canadian development. First, until the 19th century, fur and fish were staples in demand in the home countries. The Canadian economy was very dependent on this external demand. During the 19th century, other resources and staples came to be valued in international markets — mainly Britain and the British Empire. Cereals — wheat in particular — and wood were exported to Europe and elsewhere. In the 20th century, other resources found their place in the international marketplace: minerals, oil, gas, as well as the more traditional demand for cereals. Canada has thus always linked its economic development to the external demand for its natural resources.[3] A particular feature of this sort of economic growth is that, as resource demand changes, new regions of the country are opened. When fur and fish are exported, the eastern part of the country prospers. When wood and wheat are shipped to Europe, the central part of Canada — the provinces of Quebec and Ontario — develops more rapidly. And when oil, gas and minerals are in high demand, the West and northern part of Central Canada grow economically. However, when these resources are less in demand, or when their prices are too low, whole regions start to decline.

Canada entered the industrial world during the second part of the 19th century. Its industrialization was rapid, built on a nationalistic policy, centred in Quebec and Ontario and offered the peculiar feature of being autonomous, and at the same time, dependent on foreign capital and technology — first from Britain, then from the United States.[4]

Industrialization here, as elsewhere, greatly modified the environment. It also created problems, problems that were, at first, tackled timidly by some policies, measures and institutions. Two important policies characterized the growing environmental concerns following industrialization. First, following on the steps of a movement initiated in the United States, Canada, in the 1880s, adopted a policy of wilderness conservation by creating national parks. Between 1885 and 1920, 10 national parks were created; they were ecological reserves for the enjoyment of future generations of Canadians.[5] Most of them were located in the West, in regions that were not soon to be industrialized, but were nonetheless gradually incorporated, for their natural resources, in a growing and modern economy. It is worth noting that national park creation in Canada did not respond to the same pressing needs as in the United States. In the U.S., national parks were created to counter development in beautiful areas; aesthetic and spiritual reasons were given to justify the setting aside of huge tracts of land. In Canada, the newly created national parks were not left totally untouched. Although the parks were mostly created for recreation and tourism purposes, mineral and forest exploitation was authorized, though regulated — a decision that profited some developers.[6]

The second case is the creation of a *commission of conservation* by the federal government in the 1910s, to assist and advise on the proper measures to conserve natural resources. But soon, the commission became concerned not so much with conservation policies, but with the occupation of the West and the opening of the vast Prairies to European immigrants.[7] However, the idea of conservation was planted in federal institutions.

It is perhaps only with some qualification that we may speak of an environmental policy in Canada at that time. Economic development was the main concern, not environmental protection. Canadians had, in the 19th and at the beginning of the 20th century, very special attitudes towards their environment, especially if we compare them to those of our American neighbours. As one author has documented, literature and popular publications about nature in Canada were different from those widely read in the United States. The glorification, reverence and respect for nature that one finds in American literature — in the writings of Emerson, Thoreau and Fennimore Cooper — have almost no counterpart in Canadian publications.[8] Nature and wilderness have always been regarded as opposite to civilization, whereas in the United States, though such an attitude was also widespread, a counter model emerged. Protection of the wilderness was seen as one of the main contributions Americans could make to mankind. This attitude, if it existed at all in Canada, was certainly highly marginal. The town and the humanized countryside were preferred to the wilderness, the wild and wide spaces. To the extent that this attitude was deeply rooted in the Canadian psyche, and owing to its particular stage of economic development, Canadian environmental protection did not emerge until after the rapid economic expansion of the post-Second World War years, when environmental problems of a different nature emerged. The attitudes, as well as the policies, were about to change.

Environmental Problems

The Canadian economy became fully industrialized with a great variety of industries, a modern agriculture and an important mining industry. However, like so many industrial nations, it expanded without proper consideration for protecting its environment. The government policies of the 1950s and 1960s proclaimed the need to modernize the traditional economic sectors. Regional policies were modernizing ones: the forest, fisheries and agricultural sectors were to change as capital and equipment entered these sectors. But in so doing, it became more and more evident that some trade-offs had to be made, since an increase in production is usually accompanied by an increase in environmental degradation. The influence of the American environmental movement of the 1960s must be recognized here. However, indigenous concerns were well founded and environmental problems multiplied.

First, Canadians came to realize that environmental degradation was not a phenomenon that hits only the most populated nations. Pressures on land, resources, air and water are as strong here as elsewhere. Good agricultural land has been turned into urbanized and suburbanized land. The quality of the water is difficult to maintain close to urban areas, or where modern agriculture has taken root. In cities air quality is not guaranteed. Great waterways, river basins and large rivers are polluted by wastes from domestic, agricultural and industrial uses.

In a recent comprehensive report on the state of the environment in Canada, published by the federal Ministry of the Environment, Canadians can see, in great detail, how their environment is threatened.[9] It would be cumbersome to report all the statistics to describe this state. I shall, however, concentrate on three particular problems: (1) the water quality of the Great Lakes–St. Lawrence River system; (2) the state of the boreal forest, particularly in Eastern Canada; and (3) the specific problem of acid rain. For each of these questions, I shall give a brief description of the environmental degradation, the measures taken to improve the quality of the environment, and the nature of the public debate. Any identification of an environmental problem, and any approach to its solution, generate public discussion and debate. The identification of an environmental problem is often a matter of scientific and technical expertise, but the solution, complete or partial, requires a broader involvement. People are associated in the process, and there may not be better examples than the public hearings on the environment which Canadian governments recently institutionalized. I shall address this later.

One of the first environmental problems that arose in Canada on a large scale was pollution in the Great Lakes.[10] Around this huge reservoir of fresh water, one of the most important in the world, lives a population of more than 35 million.[11] It is an industrial centre of great importance. Large cities such as Chicago, Cleveland, Toronto, Detroit and Buffalo border the Great Lakes basin. Moreover, good agricultural land is located not far from it. It is thus a region that is highly industrialized, populated and is polluted by residues and wastes that its intense activities generate. But to start a programme to control pollution and clean the Great Lakes, a close collaboration and co-operation is needed between the Canadian and American governments. An International Joint Commission was created after the 1909 Boundary Waters Treaty to regulate all joint use of boundary waters. It later started to study pollution problems in the Great Lakes region.[12] But it was not until the 1960s that massive evidence had accumulated to describe the nature and the extent of the environmental problems.

Despite the commission's good intentions, governments had other pressing problems on their agenda. Also, one should not overlook the difficult, scientific and technical language in describing the problems. One of the first problems to emerge was the accelerated process of eutrophication that some lakes — Lake Erie in particular — had undergone due to pollution from high levels of phosphate from a wide use of domestic and industrial detergents.[13] The International Joint Commission had published, in the 1950s and the 1960s, a series of research results. In 1965 and 1969, it reported high concentrations of phosphorus from effluents of untreated municipal sewage systems. At first, the representatives of the detergent industry were not convinced that their product was the main cause of eutrophication. They claimed that there was not enough scientific evidence to assert that the water pollution problem of Lake Erie was primarily due to their product. But a consensus emerged, after extended public hearings, that phosphorus contents had to be controlled, and the industry regulated.[14]

The negotiations between the Canadians and Americans were not always easy. Each country had its own preoccupations. In this, as in other cases, Canada would appear to be the net loser if nothing were to be done. Owing to their greater size and number, the American cities polluted more that Canadian ones, and a great deal of the pollution ended in Canadian waters. Despite difficulties in the negotiation process, the Canadian government decided to accelerate its programme of helping the municipalities to equip themselves with proper sewage treatment facilities. In 1972, the two countries signed the Great Lakes Water Quality Agreement, setting standards and targets for controlling pollution. Even though American cities lagged behind Canadian, phosphate reduction in the Great Lakes was well under way in the 1980s. But during this time, new problems arose, and new sources of pollution were identified.

Major threats to the quality of the Great Lakes and the St. Lawrence River waters were coming from industrial plants, which release much more toxic pollutants into the environment. These new "areas of concern" focused in particular on the chemical and the pulp and paper industries that dump pollutants in the waters. Progress is being made on both sides to better control the huge production of pollutants, but industries and governments are just starting to realize that to clean up the environment necessitates huge investments. It will certainly be one of the main areas of contention in the years to come. In the 1980s, an important step was taken by the federal government and the provincial government of Quebec to control the largest industrial polluters along the St. Lawrence River, especially the pulp and paper plants that release highly polluted effluents (furans and dioxins) into waterways.[15] The pulp and paper industry has been environmentally regulated for some time by the federal, as well as the provincial, governments. However, the past record is far from brilliant. Individual companies could always threatened the governments with closure if they felt that the environmental regulations were too severe or too quickly applied. Governments were often forced to back down for fear of lost jobs, especially in areas highly dependent on a single large employer.[16] Public pressure is, however, getting stronger, and is starting to press the pulp and paper industry, which has recently issued an environmental protection statement.

The acid-rain problem is another case where environmental problems are not localized but are international.[17] Since the 1950s, Canadians had known that acid deposition — dry or humid — is causing great damage to their environment. Previous studies had shown that, close to a large nickel smelter in Sudbury, Ontario, trees had been dying and lakes had turned acidic, killing valuable varieties of fish or impeding their development.

At the end of the 1970s, the acid-rain problem entered the Canadian political agenda. In this case, it is fair to say that politicians reacted rapidly to people's concerns. For instance, in the Province of Quebec, one of the first social groups to alert the authorities were maple-syrup producers who had started to observe that their trees were yellowing and aging more rapidly, growing more slowly and producing less syrup. More generally, in Eastern Canada, a broad coalition against acid rain (The Canadian Coalition against Acid Rain) was set up in the 1970s. It represented a large diversity of interests, mostly economic. In Ontario, private residents located on the threatened lakes, and the tourist industry in these areas, pressed the government to intervene. Scientific evidence was accumulating, showing that Canada received more acid deposition than it produced. Although some acid rain is produced in Canada, about two-thirds of acid deposition that lands on Canada is transported in the atmosphere from the United States, and Canadians have to clean up their own backyard. Previous government regulations to reduce the local concentration of pollution, led to the industry construction of the super chimney in Sudbury. The result did not reduce pollution, it merely spread it out over a wider area. Similar technological adoption was undertaken by American power stations and metal-smelting plants. Thus, the trans-border problem was generated. It was now time to act, in particular, in order to give credibility to Canadian government pressures on the U.S.[18] The federal government forced the provincial governments, chiefly Ontario and Quebec, to apply tougher environmental regulations. It was agreed to reduce by half (in comparison to the 1980 emission level), the amount of acid gas emitted into the atmosphere by 1994. The metal smelters and Ontario Hydro were the producers that were to cut down their emissions the most.

For more than a decade, the federal government, the governments of Ontario and Quebec, and those of New York, Vermont, Maine and New Hampshire, pressed the U.S. federal government to amend its 1972 Clean Air Act to control industries (mainly power plants) that were great producers of sulfur dioxide (SO_2), the main source of acid deposition in Canada. This particular case of trans-border pollution is not unique to Canada and the U.S. The Scandinavian countries face a similar situation with acid rain, a product of emissions in Great Britain, Germany and Eastern European countries.[19]

The American Clean Air Act was finally amended in 1990 after many years of pressure from Canada and public, as well as research and negotiations between the two governments. In this case, as in the previous one, many people, many states (mid-west states), many industries (smelting and electricity-generating) realized that the costs of controlling pollution were high. But the arguments of the environmentalists, and of those who received more pollution than they produced, were twofold: (1) the problem of externalities is also a problem of regional and social justice; (2) the costs of controlling at the point of origin are high, but may also be high, and perhaps higher, at the point of destination. It is interesting to note that during this whole debate, scientific evidence was put forward either to prove or disprove that acid deposition is a threat to the environment. Canadian and American research programmes were set up to better understand the effects of acid rain on trees, lakes, fish and buildings. If the scientific community agrees now that acid rain has a great impact on the biological vitality of lakes, it does not agree that acid rain is the *main* cause of forest problems, at least in Canada.[20]

The third problem may be defined as the decline of the boreal forest. Decline does not mean the disappearance of the forest, but the decrease in the amount of wood that can be economically taken out of the forest. Wood harvesting has had to move to more distant, and less rich, areas owing to overexploitation of more accessible areas and to the slow process of forest regeneration which may in part be the result of improper forestry practices. The boreal forest is one of Canada's main ecozones.[21] Fifteen ecozones have been identified in Canada. The boreal forest ecozone spans five provinces, but is mainly situated in the provinces of Ontario and Quebec. I shall concentrate on the problem in the Province of Quebec.

For years the traditional system of forest exploitation has been carried out under a government contract that leases the cutting and harvesting of wood on public land to private firms. In contrast to other countries, more than two-thirds of the forest land is publicly owned in Canada. The government had established rights to cut and harvest by giving long-term forest concessions to logging and pulp and paper companies. But during the seventies this system came to be questioned. It was believed that the concessions system gave rights to exploiters but did not attach any, or very few, obligations. Ecological forest management was underdeveloped, despite the existence of a small, but coherent, association of professional foresters.[22] Their skills, talent and professional knowledge were employed to better exploit the forest resources, however, less interest was paid to the proper techniques of improving and protecting the quality of the forest. Debates arose inside the professional association over what forestry is all about: should foresters behave solely as economic agents, or should they also try to protect the forest environment for purposes others than economic exploitation? Forests are key economic resources in Canada. The forest sector is a strong exporter of wood, pulp and — most importantly in Quebec — paper. A strong export-oriented industry was developed by foreign and Canadian capital. But by the end of the 1970s, reports from government, industry and experts became alarmed: the resources were diminishing; the forest sector might face a shortage of wood if nothing was done.

During the 1980s, the government, with the help of industry and the professional foresters, abolished the old leasing system, replacing it with a new regime of tenure on public land. This new forest regime was twofold: firstly, public lands were to be allocated to producers on long-term contracts (up to 25 years), but the renewal of these contracts every five years would depend on the proper management of forest resources by the producers. Forestry management should incorporate a principle a sustainable use: trees that are cut must be regenerated. Secondly, all the economic interests recognized that the forests on public land may have different uses. A three-part zoning policy was adopted: (1) ecological reserves were designated where no commercial exploitation is authorized; (2) recreation areas were also identified where some exploitation may be allowed and; (3) large tracts of forest land were left to commercial exploitation. The new regime took into account the needs of different economic and social interests, and was more respectful of the various functions of forests, one of which was the preservation of habitats for wild animals.

This new regime was negotiated among many interested parties. The great economic and professional interests took an active part in the discussions. The professional foresters were encouraged by their association to participate, whenever they could, in the public debates and hearings. Other social interests, such as local communities, ecologists, associations of hunters and anglers and local co-operatives of small producers, were important players in the process. Each had to compromise. During these negotiations, most of which were open and public, divergent interests clashed. But a compromise was reached. Though it is too early to evaluate the impact of the new legislation, most of the people involved agree, that a step in the direction leading to sustainable development, has been taken. The negotiations leading to the new regime took place at almost the same time as the Brundtland Commission held hearings and prepared its report on sustainable development. The idea of sustainable development was in the air, and it is not surprising to see how quickly the ministry responsible for forests adopted the new idea. Some professional foresters were already advocating the concept. The industry itself, concerned over the expected shortage of timber if nothing was done, was ready to change its forestry practices. Though the conclusion of the negotiations was built on a large consensus, some environmentalists remained dissatisfied. For them the new regime fell short of addressing all the problems that have plagued forest exploitation in Quebec. In particular, clear-cutting is not banished, aerial spreading of chemical pesticides to control insect populations, that cause great damage to trees, may still be allowed, and nothing much is planned to support local communities living off the forests. For them, the new regime is not so much ecological, as it is a little more control over the uninhibited capitalist development of the past.

Legislation: Structures and Process

What I have just reported are important environmental issues in Canada that developed over a period of years. It is not, however, sufficient for a country to tackle specific environmental problems as they come up, it is also important that more general legislation and policies be designed to institutionalize the protection of the environment. In our type of society, institutionalization may be attained only when governments act.

Recent steps in the institutionalization of the protection of the environment were taken by the federal and provincial governments during the 1970s and the 1980s. Firstly, the two tiers of government in Canada enacted legislation creating ministries of the environment. Secondly, one after the other, the governments passed general legislation to protect the environment. For instance, the federal government passed its Canadian Environmental Protection Act in 1988, the Province of Quebec passed its Quality of the Environment Act as early as 1972, and the Province of Ontario, which is now examining an *environmental bill of rights* — whatever that may be — passed an Environmental Protection Act in 1976. Other provinces followed suit.

A common feature of this legislation is the deliberate focus on controlling pollution, or "contaminants," as it is called in the laws. Other courses of action could have been chosen. For instance, legislation to promote alternative technologies, or to compensate people harmed by environmental degradation, are other possibilities that Canadian governments chose not to explore at the time.[23] However, more recently, the so-called federal Green Plan of 1990, is deliberately enlarging the scope of government intervention in environmental protection.[24]

Several novel features characterize this Green Plan. The main objectives are twofold: firstly, to reconcile economic development with environmental protection, and secondly, to create an environment that is sustainable for the future generations of Canadians.[25] The Green Plan does not concentrate solely on the control of contaminants. Other interesting aspects are worth noting: (1) the government wants to promote and encourage alternative and cleaner technologies; (2) important investments in scientific research, in so-called "eco-research," are announced; (3) important steps are to be taken to reduce the amount of carbon dioxide, CFCs and nitrogen and sulphur oxides in the atmosphere; (4) finally, an improvement in the process of environmental decision making — especially a greater say of the public in this process — is planned.[26]

It is an important characteristic of Canadian environmental legislation that the public should be involved in decisions concerning development projects that may affect their natural, as well as their social, environment. For close to 20 years now, environmental-impact assessment review has been, despite its shortcomings, an institutionalized forum for the expression of views about the quality of the environment. Since about 1975, some development projects, (dam construction, road and highway construction, for instance) have been subjected to environmental-impact assessment. The procedures differ from one province to another, but they all tend to include the public. When the project is expected to have great consequences, the general public is consulted in the decision-making process. In the federal process, not all projects may be subjected to such an environmental-impact assessment review, but huge ones and those benefiting from federal money, or being developed on federal land, must be environmentally assessed. Up to 1990, the federal assessment process was governed by guidelines that were not statutory. That year, the federal government submitted a bill to parliament to "establish a federal environmental assessment process" that will be more encompassing than in the past. Moreover, environmental assessment will not be limited only to development projects, but will include federal policies as well.

Despite thorny problems with the provincial governments, the bill is going through parliament and is expected to be enacted soon. This may give enormous powers to the federal government to control large-scale development projects. It is not surprising that the provincial governments did not all react positively to the federal bill. They saw this new legislation as an intrusion of the federal government into their jurisdiction. They were worried that the new legislation may be used by ecologists or other opponents to a development project, who, having failed to get a full environmental review or to stop a particular project at the provincial level, will appeal to the federal government. On two recent occasions, opponents of provincial projects went to court and, invoking the federal guidelines, won environmental reviews, with public hearings, that they had been denied previously.

On the other hand, the federal government countered the provincial arguments by saying that these tougher regulations are demanded by the Canadian population. Despite agreement on broad objectives of pursuing economic growth while protecting the environment, we may expect some trade-offs between economy and ecology. The Great Whale hydroelectric dam project is a good case in point; the Quebec government was keen on developing its northern area, hoping to export electricity to the United States and saw the project as a major lever in its economic policy. But the native people did not see the project with the same eyes: they were worried over its disastrous impacts on their environment, a fragile Arctic ecosystem, as well as on their own culture and traditions. And they pressed the federal government to review the project thoroughly, hoping that, considering the huge social and environmental impacts, the dam project would be called off.

Changing Attitudes towards the Protection of the Environment

The institutionalization of environmental protection in Canada is the result of value changes and public pressures brought about by the growing awareness of ecological problems. Some problems can be easily perceived and identified by everybody — when the water turns brown or stinks, everyone knows that something is wrong — but other forms of environmental pollution are more difficult to identify. Air and soil pollution may only be identified by experts. As Steven Yearley has emphasized, scientific authority is a key element in the development of ecological awareness.[27] It is thus no surprise that some scientists were instrumental in the evolution of the environmental movement.[28]

But, more generally, it is people's attitudes and concerns that are the starting point of interventions, private or public, to protect the environment. One can observe strong feelings about the environment in industrialized countries. The emergence and relative success of Green parties in Europe, is a good example.[29]

Recent polls have shown that Canadians are very concerned with the quality of their environment. The relationship one has with one's environment depends on many factors. Taking the environmental movement as a whole, one is struck by how environmental values evolved more rapidly in countries where, to use Max Weber's expression, the Protestant Ethic has dominated. It is in North America, Scandinavia, Germany, Britain, Australia and New Zealand that for the last 30 years ecological values have flourished. Catholic countries joined the environmental movement later on. We can define, in simple terms, a social movement as a configuration of ideas, values and attitudes that are crystallized in specific organizations that bear these new intellectual elements. A French historian has recently asked whether nature is Protestant, namely whether Protestants, despite their great diversity, have a special relationship to nature different from that of the Catholics, and if so, from what religious texts or religious figures they draw their inspiration.[30] An answer to this question is not easy. There is some evidence that it may be so, if we regard the recent history of the environmental movement. However, other factors must be taken into account, in particular, the level of economic development. Are environmental values a luxury of well or too developed countries? Is environmental protection a necessity imposed upon developed countries by their overexploitation of nature? Before replying yes to both questions, one will be well advised to remind oneself that the Brundtland Commission has insisted that poverty, as well as affluence, can lead to environmental degradation.

Recent polls in Canada show strong support for environmental action by governments and also widespread concerns with the quality of the environment.[31] However, there may be some ambiguity in people's responses. As Riley Dunlap has shown for the Americans, the "public support for environmental protection has become a truly consensual issue — commanding support from large majorities. What is less clear, however, is the strength of public commitment to environmental quality."[32]

A public opinion poll in Canada, taken before the 1988 federal election, shows very clearly that, across Canada, people are very concerned with the environment. It is one of the most important collective issues and it has been growing for the past 10 years.[33] In this poll, people were asked to respond to a question about whether job creation or the quality of the environment was most important. A majority of people attached a greater value to the quality of the environment over job creation. Slight differences between provinces, occupation and age groups, education levels and levels of income occurred, but, on the whole, there was a strong consensus. One difference is striking: whereas 68 per cent of English Canadians considered the protection of the environment more important than job creation, only 56 per cent of French Canadians agreed. Though not all English Canadians are Protestant, we may have some evidence here that, in Quebec, a province with a strong Catholic tradition, environmentalism is less rooted than in other provinces where Catholicism has not had such a strong hold. However, other factors even more strongly influence people's attitudes: respondents with postsecondary education were more strongly in favour of protecting the environment (72 per cent) than those who only had a primary education (50 per cent). Let us also remember the dominant attitudes 19th century Canadians had towards their environment.

The authors went on to consider how values influenced people's opinions. The traditional political ideologies — left/right/centre — have a lesser impact than the distinction they make — following political scientist Ronald Ingelhart — between materialist and postmaterialist values. The postmaterialist values were defined by a strong agreement with: "ideas count," "free speech," "beautiful cities," "nature."[34] But the authors are cautious in their conclusion: "even assuming that the 'greening' of the Canadian electorate continues apace, the outcome of any given election being determined by the 'green' factor will likely be limited by the state of the economy."[35]

Conclusion

The sense of global environmental crisis that has come upon us since the 1960s and that seems to have culminated in the Rio summit, has given rise to governmental actions and a widespread concern with environmental quality. However, countries differ widely in their determination to tackle the problems. Canadian corporations are also under pressure to adopt codes of conduct that enlist them in protecting the environment. They may be doing so only because they are afraid of the bad publicity that they may incur from spectacular actions by Greenpeace.[36] They are hard pressed to adhere to the so-called Valdez principles.[37]

But attitudes as well as actions vary a great deal. What we call environmentalism is a deeply divided movement. An eco-centric outlook is very different from a technocentric one.[38] The former wants to completely redefine our relations with nature and, in so doing, redesign our social institutions. Murray Bookchin has been a strong advocate of an eco-centric ideal: harmonious relations with nature go hand in hand with harmonious social relations.[39] The technocentric ideology, on the other hand, is less radical and believes that environmental protection rests, to a great extent, on a combination of technological innovations, market forces and government regulations. The Canadian Green Plan sets Canadian society clearly in this direction.

The present policies and attitudes in Canada are clearly close to the technocentric ideology: the Green Plan's belief that a strong economy can be attained with greater protection of the environment, is certainly shared by a great majority of Canadians, though there are groups that would more easily give their allegiance to a radical redesigning of our relations to nature.

Whatever the value of theoretical writings of the eco-centric perspective, they are far from the concerns and interests of those who are active in defending the environment. The thousands of Canadians who participate in some form of environmental action are more concerned with specific problems that invade their own lives, than with large issues. The Canadian Coalition Against Acid Rain, for instance, was dominated by economic interests and by local residents. It is unlikely that its members were "deep ecologists." However, one should not belittle a more radical stand, for radical groups may be at the origin of an environmental issue. Greenpeace in Canada was certainly the group that did the most to inform the Canadian population about water pollution caused by the pulp and paper plants, by carrying out their own tests and by showing that Canadian governments did not apply their declared environmental standards, which had been agreed upon with the industry.[40]

A few factors have contributed to the institutionalization of environmental protection in Canada. By institutionalization I mean the process by which environmental protection becomes a regular activity with its own objectives, procedures and organisations. This process was initiated by a growing awareness of large ecological problems. This awareness has gained from scientific information, for many environmental problems are not, so to speak, visible to the naked eye. Scientific information may certainly be necessary, but is not always a sufficient condition for action. Without new values and social attitudes borne by militant social groups and widely diffused in society, scientific information may lie at rest. The environmental movement has been instrumental in bringing about a wider change of attitudes in the public as well as in the political institutions. The members of the environmental movement have also been helpful in diffusing and popularizing, through reports or through the press, scientific information that is not easy to understand, though scientists are not always happy with the way scientific data are handled by some environmentalists, or by the press.

Environmentalists were seldom alone in defending the environment. As has been shown above, others actors played an important part. The technical and scientific staff of government departments and agencies was often at the origin of a new ecological issue. For instance, one function of the International Joint Commission was to monitor water quality of the Great Lakes, and it became one of the first to worry about the dangers of acid precipitation in Canada. And governments are not only reacting to problems brought by the outside world, though their environmental interventions are frequently forced upon them, they can also take the initiative. For example, on the acid rain problem, the federal government was very active, in Canada as well as in Washington, in building a strong case to tackle the problem. It funded new research, published summaries of the results widely, and spared no means to convince the provinces and the American government that legislative action was necessary. True, public pressure had been building for some time, but it must be admitted that federal ministers of the environment, in particular the Conservative Minister Fraser, took the matter seriously and defended it vigorously inside and outside Canada. In their dealings with other governments they were supported by their department's technical and scientific staff.

Attitudes and behaviours towards nature are moulded by practical experience and the value system, religious or otherwise, to which one is committed. As we have seen, religious values seem to have played a part in the evolution of an environmental consciousness. The link between religion and ecological awareness would need further research, although in her history of ecological thought in this century, Anna Bramwell, tries to associate, not always convincingly, ecological thinking with the Protestant Ethic.[41] The French philosopher Luc Ferry is surprised that so few theoretical writings on environmental ethics have come out of France and that the contribution of his country to ecological thought is so poor.[42] Despite different philosophical and religious attitudes towards the environment, a strong consensus has been building inside Canada and the Canadian population and government seem to agree on a moderate course of action that is well summarized in the Canadian Green Plan. But whether this consensus — or, more accurately, the commitment of the population — is strong and long lasting, especially in difficult economic times (as implied by Dunlap), is not easy to ascertain. Besides, consensus does not mean unanimity: there is no better place to see difference of values and of interests than in public hearings and environmental assessment process, where radical opinions may be expressed and win support. It is not easy to predict the outcome of public environmental reviews and the pressure these may put on government decisions.

DISCUSSION QUESTIONS

1. In this chapter Guay discusses the larger social and political forces that are shaping Canada's environmental policies. Discuss how two of these forces (e.g. population, urbanization, industrialization) result in a specific type of environmental problem.
2. Conflicts over ever-scarcer natural resources are leading to increased tension between environmentalists, industrialists, trade unionists and others. How can government agencies protect the environment and minimize these conflicts at the same time?
3. Attitudes toward protecting the environment are changing due to an increased public awareness of the consequences of continuing present industrial practices. However, these attitudes, and the extent to which they are held, vary depending on such things as education level and socio-economic status. How can these differences in attitude be fairly and equitably accommodated when implementing environmental policies?
4. The divisive forces that make protecting the environment so difficult within a country probably mean that international co-operation is even

more difficult to achieve. Since many environmental problems, like acid rain, travel between countries, how can international co-operation be maximized so as to better protect the environment?

Endnotes

1. Kates, R.W., B.L. Turner and W.C. Clark. 1990. The Great Transformation. In *The Earth as Transformed by Human Action: Global and Regional Changes in the Biosphere Over the Past 300 Years*, ed. B.L. Turner et al., 1–17. New York: Cambridge University Press.
2. Pomian, K. 1976. Les limites écologiques des civilisations. *Information sur les sciences sociales* 15(1): 7–31.
3. Innis, H.A. 1956. *Essays in Canadian Economic History*. Toronto: University of Toronto Press.
4. Laxer, G. 1989. *Open for Business. The Roots of Foreign Ownership in Canada*. Toronto: Oxford University Press.
5. Bella, L. 1986. The Politics of Preservation: Creating National Parks in Canada, and the United States, England and Wales. *Planning Perspectives* 1: 189–206.
6. Bella, op. cit.
7. Artibise, A.J. and G.A. Stelter. 1981. Conservation Planning and Urban Planning: The Canadian Commission of Conservation in Historical Perspective. In *Planning for Conservation*, ed. R. Kain, 17–36. London: Mansell.
8. Kline, M.B. 1970. *Beyond the Land itself: Views of Nature in Canada and the United States*. Cambridge, Mass.: Harvard University Press. (Kline's analysis is based on English Canadian writings. French Canadian writings are not very different, despite the French Canadian myth of the "coureur des bois" [literally "woods runner"] who represents liberty and freedom from social and institutional constraints.)
9. Gouvernement du Canada. 1991. *L'état de l'environnement au Canada*. Ottawa: Ministère des Approvisionnements et Services. *(The State of Canada's Environment — 1991*).
10. Ibid., 18.1–18.33.
11. Ibid., 18.4.
12. D. Munton and G. Castle. 1992. The Continental Dimension. Canada and the United States. In *Canadian Environmental Policy: Ecosystems, Politics, and Process*, ed. R. Boardman, 203–23. Toronto: Oxford University Press.
13. Nitrates were also responsible for lake eutrophication, but the debate focused on phosphates partly because phosphate release was more concentrated, localized and visible.
14. Ibid., 205.
15. Gouvernement du Canada, op. cit., 19.1–19.24.
16. Macdonald, D. 1991. *The Politics of Pollution. Why Canadians are Failing Their Environment*, 225–74. Toronto: McClelland & Stewart.
17. McMillan, T. 1986. Why Canadians Worry About Acid Rain. *EPA Journal* 12: 8–10.
18. Carrol, J.E. 1983. *Environmental Diplomacy. An Examination and a Prospective of Canadian U.S. Transboundary Environmental Relations*, 239–74. Ann Arbor: The University of Michigan Press.
19. Park, C. 1989. *Acid Rain: Rhetoric and Reality*. London: Routledge/Methuen; J.L. Regens and R.W. Rycroft. 1988. *The Acid Rain Controversy*. Pittsburgh: The University of Pittsburgh Press.
20. Guay, L. and H. Laverdiére. 1991. The Science and Politics of Acid Rain Research: A Canadian Case. Paper presented at the conference, Society for Social Study of Science, 7 November, Boston.
21. Gouvernement du Canada, op. cit., 1.15–1.16.

22. Guay L. and E. Gagnon. 1988. Légitimité professionnelle et reconnaissance sociale : l'exemple des ingénieurs forestiers du Québec. *Sociologie et Sociétés* 20: 141–63.
23. Cullingworth, J.B. 1987. *Urban and Regional Planning in Canada*, 375. New Brunswick, N.J.: Transaction Books.
24. Gouvernement du Canada. 1990. *Le plan vert du Canada*. Ottawa: Ministère des Approvisionnements et Services (*Canada's Green Plan*).
25. Ibid., 3–14.
26. For evaluations of the environmental assessment process, see: Lacoste, R. et al. 1988. *L'évaluation environnementale: une pratique à généraliser, une procédure d'examen à parfaire*. Québec: Gouvernement du Québec; Bureau fédéral d'évaluation et d'examen en matière d'environnement (1987), *L'examen public: ni judiciaire, ni politique. Un forum essentiel pour l'avenir de l'environnement*. Ottawa: Ministère des Approvisionnements et Services.
27. Yearley, S. 1991. *The Green Case: A Sociology of Environmental Issues, Arguments and Politics*, 113–148. London: Harper Collins Academic.
28. Paehlke, R. 1989. *Environmentalism and the Future of Progressive Politics*. New Haven, Conn.: Yale University Press.
29. Sainteny, G. 1991. *Les Verts*. Paris: PUF; W. Rüdig. 1991. Green Party Politics Around the World. *Environment* 33(8): 7–9, 25–31.
30. Viard, J. 1985. Protestante, la nature? In *Protection de la nature. Histoire et idéologie*, ed. A. Cadoret, 161–73. Paris: L'Harmattan.
31. Bakvis, H. and N. Nevitte. 1992. The Greening of the Canadian Electorate. Environmentalism, Ideology, and Partisanship. In *Canadian Environmental Policy: Ecosystems, Politics and Progress*, ed. R. Boardman, 145–48. Toronto: Oxford University Press.
32. Dunlap, R.E. 1991. Public Opinion in the 1980's. Clear Consensus, Ambiguous Commitment. *Environment* 33(8): 10–15, 32–37.
33. The question was: "Protecting the environment is more important than creating jobs," Bakvis and Nevitte, op. cit.: 151.
34. Ibid., note 14: 297. Materialist values were defined by such items as: "strong defence forces," "first rising prices," and "economic growth."
35. Ibid., 163.
36. Gouvernement du Canada. 1991. *L'état de l'environnement au Canada*. loc. cit.: 14.15–14.21.
37. Sanyal, P.N. and J.S. Neves. 1991. The Valdez Principles: Implications for Corporate Social Responsibility. *Journal of Business Ethics* 10: 883–90. The Valdez principles are defined by a series of precepts, standards, targets and sanctions to which corporations adhere voluntarily. The principles cover a wide range of dimensions, such as: "protection of the biosphere; sustainable natural resources use; reduction/disposal of waste; wise use of energy; risk reduction; marketing safe products; damage compensation; disclosure; environmental directors/managers; assessment and annual audit": 885.
38. Bookchin, M. 1982. *Towards an Ecological Society*. Montreal: Black Rose.
39. See Cotgrove, S. 1982. *Catastrophe and Cornucopia: The Environment, Politics and the Future*. Chichester, Eng.: Wiley. Cotgrove explores in depth the differences between the two worldviews. See also, M.E. Olsen et al. 1982. *Viewing the World Ecologically*. Boulder, Colo.: Westview Press.
40. Macdonald, D. 1991. op. cit.: 172–92.
41. Bramwell, A. 1989. *Ecology in the 20th Century. A History*, 13–21. New Haven, Conn.: Yale University Press.
42. Ferry, L. 1992. *Le nouvel ordre écologique*. Paris: Grasset.

THE MAKING OF A POLLUTER: A SOCIAL HISTORY OF UNIROYAL CHEMICAL IN ELMIRA

David Cameron

University of Waterloo

Introduction

Elmira is a small, quiet town of 7,000 residents located about 125 kilometres northwest of Toronto. It's not much different from other towns of its size in Ontario. During the warm weather, children play softball and ride their bicycles. People walk their dogs in the evenings, or sit inside watching television. Some tend to their gardens during the day. For something to do, teenagers drive into the larger neighbouring community, Kitchener–Waterloo, to see a movie. And during the winter months, people play hockey. Hockey is very popular in Elmira, having produced at least one NHL star player. But most of the time, people just eat, sleep and work like most other Canadians. One thing that does distinguish Elmira from other small towns, however, is the fact that nestled comfortably on the east side of town, is Uniroyal Chemical Ltd. — one of the most infamous polluters in the country.

Over the years, the company has produced a variety of chemicals. Many of them are quite useful, like plastic resins, antioxidants, which make tires last longer. Many others, however, have proven to be quite harmful, like the now-banned herbicides DDT, 2,4-D and the deadly Agent Orange. The company also produced Alar, a toxic chemical which just recently was the centre of a huge public outcry in the United States over the safety of sprayed apples. Along with more than five decades of chemical production, has come chemical pollution. And in fact, noxious air emissions, toxic run-off, buried chemical drums and contaminated waste-water effluent have all contributed to the destruction of the local aquifer.

In 1989, a carcinogenic chemical called DMNA was discovered in the town's water supply. The Ontario Ministry of the Environment ordered Uniroyal to immediately discontinue discharging waste-water and clean up its properties. The company appealed the order. What followed was the most complex set of hearings into groundwater contamination and remediation ever held in this country. As a result of these hearings, the true extent of contamination has become known. A toxic soup of cancer-causing chemicals is floating below Elmira. The town's water source has been irrevocably destroyed. In 1991, municipal engineers found another source of water, but Elmirans had been drinking that water for years.

Uniroyal has argued that it cannot be held solely responsible for the extent of the present-day contamination. It has a very valid point. While it might be tempting to place all the blame on the company, some of the chemicals now found underneath Elmira were put there over the years according to legal disposal regulations. In the 1960s, it was common, *and legal*, practice to dispose of toxic chemicals by burying them in metal drums. It's almost as if no one considered the possibility that these drums might split open and their deadly contents escape. The government must bear some responsibility for what has happened. By the end of 1994, close to $50-million will have been spent by all parties attempting to clean up the water supply and Uniroyal properties. Experts argue that sums 10 times that amount would still not cover the costs of restoring the drinking water supply to its original condition. There are also grave questions about what this pollution has done to the health of area residents.

Uniroyal Chemical is now cleaning up its act. It has little choice. The Ontario Ministry of the Environment is now watching it more closely than any other company in the province. Uniroyal is now excavating buried waste on its properties, as well as treating the groundwater at a cost of millions of dollars. But many questions remain. How was the company "allowed" to pollute for so long? Why didn't local residents become upset that they were living next to this chemical factory — or did they? How did Uniroyal come to locate in Elmira? And how, if at all, have attitudes in town changed over the years towards the company?

This chapter answers these questions within a sociological framework. It gathers evidence from a variety of sources, including: government documents; published newspaper accounts; town centennial and reunion souvenir booklets; minutes from public meetings; company documents; and personal interviews with many town residents, local and provincial politicians and company representatives. The information collected has been weaved together to create a "moving picture" of a town, from the time the company began operations more than 50 years ago, to today. The chapter examines certain social factors within Elmira, and explores how these factors have changed over time. It also explores how these changes have affected the town's relationship with the company. People once wrote glorious poems about the company in the local newspaper and celebrated "industry" as a symbol of progress and prosperity. Now, they chain themselves to the plant's front gate, and silently carry protest signs at the home of the company's general manager — all for the television news cameras. The question to be answered is simple: What has happened in Elmira over the past 50 years to account for these changes?

In order to answer this question, this chapter will explore five areas: (1) the history of Uniroyal in Elmira and its reasons for locating there; (2) the social character of the town, then and now; (3) the economic, and therefore political, relationship between the town and Uniroyal; (4) the social (non-economic) relationship between the town and the company — i.e., the company's "place" within the community; and (5) the role that the local newspaper has played in affecting the town's ongoing relationship with the company. As well, this chapter is placed within the larger context of national and international attitudes towards the environment, and other social changes over time.

History of Uniroyal in Elmira

Given Uniroyal's recent history, it is tempting to wonder if the company intentionally located in a small town in order to carry out chemical production without being bothered. There is little evidence to support this. Naugatuck (Uniroyal's original name) came to be in Elmira for two reasons: (1) the parent company already owned a perfectly suitable building; and (2) time constraints required that chemical production begin as soon as possible. That Elmira was a small town may not have been a factor at all, since many other chemical plants were locating in big cities like Toronto and Montreal. Dupont, Dow, Reichold, IKO, Domtar and Union Carbide all have had medium- and large-scale chemical plants in these two cities since the 1950s and before. As well, almost all of the large petroleum companies situated some refineries in or near large cities in the 1950s and 1960s.

In order to fully understand the process which brought Uniroyal to Elmira, we must look back to the beginning of the century. In 1900, the Elmira Felt Co. built a factory in Elmira to manufacture felt footwear. Just prior to the First World War, demand for felt dropped, and the company went out of business. In 1916, the Canadian subsidiary of the United States Rubber Company — the Dominion Rubber Company — purchased the buildings and opened its own subsidiary, called Fleet Foot, which produced tennis shoes for export. Fleet Foot went out of business in 1930. Encountering no possible buyers for the site throughout the Great Depression, Dominion Rubber continued to own the vacant buildings.

When the Liberals were elected to Ottawa in 1940, they introduced a policy of economic revitalization. In light of the rapidly emerging war industry, this spelled great relief for the decade-long depressed economy of the country. Members of Parliament were asked to point out areas needing economic revitalization. Local MP L.O. Breithaupt contacted the Elmira Board of Trade and asked what could be done to revitalize the Elmira economy. He was told that reopening the Fleet Foot plant would help. The federal government began negotiating a deal with the Dominion Rubber Co. (the buildings' owner) and its American parent company. It was agreed that the company would begin production of the much-needed explosives stabilizer — aniline — at the old Fleet Foot factory in Elmira. Since government wartime demand for aniline was immediate, it was decided to begin production in the vacant Elmira factory, rather than delaying production by building a new factory elsewhere. And since the Elmira factory was directly on the rail line, it was just as good a place as any. It was a deal that benefited everyone concerned. While it was not expected that employment levels would reach the more than 500 workers which were once employed at Fleet Foot, Breithaupt and other local politicians were delighted that some jobs would be coming to the area. And more importantly, the Canadian government was delighted that aniline would be produced in Canada. In fact, federal Development Minister C.D. Howe was so delighted, that he attended the opening ceremonies of the plant in 1942.

At the time, Dominion Rubber's parent company, United States Rubber, had recently established Naugatuck Chemicals in Naugatuck, Connecticut, and it decided to locate a branch plant of this subsidiary in Elmira. Technically, Naugatuck Chemicals in Elmira was owned by the Canadian subsidiary, Dominion Rubber. In exchange for Canadian production of aniline, the federal government provided the company with $172,400 for plant extensions and a "recoverable amount" of $182,100. (Although this implies a loan, it is unclear when, or if, it was repaid.) As well, the company was given more than a million dollars' worth of federal contracts to produce chemicals for the war effort. That was an enormous amount of money at the time.

The Elmira Town Council also granted Naugatuck certain tax concessions. When the plant began production in 1941, the town voted to allow a fixed assessment of $5,000 per year, helped in large part by local newspaper advertisements by the town council and board of trade, and various newspaper editorials supporting the vote. The reduced rate was less than one quarter the normal assessment rate. In 1943, Naugatuck gained a further assessment reduction of $3,250 per year.

Chemical production at the plant continued for years, and in 1966, the United States Rubber Company decided to change, what it considered an outdated name (it produced much more than just rubber), to "Uniroyal (66)." Within a year, the "(66)" would be dropped, and the company would be known simply as Uniroyal.

In 1986, Uniroyal's parent company in Connecticut was faced with a hostile take-over bid from another American company. In an effort to avoid the take-over, the company divided into separate corporations, and the name of Uniroyal in Elmira was changed to Uniroyal Chemical Ltd.

There is no evidence to suggest that Uniroyal originally located in Elmira for any concealed reason. It was strictly a business decision which brought the company to Elmira. However, as will be shown, evidence does exist suggesting that later, the company may have used its economic status within Elmira to pressure local politicians into relaxing certain environmental restrictions. It may have also had undue direct influence on local political decisions. As well, it will be shown that there is evidence suggesting that the company took advantage of Elmira's conservative nature, and began secret production of certain dangerous chemicals.

Social and Demographic Nature of Elmira, Then and Now

A quick perusal of the local Elmira newspaper, the *Elmira Signet*, during the 1940s, reveals a small town no different than most others in Ontario. Discussions with residents who have lived there for decades support the view that Elmira was a proud, conservative, white, working-class town. Names like Bauman, Martin, Metzger and Weber dominated town lists. Its heritage was largely German. People who lived in Elmira were usually born there, and people who worked in Elmira, lived there. People were cautious of outsiders. The town was small, about 2,000 in 1941, and everybody knew everyone. Alluding to the size of the town, Graham Dobbin, an executive at Naugatuck, said that when he was originally transferred from head office in Montreal in 1941, he was told that "Elmira was a widening in the road," and not much more.

It had a public library, two hardware stores, two general stores and a couple of five-and-dimes. It had two dentists and two lawyers, one school with seven female teachers, and a small volunteer fire department. In 1946, Elmira's first movie theatre opened.

Like many other small towns in Canada at the time, Elmira was a deeply religious town. By 1948, there were 10 churches, and they were packed every Sunday. Churchgoing was not only a religious experience; it was also a social experience. Sunday church picnics were quite popular during the summer months, and attendance was always high. There were also a number of social clubs and associations, such as the Elmira Dramatic Club, the Tennis Club, the Horticulturalist Society, as well as the men's and women's lodges.

But along with this Norman Rockwell painting came something that did make Elmira different from most other small towns. The town was often filled with bad smells. It was not uncommon for the morning wash to be put out to dry, only to be covered in soot an hour later. The soot was coming from Naugatuck. The newspaper occasionally reported such occurrences, but said that the company was working diligently to prevent them. People tended to just put up with it. After all, many of them worked at the plant.

In the early 1960s, Elmira began expanding significantly. A new housing development, called Birdland (every street in the area carries the name of a bird), began springing up. Originally, it was the town's wealthier residents who moved into these larger, more modern homes, but gradually people from the Kitchener–Waterloo area began to relocate there. City dwellers sought out the peaceful, spacious lots that Birdland had to offer. During the 1970s, even people from Toronto relocated to Elmira, many finding work in the Waterloo area. Elmira (Birdland in particular) was turning into a bedroom community for the larger Kitchener–Waterloo community. From 1960 to 1975, Elmira's population grew by more than 3,000 people to 6,500, doubling in size.

By the late 1960s, the people in Elmira, like most Canadians, were learning about the environmental movement taking shape throughout North America. The local newspaper wrote numerous editorials about the evils of pollution and the need to clean up the environment. On closer inspection, however, it is clear that these editorials were nothing more than "bandwagon" words. Good environmental practice, according to the editor, was to ensure that the local Canacagigue Creek and the Grand River, into which it flowed, had sufficiently high levels during the summer months to ensure flow downstream, washing away the pollution. Unfortunately, the Elmira Town Council, the Grand River Conservation Authority and the Province of Ontario all agreed. Never was it considered what these pollutants would do to Lake Erie once they arrived. And never did it occur to them that it might have been wiser to reduce pollution at its source. Residents of Elmira were not the biggest environmentalists. Many of them were quick to point fingers at the province, writing letters demanding clean-up of the creek, but few were willing to do much more than that. Despite numerous studies which called for reductions in local effluent, local citizens did little to press for this.

Partly due to the influx of "outsiders," and partly due to social changes all over North America, Elmira became less religious during the 1970s. Church attendance was down, as was community involvement in social groups. Television had become the most popular pastime, and people did not get together as much as they used to. People just did not take as much pride in their community as they had in the past. The band shell in the centre of town, once the gathering spot on summer evenings, is now used only sparingly, despite extensive renovations in 1986.

Today, Elmira has a population of over 7,000. Although it has kept pace with social changes through the years, it has retained its small-town conservative feel. While church attendance is down, it is probably higher than elsewhere. There are now 14 churches in town. Neighbours do not know each other as well as they did 40 years ago, probably because most people who live in Elmira work elsewhere. Names like Bauman, Martin, Metzger and Weber, while still plentiful, no longer dominate town lists. Over the past 50 years, Elmira has grown, become less insulated, less proud, more suburban and more modern. Given all of these changes, however, Elmira is still largely a white, conservative, working-class town.

Uniroyal's Economic and Political Relationship with Elmira

While Elmira is hardly a one-industry town, Uniroyal's economic contribution to the community is considerable. In 1990, Uniroyal stated that it contributed more than $30-million to Elmira's economy through salaries and local contracts. Just prior to the discovery of DMNA in 1989, the plant employed more than 400 people. Historically as well, the company has been the economic mainstay of the town.

When Dominion Rubber announced plans to reopen the old shoe factory, the people of Elmira celebrated. The promise of a new industry, bringing prosperity to this once-depressed town, was exciting. The prospect of jobs inspired economic confidence. Real-estate prices bounced back to normal values, having been at 50 per cent during the 1930s. Although national economic recovery was a factor, the guarantee of local economic activity was impressive.

Evidence of the company's economic importance to Elmira came early. In 1942, the first tax-assessment-reduction vote took place. Naugatuck's general manager, Bruce Marr, said that the company easily could have located somewhere else, but chose to locate in Elmira. He argued that what little revenues the town would lose by way of the fixed rate assessment, would be made up many times over in the economic contribution which the plant would make to the community. Local politicians were quite aware of

Naugatuck's economic contribution. One member of the Elmira Board of Trade, fearful that the company might leave Elmira and take its jobs elsewhere if the vote did not pass, said that "the council will go the limit to keep the company here." He further stated that the council was "anxious to do anything within its power to satisfy the company." In fact, the board of trade and town council, together, wrote a public endorsement of the assessment-reduction plan. In it, they stated that the company had increased real-estate values in town and "given Elmira the promise of a permanent industry with a very bright future." The town voted almost unanimously in favour of the fixed assessment, with only a single vote against.

The size of the plant grew throughout the 1940s, '50s and '60s. Countless employment advertisements for the company appeared in local newspapers. By 1953, there were 200 employees. In 1956, the annual payroll topped $1-million when a half-million-dollar expansion to the plant added another 25 permanent employees. Workers moved to Elmira from towns more than 75 kilometres away. The economic spin-offs were tremendous. Local shopkeepers were doing a booming business. Employees at the plant, secure in their jobs, did not rent accommodations: they bought houses. Demand for housing increased real-estate values. And symbolic of economic prosperity and permanence, more than 85 per cent of all Elmira residents owned their own homes.

Naugatuck knew that it had become the economic backbone of the town. In a 1960 newspaper advertisement, the company boasted that employment levels at the plant had risen to 260 employees, with an average hourly rate of $2.00. This employment growth at the plant "not only affected the population of the town, but the million dollars in payroll and taxes has certainly been reflected in the town's economic position," the ad read. In a 1961 editorial, the *Signet* noted that with 280 employees and an annual payroll well over $1-million, "the impact of such an amount of money is undeniable in the economy of the town."

These new jobs meant increased production at the plant. Unfortunately, this was not entirely beneficial to the town. With increased production came increased wastes. During the 1960s, Naugatuck began producing enough waste-water that the local Elmira sewage treatment plant was operating near full capacity. There was little room left at the treatment plant to accommodate any new wastes. Town expansion plans, therefore, had to be put on hold until a new sewage treatment plant could be built. Only after it was constructed did Elmira begin growing. And grow it did: Birdland was built, and the population doubled in less than 15 years. Before long, the sewage treatment plant had again reached near capacity levels, and once again, in 1980, the town stopped expanding. In fact, in 1994, the population (6,978) was actually lower than it was in 1980 (6,982). Uniroyal's proportional use of the sewage treatment plant continued to be quite high throughout the 1980s. Prior to 1989, when Uniroyal was ordered to reduce discharges to the treatment plant because of the contamination, the company was the single largest user, and the treatment plant was almost always operating at full capacity.

Uniroyal's close political ties with the Elmira Town Council and the board of trade were first seen in a joint venture declaring Elmira the first "Weed-Free Town in Canada." In 1946, town officials dreamed up an arrangement with Naugatuck to rid the town of all weeds. As it was reported, "Elmira is the home of the celebrated weed killer 2,4-D," so why not advertise this fact, and clean up the town at the same time. The scheme was fully intended to put Elmira, and Naugatuck, on the map. The Elmira Town Council and the Elmira Board of Trade both wholeheartedly supported the idea. Naugatuck provided the town with the chemical at cost, and residents paid a small fee ($1.50 per year) to have their lawns sprayed. Where residents were unable to spray their own lawns, local high-schools students did it for them. While participation in the programme was voluntary, more than 85 per cent of residents had their lawns sprayed. As for the remaining residents, it didn't matter: there was enough drifting 2,4-D dust nearby, that their lawns got a good dose as well. In 1946, Elmira officially became the first weed-free town in Canada. The town was so proud. Thanks to national coverage in hundreds of newspaper reports, Elmira had become famous all across the country. And it was like advertising gold for Naugatuck. The company could not buy better coverage.

The plan lasted for the next three years but eventually, people lost interest. In 1956, a town councillor tried to revive the idea, but it did not catch on again. Ironically, on January 1, 1970, a ban on DDT, the first big-selling herbicide produced at Naugatuck, came into effect in Canada. (The ban was due in large part to Rachel Carson's milestone environmental exposé, *Silent Spring*.) In the weeks prior to the deadline, Elmira residents were encouraged to drop off all unused DDT at the designated disposal sites. Within a few years, 2,4-D, the same chemical which Elmirans had so proudly sprayed all over their town, was also banned because of its links to cancer.

Up until 1962, the company's political ties with the Elmira Town Council were indirect. In 1962, however, two senior management officials at Naugatuck were elected to the council. On two separate occasions, it became apparent that the council was quite supportive of the company's concerns. It will never be known whether there was "undue" influence from the Uniroyal employees on the council, or just its concern that it did not alienate the town's single largest employer and industry.

The first incident occurred in 1965 when several cattle belonging to a farmer, through whose land ran the Canacagigue Greek, died from drinking creek water. Naugatuck had previously admitted to discharging wastes directly into the creek. A natural instinct, therefore, would be to suspect that chemicals entering the water from Naugatuck may have killed the cattle. These concerns were brought to the town council. One councillor was quick to come to Naugatuck's defence. Although the company admitted allowing chemical waste to enter the creek, the councillor stressed that "in our democratic society, a person is innocent until proven guilty." He went on to say that the company was "continually conducting experiments and tests to prove that their particular waste products do not contaminate the water, and this is to their credit." The "experiments and tests" referred to were conducted by the same people who, years earlier, had "scientifically tested" locally grown tomatoes by eating them alongside tomatoes purposefully injected with DDT! They reported that nothing was wrong with the local tomatoes because they did not taste as bad as the injected ones. Their conclusions totally contradicted what one resident today notes was "common knowledge in town that you just could not grow tomatoes in Elmira." They did not taste right. Many residents, especially those living near the plant, refused to grow them in their home gardens.

The other incident which might call the independence of the town council into question occurred in the late 1960s. The council was well aware of the numerous odour complaints against Uniroyal, but did little more than request that the company, along with most other industries in town, sign on to a plan which would reduce air pollution. The plan, like many local environmental "regulations" at the time, had little real power. Perhaps this anaemic approach was due to the, now three, Uniroyal employees on the town council. The neighbouring *Kitchener–Waterloo Record* suggested some impropriety on the Elmira Town Council, given the high representation of Uniroyal employees. The newspaper called it the "Uniroyal Town Council." One of the three "Uniroyal Councillors" was particularly upset, and he attacked the newspaper's editors, accusing them of intentionally misleading the public into thinking that the company possessed undue influence on local politics. How dare they accuse him of such impropriety! The same councillor, however, several months later, was quick to come to Uniroyal's defence during a discussion on water contamination. He stressed that the company was doing an excellent job treating wastes before they were released, despite another report which concluded that some pollution was entering the creek from Uniroyal. It is almost as if the idea never occurred to this councillor that people might wonder about a possible conflict of interest.

Since the early 1970s, the environment has become a hot political issue. Public opinion polls have consistently shown concern for the environment to be at, or near, the top of a majority of people's sociopolitical agendas. The "jobs versus the environment" debate has become daily conversation for some. And nowhere has this debate become more acute than in Elmira since 1989.

Today, local politicians are caught in the middle. Elmira Mayor Bob Waters, and other members of the town council, are well aware of public opinion on the environment. They are also well aware of local public opinion on the economic importance of the Uniroyal plant. While they express genuine concern about possible harm that Uniroyal is inflicting on the environment, they remain politically conscious of the social and economic consequences that possible closure of the area's single largest employer would have on the community. As Waters notes, "Elmira wants, first and foremost, to clean up the contamination and prevent further contamination." He is not hesitant to add, however, that "keeping Uniroyal operating is in everyone's interest." This is not surprising. Current Uniroyal general manager, Wally Ruck, stresses that the "Uniroyal plant plays an integral role in the life of the Elmira community," just as Naugatuck general manager, Bruce Marr, stressed more than 50 years ago. As already noted, the company injects quite a bit of money into the local economy and employs hundreds. Waters recognizes this and explains that, ultimately, the goal of the Elmira council is to keep Uniroyal operating, but in an environmentally friendly manner.

In 1990, the Elmira Town Council and the Elmira Chamber of Commerce lobbied the Ontario Ministry of the Environment (MOE) to relax the stringent clean-up requirements on Uniroyal, as found in a comprehensive environmental control order. While cautious not to appear to be insensitive to environmental concerns, local officials were reported to have been quite worried about the possibility of the Uniroyal plant closing as a result of the MOE clean-up order. Although the company is not known to have publicly threatened to close the plant, it was suspected by many area residents that communications of this sort had occurred with MOE officials.

A representative of the local citizens' environmental group, APT Environment (Assuring Protection for Tomorrow's Environment), wondered if local politicians were being *too* supportive of the company to the detriment of the environment. She charged that the environmental subcommittee of the Elmira council (the Citizens' Environmental Advisory Committee — CEAC) provided no opposition to a controversial solid-waste incinerator proposed by Uniroyal in mid-1989. She suggested that members of CEAC obligingly sided with the company despite, what she called, "obvious environmental concerns" about the incinerator. The committee was willing to allow the company to construct the incinerator without pressing for an environmental assessment hearing. So she resigned her seat on CEAC and, together with another former CEAC member, and a separate group of citizens simultaneously forming in town in opposition to the incinerator, formed APT Environment. It was APT Environment, in fact, that eventually secured a promise from the Ontario government that an environmental assessment hearing concerning the incinerator would occur. In the end, however, Uniroyal cancelled the controversial incinerator project because of the contamination discovery, and the assessment hearing did not take place.

In 1992, a clean-up agreement was reached between the province and Uniroyal. Local environmentalists are critical of the MOE's choice of clean-up supervisor. He is the same person who chaired CEAC throughout the 1980s, and was often accused of being more concerned with Uniroyal's well-being than with the environment. The MOE counters that he is expertly qualified (he is an environmental-studies professor), and is fully impartial. The environmental group contends that, for certain unknown reasons, the MOE must have had to appoint a pro-Uniroyal person, in order to reach the clean-up agreement with the company. Already there is controversy over clean-up activities. It is charged that the MOE is not requiring the company to excavate buried wastes deep enough. Local environmentalists contend that only a superficial level of contaminated soil is being removed. They argue that considerably more soil must be removed in order to prevent further chemical leachate into the water table.

Uniroyal's Social Relationship with Elmira

There can be no denying that Uniroyal was once a central element of the Elmira community. Its contributions to the town were both economic and social. People respected the company and were thankful to have it in Elmira. The company was a source of community pride.

In the beginning, Naugatuck sponsored a local softball team, and held annual picnics which hundreds of people would always attend. A group of Naugatuck employees also formed a little theatre group, and would entertain town residents at the Regent Theatre. Residents frequently were treated to open houses and tours of the plant. In the 1940s and 1950s, it was not uncommon for 200 to 300 locals to show up for such events. Later, Naugatuck would become the largest sponsor of the social event of the spring: the Elmira Maple Syrup Festival.

People in Elmira took pride in knowing that their own Naugatuck was helping to save lives. Sulpha drugs developed at the plant in the late 1940s were proving to be quite successful in treating pneumonia, bronchitis and other infectious diseases. In January, 1953, Elmira was delighted with one of the very first public appearances of the new "Vibrin plastic car." Made from resins developed and produced at Naugatuck, the car later would help revolutionize the automobile industry by making plastics available to manufacturers.

In the late 1940s, Naugatuck erected its 120-foot-tall smokestack. With six-foot-high shiny, white letters spelling the company's name down each side, it was the tallest structure in town and could be seen for miles. It became the focal point of the town and Elmira was proud of this sign of prosperity.

One woman, who was particularly proud of the company, was Esther Weber. In August, 1956, she sent a poem to the local newspaper with a note that read, "I see in today's *Signet*, a write-up about the Naugatuck. It inspires me to send this poem which I have composed." The paper printed her poem:

"THE NAUGATUCK"

I stand upon a windy hill
A lone star in the sky
Is twinkling with a bright good-will
The river's gliding by.

The laughing river gliding on,
Hurrying to sea or lake,
The outskirts of the busy town
The lonely hill does make.

The river makes a sudden turn
And quickly hurries on.
As myriad lights do brightly shine
A credit to the town

These lights belong to Naugatuck,
That nestles east of town
The new plant brings the town good luck
A plant of some renown.

For many men employment find,
In this plant, new and light
And leaving all their cares behind
The see a future bright.

The Naugatuck supplies the world
With products of great worth,
It is a credit to our town
The dearest spot on earth.

Nothing is known about Esther Weber. It is tempting to assume that either she, or her husband, was an employee of Naugatuck, but it is not out of the question that she was just a regular citizen expressing pride in her town's biggest industry. Reading a poem like this today might almost make a person laugh. But at the time, it would not have been seen as strange to pay homage to a company which had brought such economic good fortune to the community. Esther Weber's poem expresses genuine feelings which she, and undoubtedly others, held about Naugatuck.

As already noted earlier, there were many social and business clubs in Elmira during the 1940s and on. These associations played a large role in the Elmira community. The most influential of these clubs was the Elmira Community Service Club. Among its members were the town's business and social élite. From 1949 to 1951, Bruce Marr, general manager of Naugatuck, was president of the Elmira Community Service Club. In 1955, Marr was also elected president of the Elmira Board of Trade. For a few years during the mid-1950s, Marr was also the volunteer director for the Elmira unit of the Canadian Cancer Society, and is credited with raising thousands of dollars. Mrs. Marr was also quite active in the community. She was president of one of the women's social associations, and was joined by the wives of several other Naugatuck management employees.

The Marrs' role in the Elmira community cannot be underestimated. Both were highly respected, well-liked social leaders in the community. In fact, it can be said that Bruce Marr's business and committee work solidified his position as *the* social and economic leader of the community. When Marr was transferred to Dominion Rubber's head office in Montreal in the early 1960s, the town was deeply saddened. They were not just losing a local business executive — they were losing a leader and a friend. Although the Marrs were gone, the town kept up with them through published reports of Bruce Marr's progress in Montreal. When Marr retired from head office, he and his wife moved back to the area and bought a house in nearby Kitchener. Before his death, they would often be seen visiting friends in Elmira.

Naugatuck knew that it had influence on the social life of the town. Frequent full-page advertisements in local newspapers would publicize its contributions to the town. In one advertisement in November, 1960, the company proudly reported that it had "caused influence to be felt in municipal politics, church, school and benevolent organizations of all kinds." The ad was oddly worded, and probably not one that a company would write today, but it was very accurate.

Given the fact that Elmirans were so proud of Naugatuck, it is a bit surprising that by the mid-1950s most residents could name only two or three of the more than 50 chemicals being produced at the plant. Despite the regular plant tours explaining the company's product line, people just didn't know what was being produced at the plant. (Later on, we discover that the company took advantage of the town's ignorance and began producing chemicals that, surely, people would not have liked in their home town.)

One thing that all Elmirans did know for sure about the chemical plant, was that it smelled — and smelled bad. Throughout the first 25 years of production, noxious smells and dust often drifted over the town. Shortly after the plant started production in 1942, the odours were strong. People reported the smells, but the company always had a good reason, and the people accepted it. Soon, residents just accepted the bad smells as a fact of life (like they accepted the fact that you could not grow tomatoes in Elmira). Perhaps they accepted it as part of the trade-off for having the large chemical company in town, pouring money into the local economy. Perhaps they accepted it because they trusted Bruce Marr. As well, there was an animal processing plant at the north end of town which sometimes gave off worse-smelling odours than Uniroyal. Often, locals just blamed all smells in town on that plant.

One particularly bad incident at Naugatuck occurred in April, 1957, when "fly ash" spewed from its stack for most of one morning. The *Signet* reported that one man woke with his face completely blackened, having rolled on his pillow which had become covered in soot through an open window. Another woman hurriedly placed her dentures in her mouth without noticing that the glass on her window sill, which had been home to the teeth overnight, was thick with black soot. She was in for a big surprise when she wandered into town for the day. Frighteningly, she was a source of amusement for locals. Naugatuck officials said that they "felt very bad" about the fly-ash release, but offered no compensation. At the time, potential health concerns did not even enter most people's minds. The company assured the locals that while the smells were sometimes a bother, they were harmless. In fact, they probably were not harmless, but no one will ever know for sure what chemicals spewed from that stack, and not just on that one particularly bad day, but everyday.

Seven years later, in 1964, the local newspaper editor noted that "when conditions are just right, the wind in the right direction and pollution is high, the community gets a full dose of the high-scented smell across town. The variety and number of complaints is astounding." It was not meant as criticism, just a fact of life. What the editor did not mention, however, was that the plant is at the east end of town, and given the tendency for winds to blow from the west, the noxious emissions probably occurred all the time. It was only on those rare east-wind days that the *whole* town smelled the *strong* odours. Even today, smells are much stronger to the east of the plant than to the west, in town. People near the plant smell it all the time. Fortunately, throughout the 1970s and 1980s, the noxious smells from the plant subsided a little. Tougher provincial environmental regulations obliged Uniroyal to install better filters. The company also erected a taller stack, diffusing the fumes over a wider area.

In 1965, when the cattle died after drinking water containing effluent from the plant, it became apparent to some residents of Elmira that perhaps there was some danger posed by Naugatuck. While it was never proven that it was Naugatuck which killed the cattle, many area residents believed it was. Many more, however, reassured by the town council (which inherently had possible conflict of interest problems), did not give it a second thought.

Throughout the 1960s and 1970s, the company began to take a less prominent role in the town. Marr was gone and no one at the plant had assumed his role as charismatic community leader. Plant tours stopped, and the company picnics were no longer a summer highlight. The company began hiring new employees from outside the Elmira area. With the convenience of personal automobiles, it was not uncommon for workers to live 20 or 30 kilometres away. By the mid-1970s, many Uniroyal employees lived outside of town — contrasted with the almost 100 per cent living in Elmira throughout the early years. Even senior company officials began living outside Elmira. Today, in fact, only one senior executive lives in Elmira. In short, the bonds which Naugatuck had once shared with the community began disappearing by the 1970s.

Public concern for the natural environment was taking hold across North America. People were beginning to see that perhaps Rachel Carson was right, and that we might actually be poisoning ourselves. We began looking at our impact on the planet. The use of phosphates was phased out because of their degradation of water quality. Sulphur emissions were reduced in an attempt to slow production of acid rain. Government programmes gave rebates to people who switched from oil furnaces to efficient, clean-burning gas furnaces. And governments across the continent began running public education campaigns promoting conservation. Helped by the 1973 OPEC oil crisis, and dramatic television coverage of never-before-seen acts of environmentalism, like the Greenpeace's whaling runs, the public was becoming knowledgeable about environmental problems. And public support for the environment increased dramatically.

Public opinion of the chemical industry also changed. No longer was chemical production seen as a glorious symbol of human progress, as was believed in the 1940s and 1950s. People began to question chemical companies. They wanted to know what was being produced, and what effect it would have on the planet and on human health. They also wanted to know about wastes.

Cognizant of these social changes, Uniroyal began behaving differently. Perhaps it was also due to a change in management. Access to their property, while secure in the past, became tight. The public tours of the plant stopped. But the most striking change was the air of secrecy which descended upon the company. The company stopped advertising its product line in the local newspaper. Once proudly hailing the development of one new chemical after another, Uniroyal became less open about what it was doing. During the early years of production, Elmirans knew about DDT and 2,4-D production in their home town, and they actually celebrated it (not to mention, spraying it all over their town). During the 1960s, however, knowledge about plant activities diminished and, in particular, there was one chemical compound being produced about which no one, including most Uniroyal employees, knew. That chemical was Agent Orange.

Agent Orange was the code name given to a herbicide used by the United States military in Vietnam to destroy natural ground cover and expose enemy positions. (It got that name because it was originally shipped in large orange drums.) Today, the name Agent Orange conjures up images of poison and destruction. A myriad of serious health problems have arisen since the chemical was first used. It has been classified as highly carcinogenic, and is now completely banned. American veterans who handled the chemical during the war in Asia are today experiencing serious health effects, including cancer. Many have died. The United States government has offered a compensation package to affected veterans and their families. Exact details of the offer are unknown, but it is estimated that the settlement could be worth well over $1-billion.

This compensation package is not the only thing about Agent Orange clouded in secrecy. Uniroyal officials refuse to discuss any details of Agent Orange production in Elmira. Other than acknowledging the fact that the chemical was produced there, and is now part of the toxic soup underneath the town, company spokespersons will not comment. What is known, is that it was Uniroyal's parent company which was contracted by the United States military to produce the chemical. The decision was made to produce it in Canada. Production was very secretive. According to a former employee, the chemical was loaded onto rail cars in Elmira and transported to Montreal, where it was loaded onto boats bound for Vietnam. It is unknown if the chemical even entered the United States.

Some employees and residents began to suspect that something odd was going on at the plant. Rumours began to circulate that Agent Orange was being produced at the plant. It is impossible to say who knew for sure that the chemical was being produced, or even who heard these rumours. It is unknown whether town officials, including the Mayor of Elmira, knew about it. It was only in 1989, after DMNA was discovered, that Uniroyal admitted to having produced Agent Orange in the past. The MOE found traces of it in the town's water supply.

Whether it was the Agent Orange rumours, continued bad smells, growing concern for the environment, or a combination of factors, the company's reputation in Elmira had become tarnished by the mid- to late-1970s. At the same time, the MOE began introducing policies with some real teeth. Environmental protection was no longer just a phrase: it was action. The province began imposing orders on many chemicals companies, including Uniroyal. The infamous Love Canal situation in New York State was lesson enough for any government official. Government agencies began monitoring closely the actions of companies such as Uniroyal.

While the province began periodic testing at the plant, some residents were still upset about the smells coming from the plant. Others remembered waste-disposal methods at the plant in the early years, and began wondering if it had caused environmental damage. It was not in the nature of this small, still conservative town, to undertake large public protests. Many people in town still worked at the plant and feared losing their jobs if they were seen to be critical of the company. The majority, however, just did not seem to care: if it was not affecting them directly, they did not pay any attention. They could put up with bad smells every once in a while. After all, many people felt as if the plant had always been there. Bad smells and dirty creek water was just something you put up with if you lived in Elmira.

Others, however, did not want to put up with it. They began to complain aloud that they thought Uniroyal was polluting the environment, and that it should stop. Many of these people had recently relocated to Elmira. They were less likely than long-term residents to see the "social history" of the company: to remember the plastic car; to have toured the plant; or to have chatted with Bruce Marr at one of the many Naugatuck picnics. They were more likely to see and smell the pollution, read about the dead cattle and remember being asked to drop off chemicals that had been produced in their own town, at designated disposal sites, because they caused cancer. It was a small, but persistent, percentage of Elmirans who began getting involved. Complaints about industrial problems in town, especially at Uniroyal, became loud enough that the town council decided to create the Citizens' Environmental Advisory Committee (CEAC) in 1982.

The subcommittee of the town council was to provide a forum for local citizens to voice their concerns about industrial pollution at the plant. It would report back to the council. CEAC was the first town-sponsored environmental citizens' committee in the country. Initially, it was thought to be quite effective in raising public concerns to the council. In one case, a request from members of CEAC received endorsement from the town council and the MOE decided to act on it, preventing Uniroyal from installing lagoon waste ponds. (This is a key point in Uniroyal's legal challenge today. It contends that had it not been prevented from installing these lagoons, the extent of present-day contamination would not be so great.) Eventually though, some members of CEAC complained that the Elmira Town Council was not listening to their concerns. They wondered about the effectiveness and impartiality of CEAC's chairperson, who was appointed by the town council. The 1989 controversy over the proposed solid-waste incinerator at Uniroyal was the turning point for two members. They disagreed with CEAC's inaction on the incinerator, and resigned. A concomitant factor in their resignation was the mayor's directive to one of these members that he be allowed to approve anything she wrote about the environment before it was published. She had been asked by the local paper, the *Elmira Independent*, to write a regular environmental column. Noting that her tough environmental stance was too much for the mayor, she refused his request that she be censored, and resigned. This all happened about the time that DMNA was discovered. APT Environment immediately became involved. The environmental group attended the hearings and presented evidence concerning the dangers of the contamination. While the MOE and the company came to respect the legitimacy of APT Environment, many locals in Elmira saw the group as troublemakers.

Elmirans watched on television how APT Environment members chained themselves to the front gates of the plant to attract media attention to their concerns, and protested in front of the Waterloo home of Uniroyal's general manager. Many residents resented the group's activities, and accused them of trying to close down the plant and put them out of work. One evening, several members' houses were vandalized. Others in town simply ignored the environmental group, hoping that it would go away. One member of APT Environment explained that many people in town saw the group as disruptive, and made up of "radical environmentalists," with whom they would not want to be associated. They saw the group as "airing the town's dirty laundry in public," and that just was not the thing to do in this small, conservative town.

The hearings eventually concluded and clean-up began at Uniroyal. While many are not happy with the way that the clean-up is progressing, they are grateful that something is being done. One surprising judgement which came out of the appeal hearings, was the Environmental Appeal Board of Ontario's strongly worded decision that the public had a right to participate in hearings of this sort. Previously, the public had often been shut out of the negotiation process between polluters and government. The board stated that public participation was crucial, and that more should be done to encourage people to participate in environmental matters affecting them. In light of the recent implementation of the Ontario Environmental Bill of Rights, this bodes well for environmental activists wanting to affect change through the courts.

Today, the dust seems to have settled in Elmira. People have gone on with their lives, most having accepted that the Uniroyal plant is a reality in their community, and that they must put up with its problems (like a chlorine gas cloud accidentally released by Uniroyal in March, 1994). However, having the company in town is not without benefit: many Uniroyal workers still live in Elmira, and local contracts contribute significantly to the local economy.

But several member of APT Environment are keeping up the fight, pressing for cleaner operating practices at the plant, and ensuring that clean-up plans proceed on schedule. And today, they have the MOE on their side, monitoring and regulating Uniroyal's activities. While it is not perfect environmental protection, it is considerably better than what Elmira has had in the past.

The Role of the Local Newspapers: the *Elmira Signet* and the *Elmira Independent*

The local newspapers have also played important roles in affecting the town's relationship with Uniroyal over the years. Initially, the *Elmira Signet* was very supportive of, almost subservient to, the company. Often, it would write that it was "indebted" or "honoured" to have been taken on a tour or given an interview at Naugatuck.

At times, you would have thought that the *Signet* was owned by Naugatuck. Indeed, the editor, Mr. Davis, was an honourable man. But for years, the paper not only provided biased coverage in favour of the plant, it also incorporated activities at the plant into its editorial policies. The regular feature, "Weed of the Week," began appearing in 1945, the same year that Naugatuck went into full production of DDT. Not surprisingly, it was "revealed" each week that every one of those pesky weeds could be zapped using DDT. Glowing editorial endorsements of DDT were all too common during the late 1940s, including an incredible story about how DDT had saved the editor's pet dog, "Zero," from a flea problem which was bothering the whole family: "There seems to be one and only one relief [from the dog's flea problem] — DDT. It is also the relief of the rest of the family." (That is until poor Zero probably died from having been sprayed with poison twice a month.) When the pesticide 2,4-D was produced, it replaced DDT as the editorial preferred poison. When controversy surrounding the safety of DDT arose in the late 1950s, culminating with Rachel Carson's *Silent Spring* in 1963, the newspaper continued to side with the local manufacturer. It even published the company's response to Carson, as well as Imperial Oil's response to Carson. In 1961, 20 years after the plant opened, the local newspaper wrote: "It seems needless to point out, that the whole town is proud of its major industry [Naugatuck] that has shown such a growth and success in operations." "What Elmirans can be proud of is the fact that this progressive company is located in their home town," wrote Davis, celebrating the company's anniversary.

In 1970, a new reporter with the *Signet*, Bob Verdun, wrote a piece about Uniroyal striving to meet pollution standards. While the article was not negative, it was tough. And it was only a taste of what was to come.

Four years later, Verdun began his own paper in Elmira, the *Elmira Independent*. Throughout the late 1970s and 1980s, Verdun's editorial policy became increasingly antagonistic towards Uniroyal. It was well known within the community that Verdun had an axe to grind with the company, although his reasons are unknown. Almost every article written about the company in the mid- to late-1980s contained some negative slant. Verdun wrote countless editorials criticizing the company. Some of the editorials had some merit, like the ones which criticized the company's waste-management practices, and occasional chemical spills. Others seemed to be critical just for fun. In one piece, he slammed the company for giving out small glow sticks to local children on Hallowe'en. The sticks were intended to make Elmira's little ghosts and goblins more visible at night, thus increasing their safety. It was a genuine gesture of goodwill on the part of the company. Verdun accused the company, however, of not giving a damn about the environment or human health by distributing dangerous chemicals in plastic tubes to small children. When DMNA was discovered and Uniroyal was accused of being responsible for destroying the town's water supply, Verdun had a field day. He began writing even more scathing editorials attacking the company.

Certainly public opinion in Elmira about Uniroyal would have been affected by such negative editorials. It also would have been affected by the *Signet's* glowing editorials during the first 30 years of the company. The *Independent* lost readers during the early 1990s. People just did not want to read all the bad news that was being written about their town. While many people today do not subscribe to the local paper, they still tend to hear about what is in it. The paper has recently changed its editorial policy concerning Uniroyal: it doesn't say anything about the company. Seldom is Uniroyal even mentioned in the paper today. Perhaps this is Verdun's way of attracting back the readers he has lost.

Discussion and Conclusions

The question asked at the beginning of this chapter was: "What has happened in Elmira over the past 50 years to account for these changes?" It is tempting to answer this question fully within a political-economy framework. Here we see Uniroyal, a corporation, which for years enjoyed a comfortable relationship with a small town. The company was supported by the town council, and was respected by the community as well. Then public attitudes changed during the 1960s and 1970s: private citizens became empowered, and began to interfere with the long-established state-capital relationship. Both the state and capital resisted this interference, but in the end, the public prevailed and capital was obliged to change its ways. The state-capital relationship was modified to incorporate new concerns for the environment, and the state began regulating the capital.

While this is not an entirely inaccurate way of examining the Elmira case, it is simplistic. What it lacks is a more well-rounded, less-structural, environmental-sociological approach. This chapter has attempted to provide such an approach by incorporating other factors into the analysis.

It has been shown that there exists no single factor which explains the changes in Elmira over the past 50 years. Rather, there are a number which have, separately and in combination, produced the observed changes. The most important of these factors is probably the change, since the late 1960s, in general public attitudes towards the environment. It has stirred public involvement all across North America, and obliged governments to change the ways they view the environment.

These attitude changes have also affected change in the government-industry relationship. Attention is now paid to industry's impact on the environment, and industrial activity has become highly regulated. Along with new public attitudes towards the environment, have come new attitudes towards industry in general, and the chemical industry in particular. Having once viewed chemical production as progressive, people are now more likely to see it as dangerous. Clearly Elmira has not remained unaffected by these attitudinal changes throughout North America. There we see increased support for environmental protection and more, and louder, voices making this known.

In addition to these general changes, there have been important changes within Elmira itself. Uniroyal's evolving role within the community is undoubtedly the most important of these. Once the social and economic heart of the town, Uniroyal slowly began losing importance. There are a number of reasons for this. The social and economic bonds with the town began to weaken over time. The Elmira Town Council, once "anxious to do anything within its power to satisfy the company," later became very concerned about chemical production at Uniroyal, and established an advisory committee to watch over the company. By the 1970s and 1980s there were no more Elmira-Uniroyal joint ventures like the mid-1940s campaign to declare Elmira the "First Weed-Free Town in Canada."

Over time, Elmira's pride in the company also faded. When the charismatic Bruce Marr left town, people lost a highly respected community leader and a friend. His departure severely affected each Elmiran's personal identification with the company. Over time, Uniroyal also became more secretive about its activities. Plant tours stopped, product-line advertisement ceased, and secret production of dangerous chemicals began. Jaded by increasing reports of pollution at the plant, and reminded of the various chemicals once produced at the plant and now being banned, the company's reputation took a beating. The company began hiring workers who lived outside Elmira. As the town grew, the percentage of Elmira residents, who actually worked at the plant, decreased dramatically. Simply put, the company's ties with the town were weakening.

The town itself also changed. Younger, less conservative people began moving to Elmira. They did not work in town, and did not feel any ties to Uniroyal. They were also more outspoken. Empowered by the wave of environmentalism sweeping through North America, many of these newcomers began to question activity at Uniroyal. They attracted the support of many long-term residents as well: people who had become fed up with decades of bad smells and bad tomatoes. In 1989, when it was discovered that the town's water supply had been poisoned, and that Uniroyal was probably responsible for it, more residents became involved. Discovery of DMNA was a catalyst for further citizen involvement.

The local newspapers have also been factors in this changing relationship. While the old newspaper glorified the company, the new newspaper attacked it. Public opinion of the company was affected. Where Elmirans once wrote poems about the company, they now wrote protest letters.

While the Elmira story is about a chemical company that got caught polluting, it is also about much more. It is about a town which once took pride in its largest industry, and how it came to lose that pride. It is about a company which saw its bonds with the people and government of a small town disappear over time. It is about changing attitudes about the environment all across North America and in this small town. And most importantly, it is about a few people who, despite continued, albeit waning, economic dependence on a company, wanted a cleaner environment, and pressed for it.

DISCUSSION QUESTIONS

1. Discuss social attitudes towards the environment and how they have changed over the years. What factors have influenced these changes?
2. Identify an important company in your local area and discuss its social and economic impacts on the community. What kind of direct and indirect impact does the company have on the environment?
3. Who do you feel should be responsible for cleaning up the Elmira aquifer and why?
4. The Elmira case raises real concerns about the links between business and government. What do you think the appropriate relationship should be? Consider issues like scientific research and development, company bankruptcy and closure, and environmental protection.
5. Should the government shut down corporate polluters? Consider environmental, legal, economic and social factors in your answer.

Selected Bibliography

APT Environment materials, 1989 to 1994.

Elmira Citizens' Environmental Advisory Committee (CEAC), meeting minutes and reports, 1982 to 1994.

Elmira Independent. 1986. Elmira Centennial, especially Ron Howey, "Wartime Chemical Needs Brought the Vacant Fleet Foot Factory Back into Productive Use."

Elmira Public Library. 1948. Souvenir of the Elmira Old Boys' and Girls' Reunion.

Elmira Signet and *Elmira Independent* newspaper clippings, 1940 to 1994, on file in the Elmira Public Library, as compiled by long-time resident Esther Thur.

Kirkby, Ian. 1990. Haunted by the Past. *Exchange* 7(6).

Meyers, Anita. 1990. Survey of Elmira/St. Jacobs Residents, re: Perceived Impact of Water Contamination. Paper prepared for the Regional Medical Officer of Health — Waterloo, November 26.

Numerous Ministry of the Environment Documents, 1982 to 1994.

And many personal interviews with Elmira residents and company and government spokespeople.

ORGANIZATIONAL ANALYSIS AND ENVIRONMENTAL SOCIOLOGY: THE CASE OF GREENPEACE CANADA

Éric Ouellet

York University

INTRODUCTION

This chapter proposes to illustrate how environmental sociology and organizational analysis can enrich each other through their concepts and empirical findings. In order to provide such an illustration, a case-study of Greenpeace Canada (and particularly its Quebec chapter) is provided. The central analytical assumption of this case-study is based on looking at Greenpeace as a formal organization, rather than as a social movement.

Greenpeace, as one of the world's largest environmental organizations, has many of the characteristics of a modern bureaucracy (Courrier international 1991: 3–6; Plog 1991). Also, since Greenpeace is an organization that promotes environmental consciousness, it is likely to have an unorthodox organizational structure. In other words, often an activist culture produces unusual and innovative organizations, referred to by Mintzberg (1990) as "missionary." This potential mix of traditional and new organizational structures makes Greenpeace an interesting case. Additionally, using the perspective of environmental sociology, it is worthwhile to examine how natural or physical environments can influence the organizational structure of a major player in the environmental movement.

On a theoretical level, a study of Greenpeace has valuable inductive appeal. Organizational analysis can benefit by integrating the influence of the natural environment with an organizational setting. This has been neglected by organizational theorists (Shrivastava and Hart 1994), even while they continue to borrow many of the ideas and concepts from the biological sciences.

In order to evaluate the bureaucratic nature of Greenpeace's organizational structure, a comparative framework is proposed. It is based on the Weberian ideal-type of bureaucracy in relation to the ideal-type of new collective functioning proposed by Tixier (1983). A brief theoretical discussion about environmental/biological concepts in organizational analysis is also provided. The objective is to highlight how environmental sociology can inform organizational analysis.

THEORETICAL DISCUSSION

Bureaucracy and Its Ideal-type

Choosing an analytical framework for a case-study is a risky task; the information gathered can be too theory driven. However, without such a framework, one may risk getting lost in data. Ultimately, one must select relevant information, or at least rank them by importance.

In order to deal with the necessary bureaucratic structure of Greenpeace, it is important to examine how previous studies can help in selecting an appropriate framework. Unfortunately, there are no in-depth analyses of Greenpeace's internal organizational structure. Among the few studies, the closest thing to an organizational analysis is by Eyerman and Jamison (1989). This study focuses on Greenpeace's public strategies, and to a lesser extent, on its internal organizational tactics. Greenpeace is compared to an "organizational weapon," a concept developed by Philip Selznick (1952) to study Bolshevik revolutionary cells. According to Ouellet (1993), however, there is little evidence that Greenpeace Canada can be considered as an organizational weapon.

The organizational-weapon model is not the most appropriate framework. A more general, and less sophisticated, model should be employed. Therefore, it would be useful to compare Greenpeace's organizational structure with the well-known Weberian ideal-type of bureaucracy. This is based on: (1) jurisdictional areas governing activities according to formal rules; (2) office hierarchy; (3) management based on written documents; (4) a higher degree of specialization; (5) full working capacity demanded from the employees; and (6) general rules stressing the importance of technical expertise (Weber 1968, 2: 957–58).

As expected, the organizational structure of Greenpeace is also innovative and distinct from a classical bureaucracy. The ideal-type of bureaucracy developed by Weber would be adapted if it is contrasted with an alternative ideal-type. One of the most interesting models, simple without being simplistic, and ontologically coherent within the Weberian ideal-type, is the one proposed by Pierre-Éric Tixier (1983).

Tixier states that we are living in a world that is different in many respects to Weber's. Our world has been referred to as postindustrial, post-modern and the information society. If this perspective is right, then a new ideal-type of collective functioning is needed. This new ideal-type, to be conceptually coherent and allowing for comparison, must use Weber's meaningful understanding method (*verstehen*). Accordingly, Tixier (1983: 25–27) suggested that new organizations are structured based on the following principles:

1. authority is delegated by the group, and involves some consensus rather than a legalistic definition of roles;
2. collective functioning is characterized by low importance given to formal rules;
3. tendency toward a reduction in division of labour and specialization, leading to polyvalency;
4. behaviour of social actors is marked by very affective relationships;
5. recruitment and promotion are based on the quality of participation, rather than on seniority or a formal evaluation of knowledge;
6. the decision-making process rests on collective decision-making structures, usually involving direct participation.

These ideal-typical principles can be subjected to a critical analysis, and integrated with other publications to provide a more synthetic model. However, because this case-study is essentially exploratory, the exercise of producing a more complete organizational model would not be worthwhile, yet remains, nonetheless, useful for comparative purposes.

Organizational Analysis and the Natural Environment

Organizational analysis, like sociology, has been influenced by biological and ecological metaphors (Morgan 1986). Organizations, like societies, were analysed as functional organisms, having needs that must be fulfilled to maintain an equilibrium within the system. This perspective has been clearly influenced by the ideas of Durkheim, Parsons and Merton, and their respective followers, often called structural-functionalists. Also, the influence of general systems theory (Bertalanffy 1950) is quite obvious. The works of organizational theorists like Argyris (1964) and Emery (1969) are good examples of such influence.

Other theorists proposed that organizations, like organisms, are part of a larger environment, and to survive, must adapt to the fluctuations in their environment. However, this conception of environment and of adaptation is essentially an economic one, placing an emphasis on elements like the type of industry, technology, clientele, etc. Books written by Burns and Stalker (1961), and Lawrence and Lorsch (1967), are famous examples of this type of analysis.

More recently, a new type of ecological analysis has been proposed: It is usually called "population ecology." Organizations are divided into different "species" according to their type of activity, technology, etc. These different "species" are interdependent, competing as well as co-operating, and have different capacities to survive modifications in the economic environment. From this perspective, it is possible to see "species" "disappearing" or "evolving" into new "species." Again, this conception of organizations, although using biological/ecological ideas, remains fundamentally based on economic considerations. The best-known theorists using these approaches are Aldrich (1979) and Hannan and Freeman (1977).

The production of organizational ideas based on the ecological metaphor occurred in the 1960s and 1970s. These ideas are not "out of fashion," but rather have been integrated, and are now an integral part of most new theoretical developments in organizational analysis. These new developments have focused on the concept of organizational culture in 1980s, and tend to be oriented towards organizational strategy in the 1990s. (However, there is also a growing interest in postmodern/deconstructionist perspectives.)

From this brief history of the use of biological and ecological concepts in organizational analysis, one important question must be raised: What is the place given to the natural or physical environment? In fact, one must know that most organizational theorists are writing from a business perspective, in order to find new ideas to improve the economic condition of business organizations, and reduce the costs of operating government organizations.

Interest in the physical environment has been limited to what is often called "environmentally friendly industry" and "Green marketing." In recent years, management and marketing journals published many articles related to the above topics (for example: Westley and Vredenburg 1991; McDougall 1993; Adrian and Dupre 1994; Jones, Green, and Coombs 1994; Peattie and Ringler 1994; Zimmer et al. 1994). Most of these articles, nevertheless, suggest that business organizations should include more "environmental considerations" in how they organize themselves, in order to maintain their competitive edge.

There are, however, studies which are more critical: for instance, research on the danger to workers related to asbestos exposure (Reasons et al. 1981), and health problems related to industrial organizations in the developing world (Navarro and Berman 1983). More recently, Pasquero (1991) studied interorganizational co-operation for addressing some environmental problems; Swinth (1991) analysed organizational strategies related to hazardous waste production; and Wolfe and Howes (1993) discussed the linkage between environmental performance and Total Quality Management.

Using elements from these studies, it is possible to propose that an organization must adopt a set of co-ordinated activity within the natural/physical environment. Therefore, the environmental contingency is also "environmental" (in its natural/physical sense), even if it is mediated by the social context.

Before drawing any conclusion about how to integrate environmental sociology with organizational analysis, it is important to highlight some theoretical elements of the former. A major contribution of environmental sociology has been to try to address the neglect many social scientists have given to interaction between the social realm and the physical (or natural) environment. This neglect has been criticized, particularly by social scientists like Catton and Dunlap (1978), and Schnaiberg (1980).

Study of environmental movements has been one of the major subfields of environmental sociology. Sills (1975), in an article entitled "The Environmental Movement and Its Critics," has provided a starting point for the study of environmental movements, even though there are many earlier studies on this subject (for example: Albrecht 1972, Mandel 1972, Morrison 1973). Sills analysed several dimensions of environmental movements, but points out that the movements are élitist, in their membership and proposals for solving the environmental crisis. The movements are constituted mainly of upper middle-class activists, the decision-making process is in the hand of a few, and their environmental solutions put the weight of sacrifice on the most underprivileged, often the low-paid workers of polluting industries.

The debate about élitisms in environmental movements continued over the years. For example, Gale (1983) compared the Left and the environmental movement; Scaminaci and Dunlap (1986) studied the antinuclear movement; and Morrison and Dunlap (1986) provided a more general analysis of environmental movements. It is still an unresolved issue, that is less important now due to the increasing variety of environmental groups. This debate, even if less important now, is an old one, well-known by political scientists and organizational analysts. It can be traced back to the Michels' (1949) study on the German socialist party, from which the well-known "iron law of oligarchy" concept is derived.

Élitism, however, cannot always explain choices related to organizational strategies. An élite group, or one considered as such (Marutollo 1990), can be organized in a strictly hierarchical manner, like the U.S. Marines Corps, or in a highly decentralized way, like scientific networks (Latour 1987). Therefore, without trying to resolve the debate, it is possible to conclude that there are many forms of élitism, and these forms are often influenced by the natural/physical environment.

Greenpeace, being an international organization, deals with a large range of environmental issues, that are diverse, both qualitatively and geographically. These issues may be divided into two main categories: (a) a continuous flow of environmental destruction, like the use of CFCs, air pollution caused by automobiles and the industrial processes, and waste management problems; and (b) catastrophes, such as the Exxon Valdez oil spill, the Chernobyl nuclear accident or the St. Basil-le-Grand fire.

A relationship with the natural environment that is sporadic, such as responding to catastrophes, involves concentrated and rapid mitigative action, either to solve the problem, or to alert the public. Such relationships, like most military organizations in times of war, involve bureaucratic forms of command and a strict hierarchical chain of command, to be able to produce fast and concentrated action and reaction. It also implies less formalism, higher technical expertise, and low specialization to assure, again, quick response. This represents an organizational response to uncertainty.

On the other hand, continuous degradation of the natural environment can allow for a more decentralized decision-making process, but implies more formalism and specialization due to the routine nature of the work. Technical expertise is less important because of this routine. This represents an organizational adaptation to a more stable and predictable context.

Having this not-too-formal exploratory framework in mind, as well as the comparative ideal-types, it is possible to explore Greenpeace's organizational structure, and present the case-study.

THE CASE

Methodology

This illustrative and exploratory case-study is designed as an embedded single case (Yin 1989: 50). Two units of analysis are used — namely, Greenpeace as a national and international organization on a general level, and the Central Canada regional offices on a local level. The emphasis of this study is put, however, on the local level of analysis. Data used for the general level, due to its dispersed nature, are mainly found in documentary analysis, and complemented by interviews. Information about the local level has been found essentially in interviews and some participant observations, to get deeper insights of Greenpeace's organizational life.

The limits used to define this case-study are grounded on the formal organizational structure of the general level, as designed by Greenpeace officials. The criteria to circumscribe the local level are, however, more sociological. The local offices have been considered as a social group, where membership is defined as full-time (35 hours per week or more) participation to Greenpeace Canada activities. All interviewees were paid employees of Greenpeace Canada. In other words, membership here is conceived of as more than simply having a membership card and paying annual fees.

Data come from: (a) written documents from inside and outside the organization intended for both members and non-members; (b) semi-directed interviews dealing with personal history, views and opinions about the environment and organizational life; (c) participant observation, including the physical setting of buildings and interior decoration, and discussion occurring outside interviews.

The actual gathering of data was preceded by establishing contacts with various Greenpeace officials during the winter of 1992. It is important to mention that some officials were suspicious, but establishing contact had been greatly helped by the personal recommendation of a major Quebec ecologist. Interviews and participant observation were conducted in Quebec City, Montreal and the head office in Toronto. A total of 13 semi-directed interviews were conducted, nine in French and four in English, each three hours in length on average.

Results and Discussion

Data related to the general level will be introduced first, and then those dealing with the local level. To ease description of the data, it is suggested to refer to Figure 1.

FIGURE 1

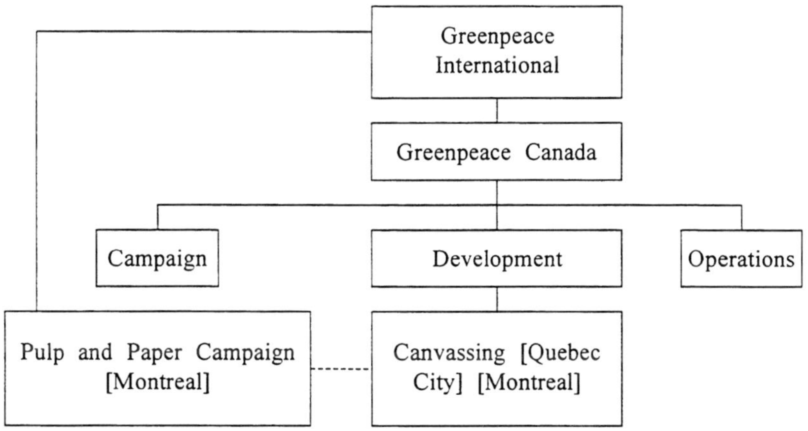

Greenpeace originated in Canada, but nowadays, Greenpeace Canada is only one of many national chapters. The Canadian section is under the direction of Greenpeace International, with its offices in Amsterdam, Netherlands.

Greenpeace Canada is formally managed by a board of seven voluntary directors. This board is elected, and its electoral population is divided into three groups, each having the same number of votes. The first group is formed by Greenpeace Canada employees, the second by Greenpeace International representatives, and the last, by members randomly selected. Candidates are selected by a joint committee composed of employees of Greenpeace Canada and Greenpeace International. Many of the members of the board of directors are prestigious scientists or community leaders.

It is clear that the ordinary member does not appear to have real access to the decision-making process. This board of directors formally gives directives to an executive director, but according to many interviewees, this power of the board is only virtual, and has limited effect on the organization's destiny. Therefore, the board should be seen as a means for enhancing the legitimacy of the organization, especially if it is constituted of prestigious individuals. Nevertheless, this structure corresponds to the jurisdictional nature of bureaucracies as described in the Weberian ideal-type.

At the time of this study, the executive director position was not filled and an executive committee was in place to manage the organization. But as soon as a new director is nominated, it will be dissolved. This committee included directors of the three functional divisions working at the Toronto head office, and all campaigners. The chairperson was the director of the campaign department. The task of this committee was to co-ordinate all actions of Greenpeace in Canada. It is important to mention that interviewees were very satisfied with this committee's decision-making process, which is more collectivist and closer to their activist culture. This constitutes a significant difference from the traditional bureaucratic model.

The three functional divisions of Greenpeace Canada are the as follows: (1) Campaigns, which co-ordinates the activities of campaigners in Canada; (2) Development, which has an objective of promoting Greenpeace's public visibility and managing its fund-raising; and (3) Operations, which takes care of all personnel and administrative matters. This division in functional tasks is typical of a bureaucratic organizational setting.

However, there is a breach in the chain of command. Campaigners are formally under the supervision of both the national director of campaigns and the international co-ordinator of campaigns. All interviewees stated that campaigners are autonomous agents who manage their respective campaigns. Two major constraints, which are widely accepted, are the non-violent character of protest action, and ensuring that they are co-ordinated with other campaigns at both national and international levels.

The co-ordinating power at the national and international levels is reinforced by the funding of campaigns. The campaigners are accountable for the use of their resources according to the level from where the funds originate. All money from canvassing, and fund-raising, is sent to Toronto, and then part of this money is sent to Greenpeace International, which redistributes these funds to various campaigns in the world, according to priority. The national headquarters does the same for national campaigns. It is important to mention that at the time of this study, Greenpeace Canada was not able to cover its operating costs, and was subsidized by Greenpeace International. The financial position of the Canadian constituent was precarious. In a sense, this predicament reinforces the situation in which employees of a bureaucracy are non-owners of the means of production.

Greenpeace also experienced communication problems. First, because Canada is a large country, and travel funds are scarce, it is difficult to organize meetings with all constituent leaders. There are phone forums, but many interviewees mentioned that this form of communication is in many ways ineffective for dealing with "real" problems. This is because telecommunication technologies create more difficulties for affective relationships, that are so important in the new collective forms of functioning.

Furthermore, many interviewees complained that at times, the media knew, before they did, about some of the decisions made in Toronto or Amsterdam. Another example is that they often receive only the English version of the member's bulletin in Quebec City.

This last example demonstrates how Greenpeace, like many national and bureaucratic organizations in Canada, experiences a linguistic division. An interviewee stated:

> Co-ordination is done at the national level [from Toronto], but we have a hard time to get on with the Anglophones. There is a problem of culture.... There is a language problem; therefore we have to co-ordinate by ourselves [the Francophones]. We don't have many contact with the Anglophones. They have their own manner and we have our own.

More practically, many employees of Greenpeace in Toronto are not fluent in French, but a limited number of Quebec-based members are fluent in English. Therefore, telephone and written communication suffers due to this linguistic barrier. Additionally, it is possible to show how cultural differences aggravate these communications problems. For instance, during the time of this study, there was a great deal of criticism about the next stage of Hydro-Quebec's project — the dam construction in James Bay (see Tessier this volume). The Toronto office produced a few press releases which were highly

FIGURE 2

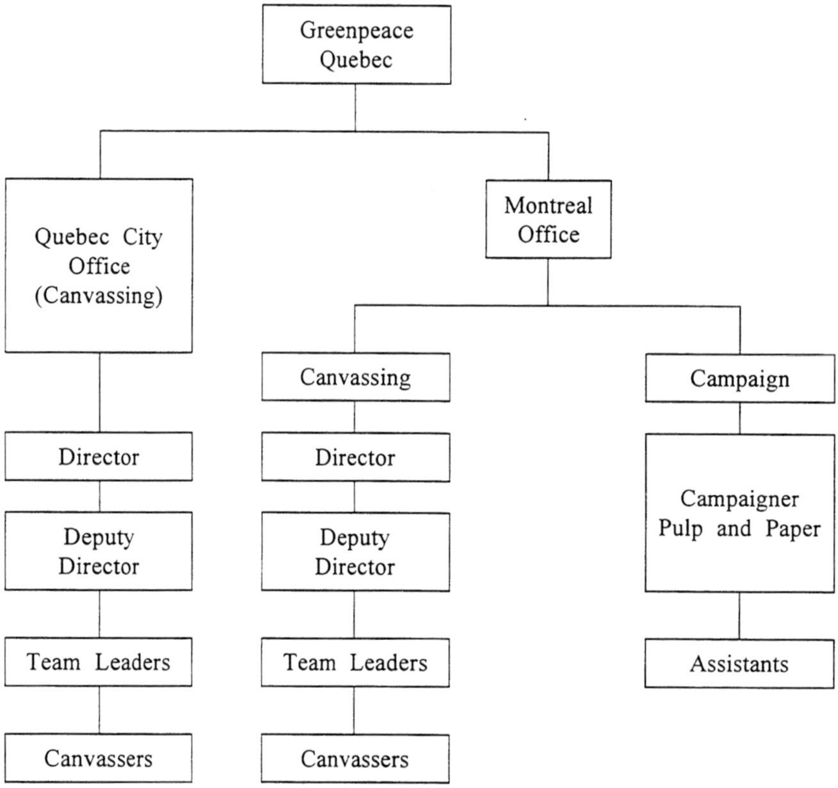

critical of Hydro-Quebec. This was well received in the English-speaking part of Canada, but was understood as an anti-Francophone statement by French-speaking Quebecers. Greenpeace was, at the time, experiencing a major backlash from the French-speaking media. Members of the Montreal office, understanding the Quebec cultural context, were well aware that such statements from Toronto would lead to upheaval, and were able to suggest a different approach, much less harmful to Greenpeace's public image.

It is important to show here how the communication problems, and their geographical dimensions, can be considered environmental issues. However, the linguistic/cultural dimension of the problem is not. This is a good example of how organizations are inexorably linked to a sociocultural context, and how such a context should not be underestimated.

The communication problems experienced by Greenpeace do not help make its various constituents financially autonomous. All offices in Canada have their rents and various invoices paid directly by the Toronto headquarters. They are entirely dependent on Toronto in matters of finance. The only discretionary resources for the local constituents are petty-cash funds, which are often less than $100. These financial controls, that are centralized in Toronto, are similar to those of classical bureaucratic organizations.

At the time of this study, the Quebec division was comprised of two offices: one in Quebec City, and the other in Montreal. Quebec's section has its own board of directors, but has not met for almost two years. This leads us to believe that the board has a very limited governing role. To get a better idea of the details of Quebec's Greenpeace constituents it can be useful to refer to the Figure 2.

The office in Quebec City is devoted essentially to canvassing — selling membership cards and receiving donations. One interviewee was very explicit in saying: "The small Quebec City office is for recruiting members, to get the cash." Due to the economic nature of these activities, the Quebec City office is mainly under the supervision of the development department. Another function is listening to complaints from members, especially problems related to the delivery of the members' bulletin.

The Montreal office is a canvassing one as well, but is also the regional headquarters for Eastern Canada. Nevertheless, this regional supervision task is not clearly defined. The regional director plays many roles: "there is an Eastern Canada director, but it is really a catch-all job," an interviewee said. These various jobs include public relations with the media, co-ordination of various campaigns in Eastern Canada, as well as being a communications link with the national office in Toronto. However, the regional director works essentially in the Province of Quebec.

One campaigner, at the time of this study, was based in Montreal, working on the pulp and paper campaign. This location was justified, since most of the pulp and paper industry in Canada is based in Quebec. The number of campaign activities and the personnel involved in each, varies accordingly. At the time of this study, the financial situation of Greenpeace was precarious, and concentration of resources was necessary.

The role of the campaigner is to co-ordinate all Greenpeace activities relevant to a specific issue. This includes contacting and dealing with the media, participating in public hearings and organizing demonstrations and non-violent civil-disobedience activities.

Campaigners do not work in isolation. In the Montreal office, for instance, the campaigners meet on a daily basis with canvassers. The canvassing director said that even if it is on an informal basis, all major decisions about the pulp and paper campaign are made in consultation with members of the canvassing section. This situation is also complemented by a very limited number of formal rules. The decision-making process is often informal for the campaign. An interviewee confirmed this by saying: “There are many decisions made in the corridors and we must react rapidly, as, for instance, when an article against Greenpeace is written in the newspapers. There is a certain spontaneity in the decision making.” This shows a clear tendency toward pluralistic decision making found in the new collective type of functioning.

Formally, the canvassing services in Montreal and Quebec City have four hierarchical positions: director, assistant director, team leader and canvasser. The functions of both the director and the assistant director are not distinct. They train and hire canvassers, keep account of the money received, perform office duty during the day, and go out with canvassers during the night (canvassing is usually done after dinner in order to reach people at home). Also, with team leaders and canvassers, both participate in the selection of neighbourhoods to be canvassed each night. Canvassers have the sole official function of convincing people to buy a membership card, and if possible, to give extra donations to Greenpeace. However, one can see that there is a low degree of specialization and of division of labour in the local constituents.

The selection of canvassing areas is the time for mutual exchanges of ideas on various topics. For instance, during this study, all members of the Montreal canvassing section were considering whether to canvass a French-speaking, socio-economically underprivileged part of Montreal. The decision was made collectively, knowing, based on their previous experience, that these types of neighbourhoods were more difficult to canvass than more privileged ones, while insisting that this fact was not a good enough reason to exclude them. They also exchanged ideas about the type of environmental issues and approaches suited for these neighbourhoods, especially in relation to the negative image of Greenpeace, due to its critiques of Hydro-Quebec. It is important to note that the director or the assistant usually propose discussion topics. This constitutes, nevertheless, a good example of authority delegated by the group based on consensus.

Campaigners, and in particular the one in charge of the pulp and paper campaign, have assistants. Their number varies according to the needs, and also to the financial capacity of Greenpeace. Campaigners can also direct those working in the canvassing section — but on a voluntary and unpaid basis — in demonstrations and non-violent acts of civil disobedience. At the time of this study, the Montreal section was organizing a demonstration. Most of the canvassers were happy, if not excited, to participate in this demonstration. The canvassing activities for the day of the demonstration were less extensive, to allow for participation in the demonstration.

Expertise in Greenpeace is rather limited. With the exception of two interviewees — one was a union leader, and the other has a B.Sc. in environmental studies — there were no canvassers that had work experience or training related to the environment or social activism. There were no specific career patterns. Most of the interviewees, wished to have a job, and to participate to the environmental movement. The campaigners, however, were usually hired according to their technical expertise about the environment, and public relations experience. The campaigner responsible for the pulp and paper campaign was a professional biologist.

The most important function of Greenpeace canvassers is to convince the public, therefore, this skill is the main criterion in hiring new personnel. However, it is not enough to simply follow orders. Assessment is based on the interviewees' very broad spectrum of interest and their perception of the environmental problems or crises. An interviewee gave a colourful and illustrative description of the situation:

> I agree with some principles. But to apply them [strictly] in your personal life ... we are not a sect. There are vegetarians [among us]; it is a personal business. We still must have the [environmental] discourse and basic beliefs, a long-term perspective. There are things that are confronting if you are not in this thought system. If you come here with a big Cadillac ... this is the non-spoken [rules].

Most of the formal organizational rules are related to canvassing, especially, about the minimum amount of money a canvasser must collect, on an average basis, after their training period which lasts about six weeks. This amount of money is calculated on the basis of salary; they must bring in at least a little more than their salary. If they do not attain that minimum level, they are asked to leave. However, according to the interviewees, there have been very few cases of firing. Extra time is often given to achieve the minimum level, or the canvasser may leave on their own accord, realizing their deficiencies as fund-raisers. Otherwise, there is much informality in the internal policies. As an interviewee described it: "It is very cool here, I don't have the feeling that I am in the army. If you're late, they will tell you so by making a joke, but if it is all the time ... there are limits. The people here are responsible, we know what we have to do."

This last quote must be understood in the larger context of conviviality and friendship, that has been observed in all offices. Many interviewees responded in the following manner: "We are a gang of chums, after work we go together for a beer." On many occasions, it has been observed that canvassers arrive for work one or even two hours before the required time. They use that time for socializing and discussing informally various news about the environment. The normative incentives (Zald and Jacobs 1978) are important, even if the economic incentives still have a large place in that organization. This illustrates the importance of the affective character of the collective function in Greenpeace's local constituents.

From this short study, it is possible to uncover clear tendencies within Greenpeace's organizational structure. First, the national level is closer to the ideal-type of a classical bureaucracy. This can be explained by the necessity for Greenpeace to convey clear and unique messages. The management of financial resources is typical of a bureaucratic organization, which can be justified by the need to create economies of scale. These elements are not from the natural environment, and in that sense, Greenpeace is not innovative. The pressure of the social/economical context of Canada cannot be totally avoided.

Second, the campaigns are between a bureaucratic and a new collective functioning type. These particular types of arrangements can be derived by both social and natural influences. Campaigners have to produce messages about major catastrophes or critical public issues' moments about a specific environmental problem (like the James Bay hydroelectric development). There is a necessity to have a unity of command in order to be socially credible, but the destruction of the natural milieu, or its threat, cannot be controlled, and demands flexibility. This flexibility is, however, based on a non-economical rationality. Greenpeace employees cannot make a fortune in working for their organization, as well as canvassers, who give freely of their time to help a campaign. Affective involvement, for fellow colleagues and for the natural environment, creates a more innovative organizational structure.

Finally, the canvassing section, in doing the daily work of raising environmental consciousness as well as raising funds, produces the most innovative structure. As presented in this case-study, Greenpeace Canada is the closest example to the new ideal-type of collective functioning. This can be partly attributed to fact that they have to deal with a continuous flow of environmental issues on a daily basis. It is more than working on a specific project that needs tight co-ordination. It is a never-ending task, driven by a sensibility toward the natural environment, that leads to organizational forms which are not based on the same foundation as Weber described in his ideal-type.

CONCLUSION

This presentation of Greenpeace Canada's organizational structure is obviously partial and limited. Many other elements and factors could have been taken into account, like the relationships with other environmental organizations, potential gender biases in the organization, etc.

This modest case-study can still show how the natural environment can influence organizational settings, in the case of an organization that is not working in the natural resources exploitation or transformation. Greenpeace succeeded in developing structures which can accommodate various types of environmental destruction. Organizational theorists, and organizations like government environmental agencies, or the nuclear industry, could learn much from Greenpeace.

Furthermore, with the globalization of environmental changes (Dunlap and Catton 1994), the natural environment will have a more deterministic influence. Therefore, organizational analysis, as well as other fields of research, have much to gain in attempting to build bridges with environmental sociology.

DISCUSSION QUESTIONS

1. Discuss how organizational analysts conceive of the concept of environment.
2. Using an environmental organization with which you are familiar, discuss the assumption that the environmental movement tends to be élitist.
3. Based on what you know about Greenpeace, discuss the assumption that bureaucratic functioning cannot be totally avoided within large organizations.
4. Based on an organization you know, discuss how the natural/physical environment can influence its internal structuring.

References

Adrian, M. and K. Dupre. 1994. The Environmental Movement: A Status Report and Implications for Pricing. *SAM Advance Management Journal* 59: 35–40.

Albrecht, S.L. 1972. Environmental Social Movements and Countermovements: An Overview and an Illustration. *Journal of Voluntary Action Research* 1(4): 2–11.

Aldrich, H. 1979. *Organizations and Environments*. Englewood Cliffs, N.J.: Prentice-Hall.

Argyris, C. 1964. *Integrating the Individual and the Organization*. New York: John Wiley.

Bertalanffy, L. 1950. The Theory of Open Systems in Physics and Biology. *Science* 3: 23–29.

Burns, T. and G.M. Stalker. 1961. *The Management of Innovation*. London: Tavistock.

Buttel, F.H. 1987. New Directions in Environmental Sociology. *American Review of Sociology* 13: 465–88.

Catton, W.R. and R.E. Dunlap. 1978. Environmental Sociology: A New Paradigm. *The American Sociologist* 13: 41–49.

Courrier international. 1991. Les dessous de Greenpeace, une grande enquête du *Spiegel*. *Courrier international* 26 Sept.–2 Oct.: 3–6.

Dunlap R.E. and W.R. Catton. 1994. Struggling with Human Exemptionalism: The Rise, Decline and Revitalization of Environmental Sociology. *American Sociologists* 25: 5–29.

Emery, F.E. 1969. *Systems Thinking*. Harmondsworth, Eng.: Penguin.

Eyerman, R. and A. Jamison. 1989. Environmental Knowledge as an Organizational Weapon: The Case of Greenpeace. *Social Science Information* 28: 99–119.

Gale, R.P. 1983. The Environmental Movement and the Left: Antagonists or Allies? *Sociological Inquiry* 53: 179–99.

Hannan, M.T. and J.H. Freeman. 1977. The Population Ecology of Organizations. *American Journal of Sociology* 2: 929–64.

Jones, O., K. Green and R. Coombs. 1994. Technology Management: Developing a Critical Perspective. *International Journal of Technology Management* 9: 156–71.

Latour, B. 1987. *Science in Action.* Cambridge, Mass.: Harvard University Press.

Lawrence, P.R. and J.W. Lorsch. 1967. *Organization and Environment.* Cambridge, Mass.: Harvard Graduate School of Business.

Mandel, W.M. 1972. The Soviet Ecology Movement. *Science & Society* 36: 385–416.

Marutollo, F. 1990. *Organizational Behavior in the Marine Corps: Three Interpretations.* New York: Praeger.

McCarthy, J.D. and M.N Zald. 1973. *The Trend of Social Movements in America: Professionalization and Resource Mobilization.* Morristown, N.J.: General Learning Press.

McDougall, G.H.G. 1993. The Green Movement in Canada: Implications for Marketing Strategy. *Journal of International Consumer Marketing* 5: 69–87.

Michels, R. 1949. *Political Parties.* New York: Free Press.

Mintzberg, H. 1989. *Mintzberg on Management: Inside Our Strange World of Organizations.* New York: Free Press.

Mitchell, R.C. 1979. Silent Spring/Solid Majorities. *Public Opinion* 2: 16–20, 50.

Mohai, P. 1985. Public Concern and Elite Involvement in Environmental-Conservation Issues. *Social Science Quarterly* 66: 820–38.

Morgan, G. 1986. *Images of Organization.* London: Sage.

Morrison, D.E. 1973. The Environmental Movement: Conflict Dynamics. *Journal of Voluntary Action Research* 2(2): 74–85.

Morrison, D.E. and R.E. Dunlap. 1986. Environmentalism and Elitism: A Conceptual and Empirical Analysis. *Environmental Management* 10: 581–89.

Navarro, V. and D.M. Berman. 1983. *Health and Work Under Capitalism.* New York: Baywood.

Ouellet, É. 1993. Étude de cas sur les structures de gouverne d'une organisation écologiste. Master's thesis, Université Laval, Quebec City.

Pasquero, J. 1991. Supraorganizational Collaboration: The Canadian Environmental Experiment. *Journal of Applied Behavioral Science* 27: 38–64.

Peattie, K. and A. Ringler. 1994. Management and the Environment in the UK and Germany: A Comparison. *European Management Journal* 12: 216–25.

Plog, K. 1991. Greenpeace n'est pas une association de randonneurs. *Courrier international* 26 Sept.–2 Oct.: 6.

Reasons, C.E. et al. 1991. *Assault on the Worker.* Toronto: Butterworths.

Scaminaci, J. and R.E. Dunlap. 1986. No Nukes! A Comparison of Participants in Two National Antinuclear Demonstrations. *Sociological Inquiry* 56: 272–82.

Schnaiberg, A. 1980. *The Environment: From Surplus to Scarcity.* New York: Oxford University Press.

Selznick, P. 1952. *The Organizational Weapon: A Study of Bolshevik Strategy and Tactics.* New York: McGraw-Hill.

Shrivastava, P. and S. Hart. 1994. Greening Organizations — 2000. *International Journal of Public Administration* 17: 607–35.

Sills, D.L. 1975. The Environmental Movement and Its Critics. *Human Ecology* 3: 1–41.

Slaton, C.D. 1989. The Politics of Symbolic Consistency: Being Green While Organizing Green. Paper presented at the 1989 Annual Meeting of the American Political Science Association, 31 Aug.–3 Sept., Atlanta, Georgia.

Swinth, R. 1991. The Organization of Hazardous Waste Production in the U.S.: Social and Ecological Performance. *Human Relations* 44(2): 147–74.

Tixier, P.-E. 1983. Démocratie directe et organisation: Pour une théorie du fonctionnement collectif. *L'Année sociologique* 33: 21–36.

Weber, Max. 1968. *Economy and Society*. 2 vols. Berkeley: University of California Press.

Westley, F. and H. Vredenburg. 1991. Strategic Bridging: The Collaboration Between Environmentalists and Business in the Marketing of Green Products. *Journal of Applied Behavioral Science* 27: 65–90.

Wolfe, A. and H.A. Howes. 1993. Measuring Environmental Performance: Theory and Practice at Ontario Hydro. *Total Quality Environmental Management* 2: 355–66.

Yin, R.K. 1989. *Case Study Research: Design and Methods*. Newbury Park, Calif.: Sage.

Zald, M. N. and D. Jacobs. 1978. Compliance/Incentive Classification of Organizations: Underlying Dimensions. *Administration and Society* 23: 403–24.

Zimmer, M.R. et al. 1994. Green Issues: Dimensions of Environmental Concern. *Journal of Business Research* 30: 63–74.

ENVIRONMENTAL HARMONY AND HUMAN PERCEPTION OF NATURE ECOSYSTEM MANAGEMENT TOOLS FOR LAND-USE PLANNING

Paul Harpley Rob Milne

Wilderness Places Ltd., Pefferlaw, Ontario

INTRODUCTION

Modern land-use planning requires integration of the skills and perspectives of many specialists. The biophysical environment, the natural features of a place, are generally studied by natural scientists — ecologists, geologists, physical geographers and others. Human social concerns are the interest of sociologists, human geographers, psychologists and cultural anthropologists who investigate our world from a humanistic perspective.

Often natural scientists and human or social researchers work independently of each other within the same landscapes, studying places and issues that are interrelated and interdependent. Recent planning experience has shown the viewpoint to achieve rigorous, broadly based explanation, and good planning results. As in Southern Ontario, the skilful balancing of humanistic and natural considerations is vital to future land-use planning. The Toronto Waterfront (Crombie 1990), and the Oak Ridges Moraine study (1994), are current examples of new approaches. The interdisciplinary nature of the field of environmental research has not developed a theoretical base. It has integrated concepts and approaches from several disciplines, especially those of geography, psychology, sociology and anthropology (Mitchell 1989). This coming together of several disciplines, so necessary for progress in this area, has perhaps also contributed to the lack of theoretical development. Concepts from different disciplines have made conceptualization of problems difficult, and this has been further hampered by difficulty in operationalizing terms (Schiff 1971; Mitchell 1989).

Social-environmental-perception research is a valuable tool that has been used by many social scientists for some time to sample societal knowledge, perceptions, values and attitudes for a wide range of issues. Original, rigorous and extensive application of techniques is not commonly applied to typical public or private land-use planning studies, even though academic studies and survey design models are readily available. Traditional cultural/economic studies for large government projects focus on resident meetings and group workshops with identified stakeholders.

The present relationship between humans and nature in Western thought, the perception of current environments and their future possibilities, must be confidently known and understood. Accurate information and understanding of individual and societal perception of environment will lead, in time, to better information for decision making, to make accurate and desirable decisions of environmental harmony, and to direct future land-use planning with confidence.

Increasing interest in environmental issues and resource management has led to a proliferation of environmental assessment studies in Ontario, particularly from a biophysical perspective. In general, humanistic research into environmental perception's particular societal publics, has not been extensive except with the exception of socio-economic impact assessments of large government projects. A problem hindering the comparability and verification of finding (in some studies) has been the wide range of definitions of the terms "perceptions" and "attitudes" (Mitchell 1989).

ENVIRONMENTAL PERCEPTION

Environmental perception is a complex multivariate field of investigation. As every human being (and social group) is continually surrounded by environments, they can be experienced from many viewpoints and perspectives. Environmental perception is more like an exploration than a passive endeavour. It involves motoric experience through an active physical interchange with the environment, and is succinctly defined as involving the process of apprehending, through sensory input, the physical environment that is immediately present (Holahan 1982).

Human perception originates from sensation, notion or intuition — an action rooted in human cognition. Environmental cognition concerns the storage, organization, reconstruction and recall of images or environmental features that are not immediately present. Environmental attitudes are the favourable and unfavourable feelings that people have towards features of the physical environment (Holahan 1982). Attitudes are perhaps even harder to define than perceptions. General usage of perceptions implies an individual's feelings towards an object, and beliefs about an object attitude describes. Schiff (1971) defines this as "an organized set of feelings and beliefs which will influence an individual's behaviour." Together, environmental perception and attitudes form a set of expectations about the environment and our attitudes towards it (Holahan 1982).

The major problem in studying environmental perception is that of measurement, since people often have difficulty articulating the conscious or unconscious feeling, attitudes and ideas associated with perception (Saarinen 1969). Since an individual's perception is governed by a combination of past experiences and present outlook conditioned by values, moods, social circumstances and expectations, two people viewing the same stimulus may "see" different images (Mitchell 1989).

SOCIAL PERCEPTION

The neurological and physical (psychological) behavioural aspects of human-perception theory have been well studied, but are idiosyncratic to individuals, and problematic to a broad holistic understanding of societal perception and behaviour. What is more properly sought is an exploration of the way human perception is linked to social and organizational worlds — a sociological approach. Of great importance to resource management is how these relationships can influence human perception of environment, and ultimately through bureaucratic and political action, affect land-use planning and design decisions in communities.

Social perception is concerned with the impression one has of a social stimulus or set of stimuli, as that impression is modified by the perceiver's past experience, his previous experience with that same or similar stimuli, and the individual's state at the moment he is viewing the stimulus of interest. When discussing perception of the environment, it is this type of perception that is of interest (Schiff 1971). In this chapter, perception will therefore refer to social perception, the observable attitudes, values and notions of a survey cross-section of a community.

The positivistic viewpoint of scientific research is considered essential by many to the study of the objective world — the physical and biological attributes of a place. Yet human perception acts in a subjective way to influence how we as individuals view our personal and collective experience. The study of individual values and attitudes towards the objective world — the geographic places around us — is not inherently subject to the same research controls and methodologies of the natural scientist. Hence, it would not be difficult to expect a significant dysfunction between the results of typical biophysical-environmental studies and humanistic-societal research understandings of the same places. It is a tension between the natural and human sciences, the difficulty in reconciling the proper relationship between each, and the vast specialization of knowledge and viewpoint in the modern world that can make land-use planning consensus in contemporary 20th century places so difficult to achieve in Western societies.

Understanding of broad societal commonalities of knowledge, attitudes and values are difficult to ascertain. It is particularly problematic when notions of human perceptions, especially by politicians and bureaucrat planners, are rarely rigorously tested and measured. Often by default, from communication breakdown, time constraint, internal government priority pressures and so on, decisions of the political process remain the only alternative to rigorous study and design of considered land-use change.

ENVIRONMENTAL SOCIOLOGY

The field of environmental sociology represents a significant shift in traditional sociology, stressing social facts in influencing societal change. Environmental sociology involves a recognition of the fact that physical environments can influence (and in turn be influenced by) human societies and behaviour (Dunlap and Catton 1979).

It has become apparent that the great local and global assaults on our world through the activities of humans can only really be mitigated, brought to a halt, and the Earth restored and nurtured by radical social-environmental change. Research in environmental sociology has already yielded information about attitudes of environmentalism, variations in these attitudes between groups, if and how these feelings are translated into behaviour (Krause 1993).

The understanding of the relationship between humans and nature in Western thought is critical to the knowledge of how societal perceptions and attitudes towards landscape and land-use change have evolved to the present. It should be noted that in the past, these understandings were complex and confused. They require a special investigation of themselves.

HUMANS AND NATURE

The Western relationship between humans and nature, while confused, is nevertheless well known. There are many illustrations of that separation between humans and nature which occurs so frequently in Western thought, and conversely of the union of the two as parts of live and indivisible wholes (Glacken 1967). Today, even with radical environmental and economic change and uncertainty reflected in societal change and shifts in paradigms, clarity in understanding Western human/nature relationships and cosmology eludes us. Indeed, in spite of many historical and contemporary attempts to render nature and humans as one, none have been successful. Of all creatures, humans alone can carry out changes deliberately and self-consciously. Humans have control over their world that no other creature possesses, and with this, they have, to a very great extent, made provision to shape their future (Ittelson et al. 1974). The changes to human societies through industrial and technological advances, are well known to us all. The rise of the importance of nature to humans has occurred in Western thought in three previous periods of constructive cosmological thinking, when nature has become the focus of thought. Nature has become the subject of intense and protracted reflection (Hegel [1821] 1970; Collingwood 1957), perhaps resulting from an emerging realization among many today that local places in the global biosphere are teetering on the verge of environmental disaster. Recent human attention on the Earth and familiar local landscapes is reflecting the emergence of a holistic-societal outlook in Western democracies like Canada. Implicit in this current social phenomena is a desire for positive perceptions of natural and wilderness landscapes (Harpley 1992).

Although geographers, sociologists and others have undertaken research of environmental perception for many years, it is clear that with the global biosphere under attack everywhere from human activities, original research of environmental perception, and Western conceptions of nature, to understand fully reasons for our assault on nature, is of utmost importance. Our understanding of human relationships to nature in the West, in the future, must not continue to be confused as it has so often been in the past. Indeed, as Larsen has said, research is necessary into all the cultural processes which include perceptions of, attitudes to, and human actions with, nature. This difficult task is to understand how microbes and aesthetics, bacteria and folk-culture, art and rain forests, fantasy and atomic decay fit together. Only one thing is certain: the boundary between culture and nature is not dependent on only one kind of explanation or one overlapping factor (Larsen 1992). If scientific inquiry in the last few decades has yielded volumes of information about the nature and operation of almost every little corner of the objective Earth, it is nothing in comparison to all that we do not know about the wide range of human society, culture and their relation to human perception of nature.

WHAT IS NATURE? WHAT IS NATURE'S RELATIONSHIP TO CULTURE?

Albert Schweitzer gave us insight into a comprehensive holistic view of nature in his philosophy of "Reverence for Life." The understanding of how we (humans) are linked to all living things was described in metaphorical terms: "As the wave in the ocean surges forward together with all waves, so must we feel in our life, the life that is around us, with its privatations and its anguish." In the world of living creatures, he rejects none. All animals, too, are included in his concept of friendship; he feels a deep responsibility towards them, as toward all living things (Anderson 1965).

Like Goethe in the 18th century, Von Humboldt came to see all nature as a whole, and "... all branches of science were involved in his exposition, and in his section (in Cosmos) on the history of scientific discovery, he also linked the sciences to the arts and the humanities" (Botting 1973) those central aspects of human society what we call culture.

Culture takes a bite out of nature when we inquire and slowly understand how it works, tiny piece by tiny piece, through scientific inquiry. Concrete knowledge of the character of nature is woven together with taboos, rituals, fantasies and symbols, through which elementary experiences of death and how to overcome it are integrated into the culture. Nature becomes culture by having some functions ascribed to it, since by nationalism, for example, nature has been turned into specific landscapes (Larsen 1992). In Canada, symbols and icons for nature in landscape are a central part of our history. The myth of the North, the pristine wilderness of the high Arctic barren lands, the Canadian symbolic landscape paintings of Tom Thomson and the Group of Seven (Mellen 1976) are representative of historical and contemporary examples of culture inventing nature.

Our culture's conception of nature is more like a series of concepts or meanings, which are present in very different types of text and artefacts of culture — in literature, landscapes, design, political speeches, scientific dissertations, architecture and so on. There are many levels of nature conception, but the most prominent is probably the conception that nature is the physical surroundings of air and water, flora and fauna, micro-organisms and molecules. That is what we turn into landscapes, what we copy in parks, what we keep in zoological gardens, what we use in town planning as light, air, cleanliness and green areas. And it is what we paint and write poetry about, and take charter holidays to (Larsen 1992).

Nature commonly takes on different objective and observable forms for different people. Often nature is recognized in local objects, individual animals, plant and rocky prominences. On the other hand, it is perceived as generalized phenomena, such as the sky, moon, earth, water, light and darkness. Indeed, Western conceptions of nature have varied widely through time, demonstrating a Christian and human-centred bias (Harpley 1992).

Politically expedient decisions are still common in much routine resource management and land-use planning. Where possible, private-sector developers and various levels of governments seek daily to expedite initiatives within the constraints of regulation and due process to meet various political and administrative agendas, social whims and attitudes of the day. The characteristics of these societal influences can change readily over time, affecting decisions and human perception dynamically over time and space. For example, in Canada and other countries, current interest and review of approaches to resource management, land-use planning, and in particular the notion of sustainability in Canada and in much of the rest of the world, is important to an understanding of future human perception and its relationship to land-use change (Mitchell 1989; Sewell et al. 1993; Crombie 1990).

SUSTAINABILITY

The concept of sustainability, a present-day response to a worsening global environmental condition, is a recent human construct devised to manage resources and to attempt to conserve ecosystems. The goal of contemporary approaches should be a methodology for ensuring an ecological framework that rigorously includes humanistic perspectives beyond basic economic and impact considerations of traditional assessment approaches.

The current international environmental experiment in sustainability, through United Nations direction in "Our Common Future" (1987) holds great promise, but equally great challenges and formidable obstacles harmonizes increasing human populations and aspirations with nature's inherent limitations, into an indefinite and uncertain future.

As some contemporary commentators point out, sustainable development suggests a phrase full of insight and promise. However, in common usage, its meaning has been altered and is now synonymous with "economic growth," which, in turn, is frequently equated with progress. In Canada, when Prime Minister Mulroney's long-promised Green Plan finally appeared in late 1990, it lacked an ecological framework, and had neither the budget nor political clout to influence big government departments. By the time of the Brazil Earth Summit in 1992, the momentum of the environment movement had faltered as governments were preoccupied with economics (Suzuki 1994).

Large-scale human paradigms, constructs and notions reflected in societal attitudes and values, ultimately articulated in political action, can be altered over space and in time. Contemporary research in environmental sociology should consider them. For example, the importance the concept of sustainability may have to human perception of nature, is clearly dependent on social, cultural, political and economic considerations. Indeed, many important resources or commodities (water, petroleum and so on) have to some degree or another influenced the direction or rate of change in society dependent on a notion consistent with the current "sustainability." Future perspectives must better understand the concept, detail how society views it, and seek to influence further knowledge and thought to achieve a more compatible future relationship between humans and nature.

RESOURCE MANAGEMENT AND LAND-USE PLANNING

A cross-disciplinary approach to land-use planning of the development of ecosystem management tools, and strategies for the administration of watersheds, is desirable. The Lake Simcoe watershed in central-southern Ontario, a land/water ecosystem that is under increasing environmental stress, is a rural, man-influenced, natural biophysical system that is becoming rapidly urbanized. It is recognized that the development of a conceptual land-use planning model must consider the importance that human perception, reflected in human attitudes and values, plays in affecting decision making in social, cultural, political and economic spheres, impacts on land-use of human change to landscapes. It is proposed that a model to clarify present and future resource management relationships must consider important natural and human-modified biophysical processes, dependent on societal epistemological assumptions, attitudes and values. These humanistic criteria are all in need of further study and consideration, since they comprise the foundations of our understanding of collective, subjective places, quite separate from objective, biophysical characteristics of those same places, but nevertheless of at least equal importance.

Indeed, resources themselves are an expression of appraisal and represent an entirely subjective concept (Mitchell 1989). It is the capability of the environment, or parts of it, to satisfy human needs, that makes a resource. Consequently, a resource is essentially a human construct, and hence belies the importance of understanding and utilizing humanistic study and research in land-use planning. In this sense, resources are not static, but expand and contract in response to human wants and human actions; they are subjective, relative and functional (Mitchell 1989; Zimmerman [1933] 1951). Even inherent natural resources like forest, wilderness areas and wildlife may or may not be considered important resources reflected in societal values and perception.

The roles of resource analyst, manger or developer are distinct but interrelated. In traditional resource analysis, approaches to land-use change, various perspectives emerge: biophysical, economic, social, political, legal, institutional and technological. Add to this temporal and spatial components, and the familiar matrix is created which emphasizes the importance of achieving in-depth knowledge about specific perspectives, time periods and spatial geographic scales, as well as an appreciation of their interaction (Mitchell 1989; Krueger and Mitchell 1977).

Perspectives are particularly complex, and because there are humanistic concerns and processes other than the biophysical component, the importance of understandings and analysis of human environmental perceptions is critical to resource-management study. Consequently, study of the human component and integration with the biophysical perspective will become increasingly important in the future, particularly in rural watersheds rapidly urbanizing with accelerating social change.

Nash (1973) has argued that studies of conservation history deserve more attention. He suggested that analysis of the history of resource management allowed opportunity to investigate such fundamental issues as the tension between individual freedom and social purposes; the relationship between the expert and the people; the conflict between utilitarian and aesthetic interests; and the advantages and limitations of alternative custodianship strategies for the environment (Mitchell 1989).

The central understanding of the relationship between perception and human actions in their environment has been suggested by many. Tuan (1976) has called for a "humanistic" geography whose purpose would be "... an understanding of the human world by studying people's relations with nature, their geographic behaviour, as well as their feelings and ideas in regard to space and place" (Mitchell 1989). It is from this perspective that this chapter is centred, and that environmental perception research in the Lake Simcoe watershed has been undertaken in recent years.

The harmonious comparison, evaluation and integration of biophysical and humanistic considerations into solutions of successful designs in landscape planning is the great challenge of future landscape philosophy and land-use planning methodology. In this regard, the moulding of good land-use change is best undertaken with rigorous knowledge of human perception specific to local geographic areas.

LANDSCAPE APPRAISAL AND THE CONCEPT OF HARMONY

Tuan's notion of humanistic-centred investigation of the human world by studying people's relations with nature, is a convenient point of departure from which to search for new means of decision making on landscape design. Environmental, economic, political and social measures of criteria considered in land-use-planning design decisions are typical. Ways of studying and measuring them vary. In general, however, methods of landscape appraisal fall into a number of categories:

1. Intuitive assessment, entirely subjective by the appraiser
2. Database surveys of landscape features, to aid judgement by the appraiser
3. Subjective scoring of landscape components by the general public
4. Other public preference techniques, such as landscape ranking and contingent valuation techniques (CVTs) (Willis and Garrod 1993)

Varying success has been achieved utilizing all of these approaches. Research methods comprise extension of traditional attitude and preference techniques employed by psychologists and behavioural scientists (Willis and Garrod 1993).

It must be pointed out that empirical science operates as a positive — that is to say positing — science by setting up (or positing) a framework of meaning within which the world is seen in a particular way, for example, as physical, biological or cultura. The methodological approach to all science study, whether by physicists, biologists, anthropologists and others, involves procedure entailing, to a greater or lesser extent, the distancing of the observer from the observed (Pickles 1988).

Today, our modern Western world faces a paradoxical situation. At the height of our technological mastery, we often find ourselves separated from both the Earth and our own human selves. After many centuries of building our world, we meet an unsettling nexus of domination and homelessness. Confronting the powerful assumption that what matters is construction in space with our planet as raw material, we find an emerging need to take seriously what is ignored or discarded, not merely technological construction, but dwelling, not merely homogenous and mathematical space, but place, not merely planetary raw material, but environment. Phenomenology — immersion of the self grounded in wonder, a compassionate intelligence that seeks to see things in themselves — offers a way of thinking rigorously and describing accurately the complex relation between person and world (Seamon and Mugerauer 1985).

Unfortunately, almost all contemporary planning methodology and practice, in, for example, Southern Ontario, ignores this approach, continuing in the path of past positivistic traditions of resource management.

It is proposed that in recent decades in Western democracies, with burgeoning populations, economic freedom and mobility, elevated human expectations, economic convenience, and political considerations have dominated much decision making. Aesthetic contemplation with regard to land-use planning has not been paramount. As a consequence, a more equitable balance between human and nature considerations among expedience, economics and the artistic, must be found in the future. A comprehensive methodology beyond the Science-Beauty-Estimation (SBE) technique, developed by Daniel and Boster (1989), should be established.

Indeed, who is responsible for the aesthetic in contemporary planning practice and methodology: the architect, the planner or anyone of many specialist consultants? Hardly. Yet designers and architects have long known that the form and appearance of a building influence certain behaviour that takes place within it. Historians of architecture have noted that man's dwelling places in every land reflect the sociocultural values of his time and place, expressing needs that go beyond those of physical shelter and comfort, to include psychological needs of identity, creativeness and harmony with the world (Ittelsonet al. 1974).

In recent years, there has been a tendency, in environmental psychology, to focus investigation on the relationship between individual and sociophysical environment on typically "molar" units of analysis, in particular that of "place" (Bonnes et al. 1991), an old geographical concept.

The identity of a place is based on its physical attributes and the mental images of its inhabitants' "sense of place," a concept that takes into account both the mental and physical aspects of place perception and interaction (Jones 1992).

Phenomenologists provide some guidelines for bringing philosophy into the arena of actual pragmatic study of situations. They focus on the investigation of the preconsciously given aspects of behaviour and perception, residing in the "lifeworld" culturally defined spatio-temporal setting or horizon of everyday life (Buttimer 1976).

Harmony is defined as harmonious, concordant, forming a consistent or agreeable whole, free from dissent (Webster 1975). To harmonize, bring into, be in harmony (with), make, be agreeable in artistic effect (Fowler and Fowler 1951), is a verbal description. Harmony is an ideal to be sought after in landscape planning and design, that consistently eludes contemporary efforts of emulation. Indeed, of the many intellectual currents in this long and confused century, perhaps the most profound and far-reaching is the shift in the order of knowledge away from a presumption of objectivity. Challenges have come from a variety of sources — some political, some ideological, some religious, and most interesting of all, some scientific. In place of objectivity, now relegated to specialized questions of narrow scope and achieving only partial answers, research has faced the unrelenting influence of the human factor. The inescapable presence of the knower in the known has forced us to recognize ourselves mirrored in what we see (Porteous 1990).

Olwig (1992) argues that the concepts of land and nature, defined in counter distinction to that of the city, are vital to understanding the ideological importance of the concepts of landscape and nature in artistic and scientific discourse.

The value of understanding these distinctions in an urbanizing, predominantly rural landscape, like Lake Simcoe watershed, are essential to future landscape planning there. What is needed is a new approach, a broadly comprehensive viewpoint to planning of landscapes, including their maintenance, conservation and rejuvenation where necessary, but that comes forth from an artistic, aesthetic, more subjective direction, based on rigorous knowledge, attitude and values gathered. Environmental-perception survey research can lead this quest for alternative ways of bringing together a more equitable balance between humans and nature in land-use development. Public response to questioning about myriad landscape evaluation preferences, if rigorous and comprehensive, would provide a basis for establishment of a series of appropriate design-aesthetic guidelines, integrating within watersheds conflicting aspects of nature, human influence, and use and resource.

HOLISM AND WATERSHED PLANNING — ESAs

Good land-use planning is a relative notion and one that is variable in time, space and any number of human or environmental parameters. However, in the interest of providing planning frameworks and methodologies to permit the implementation of actual completed plans and projects, ways of looking at land-use planning, solutions to typical design problems, and methodologies of the planning process, have occurred for many decades and changed in society. Presently, the notions of holism, ecosystem and watershed planning have come to the fore in much current planning, especially with regard to land-use-planning studies in suburban, ex-urban and rural areas.

The holistic viewpoint is recognized by others as a prerequisite to human intervention in natural systems, if undesirable repercussions of management decisions are to be avoided (Mitchell 1989). The regional study of a particular place — for example, the Lake Simcoe watershed — reinforces the applied nature of results of research in resource management. The holistic approach recognizes the value of predicting consequences of resource development, or appraising the impact of earlier development. It is essential to recognize a host of variables and relationships (Mitchell 1989).

The ecosystem concept is not new. The word "ecosystem" was coined in 1935 by scientist Arthur Tansley, who defined it as a whole system that included not only the community of living organisms, but also the complex of physical factors forming the environment. Simply put, an ecosystem is composed of air, land, water and living organisms, including humans and their interactions. The concept has been applied to many types of interacting systems, including lakes, watersheds, cities and the biosphere (Crombie 1990). It is the watershed approach to ecosystem planning, from a holistic perspective, that is desired in much modern land-use planning.

QUESTIONNAIRE SURVEY RESEARCH

Questionnaire survey research is an instrument widely used in social science. Environmental perception research involves questionnaire survey application. It is a rapid and effective way of collecting large amounts of data in spite of some well-known disadvantages (attitude creation, cognitive dissonance and so on). Many problems of survey research are overcome through good instrument design. Abstract concepts or notions, when converted into questions in a questionnaire to collect relevant empirical data in a logical and skilful way, reduce survey problems (Babbie 1973). A broad range of survey techniques, beyond verbal questions, such as the employment of projective-photographic methods, mapping exercises and so on, are possible, and have been utilized in some instances (Saarinen 1976; Sonnenfeld 1967). These techniques serve to provide the researcher with a wide range of alternative instruments for measurement of human perceptions and attitudes.

Perception and attitude studies of the general public may be pursued with a number of interests in mind: to identify problems, opportunities or alternative strategies; to determine how well perceptions relate to reality; or to ascertain how stable perceptions and attitudes are. Studies of perceptions and attitudes have also revealed a wide variation in the ability of the general public to understand the manner in which natural systems or human interventions in such systems function. There is a real need for more work on the perceptions of the general public, since their views are needed in public-involvement programmes (Mitchell 1989). Nowhere is this requirement greater than in land-use planning.

LAKE SIMCOE WATERSHED — LAND-USE MANAGEMENT

Rapid urbanization of predominantly rural landscapes best reflects the planning dilemma of achieving appropriate landscape change, at the watershed scale, yet remaining consistent with a holistic perspective. The Lake Simcoe watershed in Southern Ontario is a contemporary example of such a geographic area. Environmental perception research has recently been undertaken in the Lake Simcoe watershed, and analysis of the results has resulted in the development of a conceptual environmental watershed management model stressing the concept of harmony, based on results of the original social-science research. It is proposed that a conception of understanding and testing human perception of nature, and seeking a proper contemporary balance or harmony between nature and culture, is sought in land-use-planning application.

Today, although much resource management work has been undertaken, problems of the place of environmental and behaviour research in the broad field of environmental-impact assessment, remain nebulous. The application of the ecosystem-management methodology, based on results of rigorous environmental perception research at the social scale presented in this chapter, will be adopted as a result of research findings in the Lake Simcoe watershed. It is an attempt to address some of the shortcomings of the proper use and balance of humanistic environmental research, and objective scientific study for land-use planning.

Lake Simcoe Watershed

The Lake Simcoe Basin covers 3,707 km^2 and is dominated by Lake Simcoe (745 km^2), which drains through Lake Couchiching into the Severn River, to Georgian Bay and Lake Huron. The current population exceeds 200,000 and the Lakes Simcoe-Couchiching study (1979) estimated that this number would grow to 450,000 by the year 2030. Recent planning studies of regional municipalities within the Greater Toronto Area (GTA) in the Lake Simcoe basin (York and Durham) continue to document significant population increases (York Region 1994).

For decades, Lake Simcoe has served as an important recreational and commercial resource in Southern Ontario, particularly the Greater Toronto Area. Due to the demands being placed on Lake Simcoe and the lands that form its drainage basin, the water quality of the lake is deteriorating. In particular, eutrophication involving the addition of nutrients, specifically phosphorous, promotes increased plant growth in the lake, resulting in a decrease in the water's oxygen supply. The Lake Simcoe Environmental Management Strategy (LSEMS) was initiated after a steering committee report in 1985 recommended that a phosphorus-control strategy be designed to reduce phosphorus inputs from rural and urban sources. LSEMS studies and Implementation Program are co-operative efforts between the Ministries of Agriculture and Food, Environment and Natural Resources, and the Lake Simcoe Region Conservation Authority (LSRCA). The goal of the LSEMS Implementation Program is to restore a naturally reproducing cold-water fishery by improving water quality of the lake (LSRCA 1992).

Recent actual projections of population indicate that by the year 2011, population and employment levels in York and Durham Regions (much of which are in the watershed) are expected to double to approximately 1.9 million residents and 810,000 jobs (Cole 1994). The increased stress on the natural ecosystems, and the consequent impact on strategies of watershed management and land use from urbanization and associated changes, are of major concern.

It has been suggested that two aspects of environmental change — phosphorus loadings and accelerated erosion on farmland — are of critical importance to Lake Simcoe watershed management in the future. But it can be argued convincingly that agricultural land use and urbanization (human activities) are close to the core of the larger issue — that the early days of settlement are nearly always destructive, and that the Lake Simcoe basin has undergone huge environmental changes without seeing the beginnings of a more permanent and viable system of land use during the past two centuries. The future design of land-use planning in this watershed will be critical to the health and sustainability of Lake Simcoe. It is clear that human use of the Lake Simcoe watershed is being reflected in changing natural, physical processes and characteristics of the watershed and lake. Hence, the importance of understanding human, environmental and social perception research, to direct future land use in the watershed, is clear.

Survey Research — Lake Simcoe

Research of social and environmental perception in the Lake Simcoe watershed was designed and undertaken by the South Lake Simcoe Naturalist's, a community research and recreation organization located in the central-southern part of the watershed. The recognition of the importance of public understanding of significant resource-management issues and their relationship to human environmental and social perception and subsequent land-use planning, was the motivation for the study. Research grant assistance was provided by the Ontario Ministry of the Environment under the Environmental Youth Core Program.

The research instrument, an original random-sample survey of residents and visitors to the watershed, was undertaken between June and August, in 1993 and 1994. Sampling occurred in several communities in the watershed. Some surveys were administered outside the watershed (suburban Metropolitan Toronto) to provide a measure of public perception, attitudes and knowledge of the region from outside the watershed. Figure 1 maps major survey locations.

The 1993 survey was a foundation instrument questionnaire. It was six pages in length, seeking responses on wilderness, wildlife, land use, environmental issues and photo tests, and mapping exercises of various landscapes. It had 123 respondents. Direction gained from 1993 survey results led to the creation of three short, focused surveys on watershed knowledge and funding, policy making and human perceptions of nature in the 1994 field season. Each of these surveys, administered randomly at similar locations, had a sample size of slightly over 100.

FIGURE 1
The Lake Simcoe Drainage Basin and Subwatersheds with Locations of Interview Sites

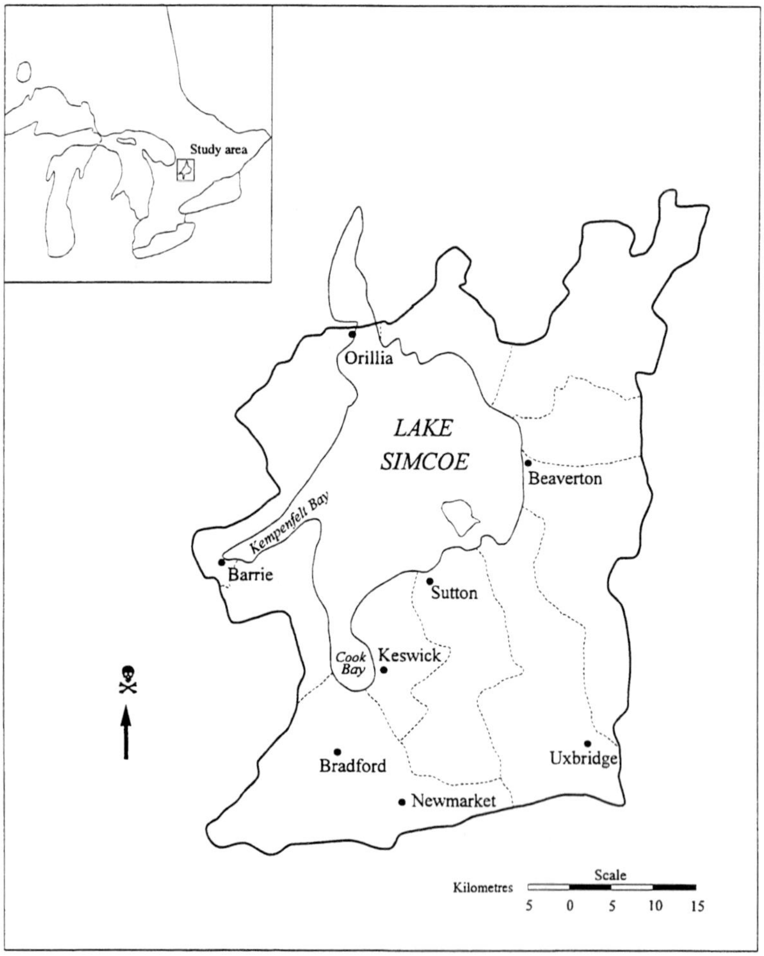

Survey Results

Discussion

The purpose and space available in this chapter allow only general reporting of research results. For more details, refer to the referenced material.

Many professional specialists — planners, architects and landscape architects — study and design changes to landscapes. Often it is thought that expert evaluations for landscape issues are definitive, and must be rigorous and consistent and valued above others. However, through recent landscape evaluation work, there is growing evidence that experts can lack consensus in rating landscape features. Research results suggest that expert or professional judgement cannot be a substitute for rigorous survey data-based aids and quantitative techniques (Willis and Garrod 1993). Indeed, there is some question about the lack of objectivity in, for example, the traditional architectural design process. How well are the attitudes, values and perceptions of the public being incorporated into design decisions of landscape and architectural space? This concern is warranted when one considers that numerous commentators have stressed the lack of rigorous psychological support for many routine architectural design decisions (Sommer 1966).

Visitor/Resident Perceptions

The notion that perception is related to residence was tested in the Lake Simcoe survey research. Results of the 1993 survey relating to perception of nature and visitor/resident perceptions, showed that overall, residents were much more accurate in identification and mapping questions; they were also more accurate in their assessments of the current state of the lake and the cold-water fishery (Evans et al. 1993). Indeed, visitors and residents often focus on very different aspects of the environment.

Often, it seems that only visitors, and particularly tourists, have a viewpoint. They use their eyes to compose pictures and unhesitatingly express themselves. By contrast, residents have a complex attitude derived from total immersion in their environment, which can be expressed only with difficulty, and indirectly through behaviour, local tradition, lore and myth (Tuan 1974). These results reinforce those who argue for more control and authority for local government with regard to land-use planning in watersheds. Results of survey data can be used to rigorously suggest and support appropriate land-use change and design of potentially unique and watershed-specific communities and facilities. These decisions are testable and verifiable through original social-perception data analysis.

In general, little work has been done in municipalities in Ontario or Canada in this regard. However, in Britain, for example, Yorkshire Dales National Park recently undertook rigorous research of this nature for resource-management use (Willis and Garrod 1993). Also, in the United States, research, similar to the Lake Simcoe study, is being undertaken at Ojai, California. Ojai, population 7,000, located in a small valley in the Transverse Range, is essentially rural, only 80 miles from Los Angeles. It is unique socially due to its high concentration of private schools, artists, various religious groups and anti-growth policies in rapidly urbanizing Southern California (Jones 1992).

Knowledge of Plans and Issues

Knowledge of important subjects of environmental concern in the watershed were found to be much better understood by the residents. For example, responses to questions about the state of the Lake Simcoe cold-water fishery and the LSEMS programme were more accurately answered by residents than visitors, although, knowledge on the whole was low (Evans et al. 1993). This is particularly distressing because political support for conservation and the improvement to the lake is more likely to occur with increased public knowledge and awareness. Results of this research suggest there is still a very long way to go to achieve a broad knowledge base, accurate perceptions, informed attitudes and accurate perceptions leading to social environmental change. The limited public knowledge and accuracy of human perception of issues in the watershed is instructive in explaining current land-use planning and resource-management lethargy in area municipalities. The LSEMS programme has made possible some direct conservation projects focused on erosion control and limiting phosphorus loading. They are, however, small in comparison to the need in this rapidly urbanizing landscape. Consequently, the results of questionnaire-survey environmental-perception research explanations of current situations, can be quantified and studied, and recommendations suggested to direct future initiatives.

With regard to the relationship between humans and nature, most respondents in the 1993 survey recognized nature as biotic and abiotic, but considered it quite divorced from humans. A larger proportion of visitors and urban residents surveyed saw themselves separated from nature, while residents more frequently considered themselves part of nature because it is part of their everyday lives (Evans et al. 1993).

Awareness is central to a discussion of perception. In order to perceive something, one must first be aware of it. Thus, awareness may be considered an aspect of perception, perhaps the most basic and preliminary one (Schiff 1971). In the 1994 survey research, respondents were asked if they were aware of the Oak Ridges Moraine, a prominent glacial, geomorphological feature in Southern Ontario of provincial conservation interest. They were also asked to place it on the map of Southern Ontario and to name any current studies being conducted on the moraine. The results showed that 30 per cent of those surveyed were actually aware of the moraine, however, only 19 per cent (20 people) of the survey group could locate it on the map. And, of those 20 people, only 14 (or 14 per cent) said they were aware of the current Ontario government study of land use and planning on the moraine. Only one individual of all those surveyed, was able to correctly identify details of the study.

These results are probably typical of awareness of many resource-management and land-use issues in the Lake Simcoe watershed and elsewhere across Canada. Many people are simply not informed or aware of issues, and therefore, although they may want to contribute, cannot participate in meaningful discourse, or significantly influence decision making.

As a consequence, much study, research and decision making effectively falls solely to various levels of government and specialist consultants, as the general public is relatively uninformed in spite of recent planning studies that claim wide public input and participation. In practice, small well-connected, special-interest groups undoubtedly dominate public consultation processes. It is not surprising that much policy and individual decisions on land-use matters are contested later at the Ontario Municipal Board (OMB). It is the rigorous testing of perception at an early stage that could influence policy makers and politicians to champion more appropriate directions for land-use planning in a future methodological model that considers public environmental perception, attitudes, knowledge and so on, in study and direction of land-use planning. From this research, production and dissemination of accurate knowledge to the general public, allowing ample time and opportunity for review, would undoubtedly make for much better comment later in the planning process by the public. Government and consultant resources spent on public contact would be better spent, at least initially, by concentrating efforts on sending out sound information to the public.

More research is needed into characteristics and perceptions of the general public, and definition of various special publics in resource-management planning. Unfortunately, research to date has overrelied upon the personal interview or mailed questionnaire. Other techniques, such as direct observation, participant observation, analysis of written material and simulation, need to be experimented with. Referring to Edward de Bono's notion of lateral thinking, perception and attitude studies represent a fertile hole for digging (Mitchell 1989). Techniques utilized in the Lake Simcoe study represent a move in this direction.

Societal changes in perception can be quite dynamic, and if not well understood and acted on, can doom land-use-planning exercises to certain failure. The case of wilderness-fire management in Montana, U.S.A. is well known, whereby public perception of policy direction changed radically ahead of government policy (McCool and Stankey 1989).

The City of Seattle undertook a long-term, progressive, development programme called Seattle City Light, constructing various dam projects on the Skagit River. By the 1960s, when the High Ross Dam was proposed, considerable opposition had arisen. This opposition was triggered by several factors, one of which was a heightened concern for environmental quality. Ross and Marts (1975) noted that "groups supporting projects such as Ross Dam fail to appreciate these changed attitudes towards nature and view the particular project as an isolated problem." The result was confrontation (Mitchell 1989). The vital need for study and understanding of human perception and societal change towards nature is clearly evident in other jurisdictions. Study of characteristics of the Lake Simcoe area could lead to predictive and good strategies in the future.

Policy Making

Public policy is a direct result of decision making, ultimately at the political level, in resource-management practice. By studying policy making, its context, process and outcome, research provides understanding of the manner in which resources are allocated. By knowing something about the variables that shape policy, analysts should be able to predict future policies, as well as to suggest how current ones might be improved (Mitchell 1989).

Sampling of societal environmental perception of resource-management issues in a geographic area would be an invaluable indicator tool for policy issues. Original research in the Lake Simcoe watershed sheds some light on perception of resource-management issues, and a policy on how the public feels should be developed.

It was also found that knowledge of studies in the Lake Simcoe area was rather poor. Only 31 per cent of the people surveyed said they were aware of any studies going on in the lake region. An interesting conclusion from a list of studies, and the researchers' feelings about respondents, was that people were not really aware of the specific study, but merely listed what kind of study they thought should be done.

Another question which probed the Lake Simcoe environmental situation, concerned the water quality of the lake. Results show that 75 per cent of the people surveyed felt that the water quality was deteriorating. The respondents were asked to list any factors they thought might be the cause. Respondents felt the top three factors were pollution/garbage, people, and sewage/septic systems. Therefore, they felt the water quality was deteriorating due to pollution caused by people. In comparison, of the people who felt the water quality was deteriorating, 45 per cent believed that government and taxes should pay for the rehabilitation and maintenance of the lake. Therefore, even though people felt the general public was causing the problem, they wanted government to fund the clean-up and long-term maintenance of the lake. Of the total population surveyed, 59 per cent felt government and taxes should provide the funds to aid in the maintenance or clean-up of the lake.

The knowledge of the Lake Simcoe Environmental Management Strategy was even lower than the knowledge of the Oak Ridges Moraine Strategy (ORM). Only 21 per cent knew of the LSEMS strategy, and of those, only two were visitors.

It is interesting to point out that when asked if it was important to save Lake Simcoe, 85 per cent said it was very important, 15 per cent said it was important, and not one person said it was not important. Undoubtedly this question was answered this way because the idea of letting the lake deteriorate and be destroyed is not very appealing to anyone. This result was expected.

Further analysis of the results of the study, and a larger, comprehensive, future study could better detail future resource-management policy.

Perceptions of Nature

In the 1993 Lake Simcoe survey, respondents were asked to participate in a photo test. They looked at five photos of different types of landscapes within the watershed, and ranked them in decreasing order of preference. The old-growth-forest photo was ranked highest by both residents and visitors, and agriculture was ranked lowest. The photos ranged from mature forest, through early successional and tree plantations, to open productive farm landscape.

The notion of nature as being wild, uncontrolled and very green is born out in the Lake Simcoe survey results. Other survey data showed that residents, asked to define nature by category, most frequently fit the category defining nature as being biotic or abiotic, with no mention of humans.

With regard to land-use planning, the still prevalent separation between humans and nature can be reflected in a persistent social distinction between the natural and the cultural landscape (Glacken 1967). Attitudes and values fall naturally out of this condition, and lead at some point to social and political decision making affecting land-use planning and design outcomes.

Indeed it is a recognition of the fact that the separation between humans and nature must not persist, and the consideration of the two as integrated wholes in future watershed-planning methodology must occur to improve land-use planning.

The Lake Simcoe perception of environment research results have shown where attitude problems exist, and identified suggested areas where societal perception must move to in the future. Efforts by government, the general public and others must continue in this direction in order to effect tangible, physical improvements to land-use planning, and influence the entire natural and human environment. Nowhere is this more the case than in the Lake Simcoe watershed, where perhaps only radical human-social change regarding new attitudes and values for the use of the landscape will ensure the goals of the LSEMS programme.

Wilderness

The recent rise in importance of nature to humans in the West, as previously detailed, is essential for leading to emerging societal values and attitudes, reflected in increasing interest in identifying, conserving, studying and enhancing wilderness areas.

The concept of wilderness is complex in itself, and a comprehensive study of it is beyond the scope of this discussion. However, perhaps the most important distinction to be made is the notion of wilderness as pristine nature. The idea of pristine is closely related to wildlife and wilderness, and is used generally to denote the extent to which non-human entities and landscapes have remained separate and untouched by nature (Bell, Russell, and Plotkin 1994).

Recent social-perception research in Southern Ontario has demonstrated the importance of modern media in interpreting landscape and place to the Western public. It is accepted that we perceive the present through the eyes of the past (Harpley 1992). Consequently, the notion of a true wilderness place in our increasingly human-dominated world, has seemed farther and farther removed from daily human experience. Indeed, few people, who understand and make use of the expression pristine nature, are likely to consider themselves living in, or being part of it. By definition, it is that which partakes of an earlier, original state (Bell, Russell, and Plotkin 1994).

The Lake Simcoe environmental-perception research challenges many of the notions about wilderness suggested above. Respondents were asked a number of direct questions, including: "Are there any wilderness areas in the watershed?"; and "Is there such a thing as pristine wilderness in Canada?"

Forty-three per cent of the people surveyed in 1994 thought that wilderness exists in the Lake Simcoe watershed. Response to this question was completely dependent on the respondents' personal definition of "wilderness," as no formal definition was given. For one person, wilderness may have been a provincial park, while for others, the northernmost part of Alaska. Surprisingly, areas such as Cooke's Bay, Willow Creek and Scanlon Creek, local areas within the Lake Simcoe watershed, were cited as examples of wilderness. When asked if "pristine" wilderness still exists in Canada, 86 per cent responded that they believed it still existed. Out of this 86 per cent, four people suggested areas within the watershed, such as Pefferlaw, the Black River and Sibbald Point Provincial Park. Others mostly picked the general direction of "north," as where they envisioned pristine wilderness to be. Responses such as northern areas of the provinces, the Yukon, the Northwest Territories and Algonquin Park were common. The myth of the North in Canada would still seem to be commonly held.

Another area in which differing attitudes of visitors and residents became evident was on the question of "wilderness" in the Lake Simcoe area. More residents than visitors (46 per cent and 33 per cent, respectively) believed that wilderness (by their definition) existed in their area. This did not necessarily mean that residents were correct in their assumption that much wilderness still existed in the Lake Simcoe area, but simply that they believed it did. Residents of the watershed may have simply wanted to believe that wilderness existed there, because they did not like to think that it was fast disappearing due to urban development, extractive industries and so on.

Clearly, although much is thought and written about concepts like wilderness, the pristine, rigorous, social-perception research may show us how little we actually know about these ideas.

THE LAKE SIMCOE RESEARCH — HARMONY MODEL OF WATERSHED MANAGEMENT

The environmental perception studies in the Lake Simcoe area, when viewed from the perspective of the broader context of environmental socialism, emphasizes the need for its application in land-use planning, and points out that it must be more widely considered in the future. Quantitative results of the case-study presented here allow for a synthesis of results of future management and planning.

In much traditional land-use planning, much attention is paid to the identification and study of the biophysical environment (soils, hydrology, wildlife and so on). This work is important. The humanistic aspects — social, cultural, political and economic impacts of land-use change — are also studied, but the investigation of the complex interrelationships between each influencing human values, attitudes and perceptions is little studied. The Lake Simcoe research presented is an attempt to concentrate study on environmental-perception research as an important component in watershed-management study.

A proposed conceptual watershed-management model — the Environmental Harmony Model, focused on environmental perception as the active resource-management component of integration, based on the Lake Simcoe area research — is presented in Figure 2. This framework proposes an expansion to typical approaches, developing an integration of human perception of nature research with traditional biophysical, social, cultural, political and economic components of typical study. The understanding of perception and action taken resulting

FIGURE 2
Harmony Model of Watershed Management

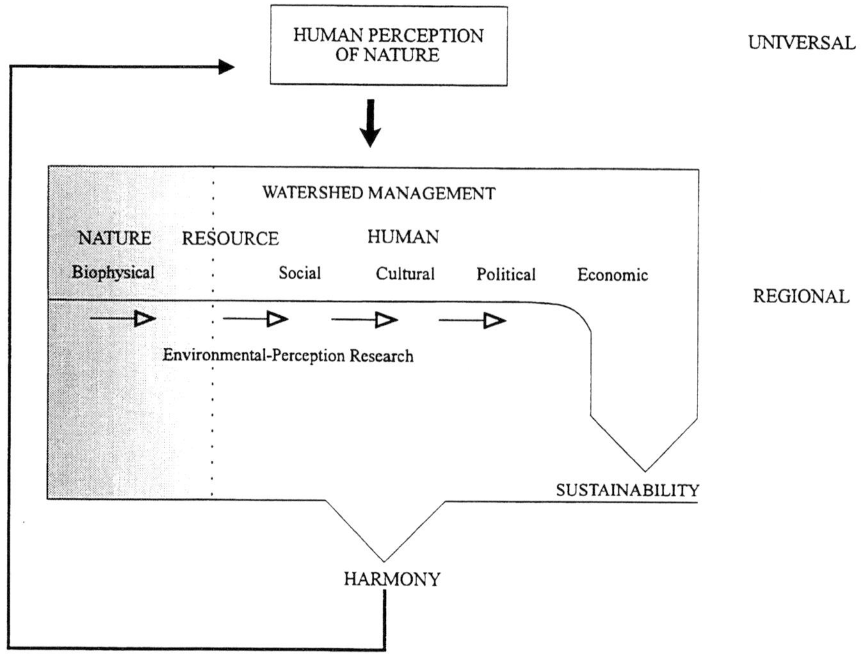

from interpretation can lead to societal movement towards sustainability, and ultimately a harmonization of attitudes of human activities towards nature.

Application of methodology, consistent with the structure of this model, provides the possibility for planners to address specific land-use conflicts within a watershed context, to arrive at a state of harmony, balancing biophysical and humanistic considerations through rigorous censuring, understanding of, and acting on results of original social-perception research findings.

DISCUSSION QUESTIONS

1. How does the increasing multicultural composition of Canadian society affect our society's knowledge, perception, and attitudes towards nature?
2. Are there significant regional differences in the perception of nature between individuals in different provinces, urban and rural areas etc.? How could these perceptions be surveyed and utilized in the planning and design process?
3. Is "real" ecosystem planning possible given the constraints discussed in this chapter? Are there other ways of harmonizing human/nature relations for future land-use planning?
4. For Harpley and Milne, the notion of a sense of place in our perceptions of nature is important for good land-use planning. Suggest ways of understanding and sampling communities to best accommodate these perceptions.

References

Anderson, Erica. 1965. *The Schweitzer Album — A Portrait in Words and Pictures*. New York: Harper & Row.

Babbie, Earl. 1973. *Survey Research Methods*. Belmont, Calif.: Wadsworth Publishing Co.

Bell, A., C.L. Russell and R. Plotkin. 1994. Searching for the Pristine. Paper presented at conference, Animals: A Reappraisal, The Post-Cartesian Vision of Animals in Art, Science, Education, Ethics and the Law, 19–22 May, at University of Toronto, Toronto, Ontario.

Bonnes, M., M. Bonaiuto, A. Ercolani. 1991. Crowding and Residential Satisfaction in the Urban Environment: A Contextual Approach. *Environment and Behaviour* 23(5): 531–52.

Botting, D. 1973. *Humboldt and the Cosmos*. London: Michael Joseph.

Buttimer, Anne. 1976. Grasping the Dynamism of Lifeworld. *Annals, American Association of Geographers* 66: 277–92.

Cole, Sherman. 1994. *Highway 404 Extension Davis Drive to Highway 12 Environmental Assessment Study*. Toronto, Ont.: Ministry of Transportation.

Collingwood, R.G. 1957. *The Idea of Nature*. Oxford: The Clarendon Press.

Crombie, David. 1990. Watershed, Interim Report. Royal Commission on the Future of the Toronto Waterfront, Toronto, Ontario.

Daniel, T.C. and R.S. Boster. 1989. Measuring Landscape Esthetics: The Scenic Beauty Estimation Model. In *Geography and Resource Management*, ed. Bruce Mitchell. New York: John Wiley and Sons.

Dunlap, R.E. and W.R. Catton Jr. 1979. Environmental Sociology. *Annual Review of Sociology* 5: 243–73.

Evans, N., K. Hynes, P. Harpley and R. Milne. 1993. Environmental Perception Survey in the Lake Simcoe Watershed. Lake Simcoe Technical Paper, vol. 3, no. 4. South Lake Simcoe Naturalists, Ontario, Canada.

Fowler, H.W., and F.G. Fowler. 1951. *The Concise Oxford Dictionary of Current English*. 5th ed. Revised by E. McIntosh. London: Clarendon Press.

Glacken, Clarence J. 1967. *Traces on the Rhodian Shore*. Los Angeles: University of California Press.

Harpley, P.J. 1978. Response to Wildlife Hazard in Ontario. B.Sc. Honours thesis, University of Toronto, Scarborough, Ontario.

Harpley, P.J. 1992. Western Perception of Nature: Perspectives From Visitors to the Toronto Zoo. Master's thesis, York University, North York, Ontario.

Hegel, G.W.F. [1821] 1970. *Werke 7, Grundlinien der Philosophie des Rechts oder Naturrecht und Staatswissenschaft im Grundrisse*. Frankfurt: Suhrkamp.

Holahan, C.J. 1982. *Environmental Psychology*. New York: Random House.

Ittelson, W.H., L.G. Rivlin, H.M. Proshansky and G.H. Winkel. 1974. *An Introduction to Environmental Psychology*. New York: Holt, Rhinehart and Winston.

Jones, Kris. 1992. *The Sense of Place for Ojai, California, in Technical Program*. Abstracts. 27th International Geographical Congress, 9–14 August, Washington, D.C.

Krause, Daniel. 1993. Environmental Consciousness, An Empirical Study. *Environment and Behaviour* 25(1): 126–42.

Krueger, R.R. and B. Mitchell, eds. 1977. *Managing Canada's Renewable Resources*. Toronto: Methuen.

Larsen, Erik S. 1992. Is Nature Really Natural? *Landscape Research* 17(3): 116–22.

Lake Simcoe Region Conservation Authority. 1992. *Lake Simcoe Environmental Management Strategy, What is LSEMS?* Newmarket, Ont.: Public Information Circular.

McCool, S.F. and G.H. Stankey. 1989. Visitor Attitudes Toward Wilderness Fire Management Policy — 1971–1984. USDA Forest Service Research Paper. INT-357, Intermountain Research Station, Ogden, Utah.

Mellen, Peter. 1976. *The Group of Seven*. Toronto, Ont.: McClelland & Stewart Ltd.

Mitchell, Bruce. 1989. *Geography and Resource Analysis*. New York: Longman Scientific and Technical U.K. and John Wiley and Sons.

Nash, Roderick. 1973. *Wilderness and the American Mind*. New Haven, Conn.: Yale University Press.

Oak Ridges Moraine Technical Working Committee. 1994. *The Oak Ridges Moraine Area Strategy for the Greater Toronto Area*. Toronto: Ontario Ministry of Natural Resources.

Olwig, Kenneth R. 1992. Space, Place, and the Ideological Construction of Nature as Landscape. *Abstracts*. 27th International Geographical Congress, Washington, U.S.A., Aug. 9–12.

Pickles, John. 1988. From Fact-World to Life-World, The Phenomenological Method and Social Science Research. In *Quantitative Methods in Human Geography*, ed. John Eyles and David M. Smith, 233–54. Cambridge: Policy Press.

Porteous, J.D. 1990. *Landscapes of the Mind: Worlds of Sense and Metaphor*. Toronto, Ont.: University of Toronto.

Ross, W.M. and E.M. Marts. 1975. The High Ross Dam Project: Environmental Decisions and Changing Environment Attitudes. *Canadian Geographer* 19: 221–34.

Saarinen, Thomas F. 1976. *Environmental Planning — Perception and Behaviour*. Boston: Houghton Mifflin Co.

———. 1969. *Perception of Environment*. Association of American Geographers Resource Paper no. 5. Washington, D.C.: Association of American Geographers.

Schiff, M.R. 1971. The Definition of Perceptions and Attitudes. In *Perceptions and Attitudes in Resource Management*, ed. W.R. Derrick Sewell and Ian Burton, 7–12. Ottawa, Ont.: Department of Energy, Mines and Resources.

Seamon, David Mugerauer. ed. 1985. *Dwelling, Place and Environment, Towards a Phenomenology of Person and World*. Boston: Martinus Nijhoff Publishers.

Sewell, J., G. Penfold and T. Vigod. 1993. *New Planning for Ontario, Final Report*. Toronto, Ont.: Commission on Planning and Development Reform in Ontario.

Sommer, Robert. 1966. Man's Proximate Environment. *Journal of Social Issues* 22(4): 59–89.

Sonnenfeld, J. 1967. Environmental Perception and Adaptation Level in the Arctic. In *Environmental Perception and Behaviour*, ed. D. Lowenthal. Chicago: University of Chicago Press.

Suzuki, David. 1994. Rethinking "Sustainable" Development. *The Toronto Star*, 17 September, E5.

Tuan, Yi-Fu. 1974. *Topophilia, A Study of Environmental Perception, Attitudes, and Values.* Englewood Cliffs, N.J.: Prentice Hall.

———. 1976. Humanistic Geography. *Annals of the Association of American Geographers* 66: 266–76.

The Merriam Webster Dictionary. 1975. New York, N.Y.: Merrriam-Webster Inc.

Willis, K.G., and G.D. Garrod. 1993. Valuing Landscape, a Contingent Valuation Approach. *Journal of Environmental Management* 37: 1–22.

York Region. 1994. Revised Draft Official Plan. Newmarket, Ontario.

Zimmermann, E.W. [1933] 1951. *World Resources and Industries*. New York: Harper and Brothers.

PROVINCIAL INNOVATION IN POLICY MAKING: ONTARIO MOVES TOWARDS PUTTING THE PUBLIC INTO PUBLIC TRANSPORTATION

Anton H. Turrittin

York University

Introduction

Canada's national transport policy has traditionally been played out in the context of Canada's élite politics where the decision makers are a combination of cabinet, the federal bureaucracy, provincial premiers and their ministers of transportation and commercial interest groups.[1] By contrast, Parliament, public interest advocates and public opinion play a much diminished role in transport policy making. While the regulatory framework has

---

The author has been involved with transportation activism since 1981 and is currently serving on the board of directors of both Transport 2000 Ontario and the Better Transportation Coalition. The author wishes to thank Dale Martin, John B. McCullum, David McRobert and Joe Murray, for granting interviews in connection with this paper. Much of the data for this paper come from direct participation in meetings (as an organization representative) with the minister's staff in the Ontario Ministry of Transportation, and with staff persons in various branches of the Ministry of Transportation. The helpful comments on this paper by Anthony Perl of the Political Science Department of the University of Calgary are very much appreciated. A general thanks must also be extended to all of the many marvellous people who collectively constitute the transportation movement described in these pages, from whom this author has learned a great deal, and whose opinions and efforts he values greatly. Needless to say, as author, I am solely responsible for the interpretations made in this paper.

been altered, neither the mode of policy making nor the direction of public finance has changed appreciably over the last several decades.[2]

Evidence from Ontario suggests that changing the mode of policy making to include a broader range of public input can dispel conventional wisdom and revitalize unresponsive decision-making arrangements, thereby facilitating more innovative policy outcomes. Policies that facilitate the shift to a more balanced transportation system, in which the technical and economic attributes of each transport mode can be applied most productively, thus appear correlated with the inclusion of additional social actors in the policy-making process. In particular, social involvement based on concerns for the environment and energy, has played a key role in the development of new transportation priorities.

Since the 1970s, there has been significant growth in the number of social movements in Canada — the women's movement, the peace movement, the environmental movement and others. Scholars have subjected such movements to analysis focusing on the role of these new movements in the political process.[3] But to say that there is now a transportation movement at work in Ontario would sound strange and even bizarre to many people. Yet, such a movement has emerged in the U.S. and in Europe.[4]

Social movements are publicly perceived as major forces for social change, especially through their demands on politicians and governments. But the obverse ought also to be entertained — that politicians can move to seek the support of movements in order to advance their political agendas against political opponents and hostile bureaucrats. While responding to social movement constituencies is not unknown to politicians, it is certainly novel in the area of public transportation.

This paper argues that a fundamental redirection in transportation policy, as it relates to intercity rail and urban transit in Canada, will likely require just such a new kind of fraternization between politicians and social movements. The argument is that letting the public play a more significant role in the long-standing élite game of transport policy making, has the potential to give the rail mode a more significant place in Canada's transportation mix, building on the inherent environmental and energy benefits of rail-based and electricity-powered transportation technologies.

While the resources of social movements for transport policy change at the national level are only now just building, this paper offers an assessment of three case-studies of transport/transit policy initiatives that illustrate how the interplay of ideas, interest, advocacy and institutions in a particular urban area may be fostering a different transport policy direction.

The first case-study concerns a provincial consultation exercise at the level of the public bureaucracy. It illustrates the pitfalls of an exercise of nebulous political commitment. The second case-study illustrates citizen activism to contain and reverse a decision by a transit agency to eliminate a significant portion of its electric transit infrastructure. This case and, flowing out of it, the third case — the forming of a publicly funded coalition of advocates for better public transportation — illustrate a linking between politics and a social movement, that has the potential of institutionalizing a new mode of transport policy making. These three examples raise the possibility that innovation can flow upward through the federal system, so that transport policy decision making at the federal level might one day encompass a broader range of social actors — the claim being that broader public input would lead away from the fly-drive fixation of Canada's current transport policy élite.

Normative Assumptions

This paper has a set of specific normative reference points which should be made explicit at the outset. Given global warming, the urgency of energy conservation, and my commitment to the compact urban form, in my advocacy I have worked towards promoting public transportation and the reduction of auto-dependency. No large nation in the world is as auto- and as air-dependent as the United States. One consequence of good roads and autos is the virtual disappearance of the compact city as a species. Canada, in my view, may be heading in that direction.[5]

At minimum, a transportation policy would attempt to link land use to transportation, would begin to reduce auto-dependency, would promote the most energy-efficient modes for the movement of people and goods, and, in the Canadian context, where many cities still have a compact urban form, would limit urban sprawl. In the context of this paper, the argument is that including the public in transportation decision making can enhance the likelihood of achieving the normative goals just mentioned.[6]

The Ontario Round Table on Environment and Economy

Our first case-study of citizen involvement is the Ontario Round Table on Environment and Economy. A round table is a first cousin to a royal commission and other bodies that are designed to give non-partisan advice to government on policy — advice to be followed, in theory, by legislative actions. Such bodies generally start with high hopes on the part of the public. High hopes evaporate when such bodies are compromised as exercises in delay and/or are tainted by commissioners and an executive with gross biases that undermine their credibility.

The Ontario Round Table seems to have had a credible enough start. It was established in 1988 by the Government of Ontario "... to create a long-term strategy that will support both a vigorous economy and a healthy environment."[7] That strategy was to have emerged from a consultation process aimed at consensus building. Consultation here means bringing together selected key representatives of government, industry and public advocacy — groups that are often adversaries; consensus was a hope, if not a prayer. In the process of achieving its mandate, the Round Table held public meetings and solicited briefs from various sources, but it also embarked on convening six sectoral task forces, one of which was devoted exclusively to transportation. This one is of particular interest here.[8]

It is fair to say that the advocacy community is generally sceptical of exercises such as round tables. Quite frankly, transportation and environmental advocacy groups have not paid much attention to the Ontario Round Table, in part because participation was highly selective. There is a strong belief in the advocacy community that such exercises accomplish little, especially with regard to policy change and its implementation.

The Sectoral Task Force on Transportation was, in effect, a mini-round table. It consisted of 10 members including an assistant deputy minister from the Ontario Ministry of Transportation (MTO), one representative each from General Motors of Canada, Imperial Oil Limited, and the Canadian Auto Workers, a leading activist concerned with automobile emissions, a representative from the Toronto Transit Commission, an environmental consultant, a representative of a public transportation advocacy group with a rail orientation, and a Native Circle representative. The task force was chaired by a Metro Toronto councillor with a reputation for concerns regarding urban design and transportation.

On balance, I believe that the advocacy sector was correct in its scepticism of the Round Table. Still, the work of the transportation task force, a rather select and divergent group, came up with a quite comprehensive view of the problem (transportation's contribution to environmental degradation), and clearly identified many key steps that could change the present situation.[9] However, a reader of the report with any knowledge of the issues and the players, can readily grasp the political nature of the document which reflects the multiple interests brought to the table or presenting briefs.

The auto, with an average of 1.2 to 1.8 occupants (depending on whether the trip is intraurban or intercity), takes a number of hits. It was recommended that automobile pollution emission standards be raised (some objections here from auto manufacturers); that a gas guzzler/sipper "feebate" programme be adopted (but that battle was already lost when the New Democratic Party [NDP] government in Ontario withdrew such legislation after a lobbying effort by General Motors through its employees); that high-occupancy vehicle (HOV) lanes be built on major urban roads; that land-use planning promote higher urban population densities; and that efforts be directed to promote greater use of public transit. There were many other useful recommendations

— for example, that intercity goods movement be switched from trucks to the more energy efficient railways.[10] Oddly missing from the sectoral report is any discussion of alternatives to flying and driving for the intercity movement of people, though there is one recommendation about negotiating the development of a high-speed train system between Windsor and Quebec City.[11]

Ironically, the report also plumps for Intelligent Vehicle Highway Systems, which aim to increase road capacity, as do HOV lanes. Highway planners (or should we say developers) at the provincial and municipal levels have been quick to utilize the call for HOV lanes to widen roads in order to add such lanes. The report makes a feeble attempt to come to grips with the road capacity issue by noting that providing more capacity leads to greater auto dependency, but warns that "limiting new road supply may encourage people to switch to public transit, but may also drive away business."[12]

The former chair of the transportation sectoral task force indicated to me that the Round Table was all about a "paradigm shift," an effort to change the thinking of decision makers to give more weight to environmental concerns in all of their actions. For this to have an impact on government, the former chair believed that the Round Table had to have the sponsorship of the highest level of government, the premier himself. But this is, in fact, the case with the ongoing Round Table, and with little discernable effect.[13]

In speaking with one political insider regarding the sectoral task force, I was bluntly told that the report of the transportation task force had many good ideas, but, where there would be any major opposition to a recommendation, it wouldn't happen, to use plain language. I take this to mean that, with regard to any of the task force's "good ideas," MTO staff would not recommend implementation if a good idea was controversial.

The Round Table's final report, *Restructuring for Sustainability*, was presented in September of 1992 to the premier of the province and a government response is due shortly. The board of the Round Table has been repopulated with new appointees, and the mandate of the Round Table has been altered. It now has three working groups — one dealing with government accountability, one with greening industry and one to stimulate local community initiatives. A number of the transportation sectoral recommendations made it into the final report, in rather general language. With respect to governmental actions, the Round Table's current strategy is to recommend better economic and environmental reporting, and instituting statements of environmental values in ministries that will guide future legislation. When I described the current role of the Round Table to a former leading campaigner in the environmental movement, I got a two-word response: "milk-toast."

My overall interest in this paper is with the role of advocacy groups in changing transportation policy. There are two positive aspects to the kinds of activities such as the Ontario Round Table that I have just described.

Firstly, these kinds of endeavours do bring together the players, perhaps for the first time, permitting them to explore areas of agreement and divergence, to assess each other, to dispel stereotypes and misapprehensions, to bargain and negotiate, etc. Secondly, the reasonably comprehensive list of recommendations of the Sectoral Task Force is a useful compilation of possible initiatives carrying the imprimatur of the Round Table and thus some modest authority. Still, the advocacy critique holds: nothing much will happen because of the Round Table. Where an advocacy group has any resource base at all — and a resource base is absolutely critical for sustained and effective advocacy — it would likely choose to do something very different, such as going to government directly with very specific demands. Our second and third cases are examples of this direct approach.

The Trolley-Bus Fight and the Revival of Transit Advocacy in Toronto

Metropolitan Toronto, Canada's largest municipality, is strongly public-transit oriented. It has a dense system of transit routes, a commuter rail system, high ridership per capita, and makes extensive use of electric transit technologies, with subways being the most significant of the electric transit modes.

Transit is especially important to the central portion of Metro, the City of Toronto itself, with its streetcar lines and, until recently, trolley buses. The transit-oriented character of the City of Toronto is in part a consequence of a major citizens' movement of the late 1960s which fought the infamous Spadina Expressway, a battle that was won in 1972 through the intervention of the then Premier, Bill Davis. This expressway fight (which also lead to the cancellation of some other related urban expressways) was the first successful revolt against freeways in North America. In 1967 the transit authority undertook a streetcar phase-out programme, but citizen efforts helped to reverse that plan and a core streetcar system was accepted by the transit authority in October of 1972. The citizen activism of the period, in turn, lead to a brief period of reform municipal government in the City of Toronto.[14]

The transit agency for the Metro area is locally known as the TTC, which stands for the Toronto Transit Commission. It is governed by a board of seven commissioners now elected from the council of the metropolitan level of government. While legally separate from Metro Council, it is financially responsible to it. The province is a player in transit decision making because it pays 75 per cent of all capital projects and shares half of the operating subsidy of the TTC with Metro (approximately 68 per cent farebox recovery on operations, with the remaining 32 per cent, therefore, needed as a subsidy).

One of the key developments in transit in North America is the turn towards surface electric transit symbolized by the construction of new light rail transit (LRT) and, most recently, the renaissance of the trolley bus. Some time in the early 1980s the TTC turned against this trend in the industry. Since 1983, the TTC has proposed terminating its small trolley-bus network in favour of natural-gas buses. Trolley buses have clear environmental and energy benefits: they are highly energy efficient, have no on-street pollution, and are virtually noiseless. These features are attracting several U.S. cities to plan entirely new trolley-bus systems, and cities that still retain this mode are planning upgrading and extensions.[15]

At the beginning of this paper, I argued that transit decisions are commonly élite decisions. I would argue that this kind of decision making has typified TTC's major capital projects since the early 1980s. My case in point is the élite alliance regarding the development and promotion of the natural gas bus. The transit industry, particularly in the U.S., has for a long time been searching for an alternative to the workhorse vehicle of the transit industry, the diesel bus, as a result of increasing tough air quality standards. In Ontario, a local bus manufacturer teamed up with a branch of the Ministry of Transportation to develop a compressed natural gas (CNG) bus. This local manufacturer has a U.S. manufacturing arm and thus the province perceived some export potential for Ontario-developed technology. This alternative-fuel bus needed both prototype testing, but, more significantly, some sales and a happy operator.

The alliance between the manufacturer and the ministry eventually took in the general manager of the TTC, as well as the powerful chair of Metro Council. Because TTC diesel buses still had considerable years of life expectancy, in the mid-1980s the TTC targeted its aging trolley-coach fleet as ready for conversion to the new "clean" CNG buses, a conversion that would net a sale by the local bus manufacturer — not a competitive tender by the way — for about 150 CNG buses. TTC staff reports were issued detailing the high costs of trolleys and recommending closure of the trolley network. The electric bus technology was depicted as outmoded in comparison to the new CNG-powered bus alternative.

To its credit, the City of Toronto has always opposed elimination of trolley coaches, and this form of transit had especially strong support from its mayors. The election of a new mayor in fall of 1990 who was not particularly interested in the issue, appears to have resulted in the "alliance" making its move. The transit commission voted 4 to 3 to mothball the trolley buses. The TTC-owned trolleys were stored out-of-doors and quickly became junk through vandalism. Public protest, however, started to build. A local transportation advocacy group (Transport 2000 Ontario) started a Transit Advocacy Centre which became the focal point rallying the public to put pressure on the commission to reverse its decision. The focus was on the chair of the commission who was regarded as being the swing vote.

The Transit Advocacy Centre and its sponsoring organization began to research the range of issues that the trolley conversion encompassed — technical questions such as the alleged energy cost savings of the CNG bus, the alleged "clean" quality of the CNG bus, the alleged fact maintained by a former transit commission chair that the public did not really know or care about the kind of surface vehicle in which they rode. Ordinary citizens began gathering petitions to save the trolleys, and ratepayer groups began to voice their support for retention. Pressure was sufficient that the commission granted a one-time only public meeting on the issue for the evening of April 28, 1992 and placed the meeting in the large Toronto City Hall chamber itself.

The April 28 meeting filled the chamber. Limited to five-minute deputations each, some 30 speakers gave remarks ranging from emotional appeals to highly documented refutations of every major argument that TTC staff had presented to date to justify the conversion. In the end, the chair of the commission changed his position and voted to reinstate trolleys on two routes (out of nine) using the most modern trolleys then in Toronto, trolleys from Edmonton which had been stored indoors because they were under lease. The two routes started running again in September of 1992.

With some kind of success under their belt, many of the citizens involved in the trolley fight began meeting on a regular basis but now under the name of the Better Transit Coalition. While two routes had been restored, the revitalization of the network had still to be fought for, and, in addition, there were other transit issues to address. The meetings of this group also had the strong support of a downtown Metro Toronto councillor affiliated with the NDP and who has a long-established record of encouraging grassroots organizations of all kinds.

More Transportation Advocacy and Some Political Attention

There is a second strand to transportation advocacy in Toronto, the rise of bicycle users and of environmental transportation groups. Bicyclists in the City of Toronto achieved a critical mass in 1974, resulting in the establishment of the Toronto City Cycling Committee as an office of the planning department. Bicycle advocacy builds on the local and social character of bicycle use. BUGs, or bicycle user groups, have formed by the dozens, at workplaces, on streets or on other community basis.[16] The bicycle community has been fighting very hard to obtain recognition from municipal and provincial planners and politicians demanding their fair share of road and other public space.

Continuing urban sprawl and highway building, deteriorating air quality, and the increasing dominance of the automobile has also lead to the founding of at least two new local environmental transportation organizations, one of these being bicycle-oriented. Several established large environmental organizations have also become more interested in urban transportation issues in recent years.[17] One outcome of this bicycle and environmental activism was the Second International Conference on Auto-Free Cities held in Toronto May 22–24, 1992 organized by Transportation Options. At this conference the public, planners, politicians and activists had an opportunity to meet and hear numerous experts and generally to digest alternative visions of what a city might be like, especially if cars were subordinated to walking, bicycling and transit use.

The citizen activism represented by the Better Transit Coalition, and the Auto-Free Cities conference did not go unnoticed by government staff and politicians. In fact, the conference drew on staff and politicians from all three levels of government (local, regional, and provincial) as presenters and participants. The present provincial governing political party, the NDP, has had a considerable history of involvement with grass-roots organizations in its many years as an opposition party (the NDP is in power for the first time in Ontario). What was perceived as grass-roots transportation ferment led one of the policy advisors at the Minister of Transportation (in a political, not a civil service position) to take a strong interest in these new transportation activist groups. This advisor attended a number of their meetings and in general kept in touch with their activities. Because this advisor became known to the activist community, he quickly became the recipient of requests for funding for various projects. Rather than responding to the transportation groups separately, this policy advisor recommended that the groups form a coalition and then approach MTO for funding jointly. The idea of supporting a citizens' transportation coalition was truly novel at MTO, but it did gain staff support, in part because several new programme initiatives were seen as benefiting from having official citizen participation. The invitation to form a coalition was received favourably by the transportation groups and they began to hold meetings to form the new umbrella organization.

The Rise of the Better Transportation Coalition (BTC)

The writing of a comprehensive proposal to go forward to MTO was no easy task for the coalition's founding groups. The groups had first to agree on common objectives, and to embody these in a formal incorporation. The proposal itself had to elaborate these objectives by indicating concrete activities and how staff job descriptions would reflect the group's goals. In April of 1993 the BTC received its incorporation, held its first general meeting, and elected its first board of directors.[18] In the end, it was agreed that the coalition would be composed of member organizations, that it would first focus on Greater Toronto issues but that it would become provincial-wide as quickly as possible, and that other organizations and individuals could join as non-voting members. After MTO funding was granted to the BTC, two full-time staff persons — both experienced environmental campaigners — were hired in June, 1993. The name Better Transportation Coalition was selected because the coalition had concerns which encompassed non-motorized forms of mobility and land-use planning as well as transit.

While the BTC is just off the ground, MTO has already put it on its Transit Integration Task Force. The BTC is to provide the users' perspective. The goal of the Task Force is to co-ordinate the multiple transit agencies operating in the Greater Toronto Area (GTA) to the end of having a common-fare system and "seamless" travel in the region.[19] The BTC has also embarked on a fight to revitalize the trolley-bus network in City of Toronto since the TTC again pulled the plug on the trolleys in July, due to a major revenue shortfall. The future of the $7-billion subway expansion plan for Metro Toronto has been reopened by the BTC both because of the very poor planning of the proposed additional lines, and the changed environment that has seen a substantial loss of riders on the TTC and recognition that real-estate developments projected for the next 50 years are not likely to be built. The BTC also has its work cut out in expanding its grass-roots base so that it can credibly argue that it speaks for real transit users and other citizen transportation needs at the local level.

The funding of the BTC is a first for the MTO; furthermore, it is highly unusual, because funding is for an operating grant. The funding of public-interest groups by government is not unknown. It is done often on a project basis, often for specific services. Public-interest groups sometimes get intervenor funding, as under environmental assessments processes. In speaking with the ministry staff to which the BTC reports, it is clear that it was sold to the ministry generally as a significant "value-adding" activity. That is to say, the BTC was seen as exactly what I described earlier as the major benefit of public-interest activity by citizens — citizen input improves decision making. The BTC does not expect to obtain ministry funding forever; it has already embarked on a strategy of fund-raising so that in the long term it can become mainly financially independent, though it will likely take several years for this to happen.

The BTC may not survive in the long term. While the BTC believes that it is playing a constructive role in current transit debates, to the extent that MTO and top-level politicians at Metro and the province have a fixed agenda, the more likely they are to respond negatively to the BTC's advocacy, which could end with the cutting of its funds. The BTC could then only survive if it has developed its own funding base. Another possibility is that the organizations making up the BTC might lose interest in it or fall out among themselves. But at this point it is as likely that the BTC will become very much a success, will survive and will become a leading player in transportation decision making in the province. The brief existence of the BTC already shows that informed and energetic citizens have much to offer to policy development and programme implementation. The BTC also demonstrates that citizens can only make such contributions when they have resources including, at minimum, some kind of paid staff. The claim here that a new transportation movement is taking shape will be tested maximally when an organization such as the BTC goes to the public for its resources.

The BTC will prove worthwhile if it can help surmount two significant and related barriers to effective transit planning and transit investment in the Greater Toronto Area. The first significant barrier — and readers may regard this as controversial, but that's the point — is the TTC's subway building programme known as Let's Move, recently renamed the Rapid Transit Expansion Program. This programme has been highly political from its inception — it was an election promise made by the previous provincial government. When the NDP government was elected in the fall of 1990, it quickly endorsed continuing the Let's Move programme without making a critical evaluation of it.

As presently constituted, four of the seven components in the Let's Move programme are subways and a fifth is an elevated rail-rapid transit technology.[20] Environmental assessments of these components, have been, or are just now, nearing completion. Final engineering design and formal funding are next. The premier of the province has endorsed these projects to achieve the priority of generating construction jobs. Originally, the jobs were to come from two sources — from the subway construction itself, and from the building boom that was expected to occur along the new lines.

Critics point out that the TTC is now in a dangerous downward spiral. The recession-caused ridership losses are leading to revenue declines which in turn lead to cuts in service, followed by further ridership losses, and the downward spiral repeats. The subway projects were predicated on a continuing boom in development along the new extensions. But the boom is now a bust and, moreover, recent data show that even in good times subways no longer necessarily attract development in the inner and outer suburbs. Even in the most optimistic scenario, ridership on the new subways extensions would only be at levels of the best bus and streetcar routes on the TTC system. That is to say, even long-term ridership projections suggest the use of less expensive alternative technology, such as light rail, instead of very expensive subways. The Let's Move programme could very well bankrupt the TTC. Increased operating costs associated with the low ridership on the subway extensions, could cause cuts to service on the surface system or force it to turn to government for a substantially higher operating grant. The whole programme would eat up capital and operating funds to such an extent that it would likely preempt many other very much needed regional transit capital projects. Let's Move is a classic mega-project based on the assumption that transit can lead development where development is assumed to be exponential and never-ending.[21]

The BTC (and others) are suggesting that Let's Move be scaled down. The pitch is to still develop rapid transit (public-works jobs are important), but to change the technologies to light rail, and to upgrade the GO commuter-rail system using the many under-utilized railway corridors of the GTA. There would be a threefold benefit. Not only would Metro get more rapid transit from this modified investment strategy, it would also gain a completed region-wide rapid transit network by a much earlier date, and there would be funds left over for improvements to the bus system as well. It remains to be seen whether citizen pressure and alternative proposals will change the course of the provincial government regarding the TTC's and Metro's mega-project plans.[22] To summarize, a key challenge for the BTC is to obtain the replacement of the planned expensive subway extensions with an affordable rapid-transit network.

The second barrier concerns transit planning for the GTA. As a result of its participation on the provincial Transit Integration Task Force, the BTC has come to realize that GTA-wide transit planning is not taking place. The proposal by the task force to integrate transit services does not envisage any larger planning authority for the region. Funding the Let's Move programme seals the fate of transit investments for a generation; it shortchanges the whole region in favour a few pieces of subway to be built over a 25-year period.[23] While the province plans its road system on a regional basis, it appears unable to do so for transit even though it has the authority to do so, especially as it is the senior funder of transit capital projects. At the task force meetings, the BTC is urging the province to rectify this situation with a new GTA planning agency. The BTC is banking on region-wide planning being able to curb the kind of excesses that the Let's Move programme represents.

Conclusions

Let us now review our case-studies in the light of the claims that I made at the beginning of this paper.

1. Transit planning is too often élite decision making.

   Occasionally, the public's nose gets into the decision-making tent. The Round Table was a form of carefully controlled access to that tent, an exercise that has proven to be more of a side show than a main attraction.

With respect to tangible vested interests, one cannot beat the trolley-bus caper. Here was a real alliance of insiders that planned and plotted the replacement of Toronto's trolley-bus network to the end of obtaining a quick purchase of natural-gas buses conveniently made by a local manufacturer. The public interest clearly showed up the strategy, but the public has not yet won this one.

2. An inclusionary decision-making process facilitates innovative policy outcomes.

   The Round Table demonstrated that even a modest "brainstorming" activity can produce many serious, responsible and innovative ideas. The vested interests tied to auto domination — industrial and governmental — are clearly not anxious to move quickly here, in spite of all the evidence of environmental damage and all the evidence of real economic and social costs of the automobile-driven society.

   The case of the trolley bus shows that it is even extremely difficult to keep decision makers honest let alone innovative. The last card of the alliance was the economic one — electric transit costs too much even though it might be good for the environment. (Note that that argument never applies to subways in the TTC's culture.) The TTC has done all that it can to avoid and prevent a truly open and fair study of the costs and benefits of trolley retention. The BTC is now in the process of demanding that such a study be done, and demanding that it has citizen and peer participation and review at each stage of the study. It still has not got that study.

   The case of the BTC turns out to be a natural experiment. Sold internally within MTO as adding value to decision making, the BTC has the heavy burden of proving its sponsors right. The BTC has taken up this challenge with relish. With respect to the Transit Integration Task Force, the BTC has entered the process not with some fully developed alternative blueprint in mind, but rather to help the task force participants think through the complicated and interconnected issues, and, through hard questioning and knowledge of what has been done in other jurisdictions, to move the process to the ephemeral "best" solution.

3. An inclusionary decision-making process may revitalize an unresponsive decision-making process.

   Must governmental organizations be forever unresponsive and uncreative as the stereotype of bureaucracy insinuates? I would suggest that one consequence of opening up governmental decision making to the public is that it may simultaneously loosen up decision making within government itself. We are all familiar with situations in organizations where a few people come to call the shots and everyone else toes the line, to use colloquial language. Let us examine our case-studies in terms of the potential for organizational change.

I wish the Round Table all the best in its quest to reform governmental processes. One key problem is the issue that crosses ministerial boundaries. It appears to take a very highly placed champion to get an interministerial initiative underway, given the unwritten law which prohibits one ministry from stepping on another's turf. While the minister of the environment did pronounce in favour of trolley-bus retention in Toronto, there was no interest in this ministry that I could detect, to work on a policy that related transit technology to air quality, either through an environmental assessment process (after all, the decommissioning of the trolleys would make Toronto air much worse than it already is), or through a more proactive process of requiring transit agencies to participate in achieving air-quality standards (this happens in the U.S., but only where a jurisdiction is a non-attainment area).

The BTC case offers more hope. As noted earlier, the intent of the sponsors of the BTC at MTO is that the BTC would provide the point of view of transit users to ministry decision-making exercises. The BTC now has close to a dozen projects of varying scope underway and is thus moving quickly to develop a transportation and transit agenda reflecting the interests of its component groups. It is beginning to clash with the conventional way transit decisions get made by Metro and the TTC, on one side, and the province, on the other, as evidenced in particular by the positions that it is taking at the Transit Integration Task Force. It remains to be seen whether the BTC's strong and well-argued positions will also be a catalyst to augment debate and exploration of alternatives within the ministry itself.

4. An inclusionary decision-making process will lead to a more balanced transportation system

   I am assuming here that present transportation decisions at all levels of government are dominated by a fly-drive mentality, fly-drive interests, and the mega-project disease. For the asserted proposition to be true, I must show that the public in general is ready for more change in the area of transportation than politicians and planners are prepared to imagine or give them credit for, let alone what fly-drive vested interests are ready to give up. I believe that, indeed, the public is ahead of both planners and politicians in their willingness to be less auto-dependent. It is not possible in this paper to present a full argument in this regard, but the emergence of a transportation movement would certainly be some support for this position.[24] In this regard, some tentative remarks may be made with respect to our case-studies.

The Round Table demonstrated that bringing diverse groups together can lead to a broader transportation vision than the dominant one, but speaking or writing down a broader vision is only a first step towards change. Real alternatives to cars must be in place — more compact urban forms, public transportation, and streets and sidewalks inviting to non-motorized transportation. The BTC is striving to build an organizational base across Toronto, indeed across Ontario, to offer an alternative vision, a vision that has behind it some political clout. That does not mean that the BTC will necessarily be the key actor in every transportation fracas. Rather, the BTC aims to become a clearing house or network centre that can provide knowledge, motivation, and some resources to the many groups that need to spring up in many different locations to push for change at their local level, assuming that such change will reflect an alternative vision.

5. Citizen activism can result in politicians seeking the support of movements.

   To this proposition, the quick answer, of course, is "I should hope so," given that politicians ought to "represent," in some sense, beyond the day of their election. As mentioned before, the NDP as a party has a history of links with community movements and community-based groups. The City of Toronto now has a concentration of transportation advocacy groups. With many of the NDP's local, Metro, and provincial elected officials from the City of Toronto, one would expect some attention, on their part, to the interests of these new social movements that are also their constituents. As I have shown, the rise of the BTC was importantly linked to an interest in it by the NDP and MTO. We can also ask whether or not the other parties (Liberals, and Progressive Conservatives) might also show some interest in the emerging transportation movement? That may also happen if the BTC continues to grow and manages its resources well, both in actions and in developing greater public consciousness of transportation as an important community and social issue.

6. The redirection of intercity-rail and urban-transit policy in Canada to promote these modes would require a new federal politics that links politicians to transportation movements.

   I would argue that for decades the federal government's transportation policy has been mired in the fly-drive syndrome, so much so that it will take major public pressure to obtain recognition of the importance of the rail mode for its environmental and cost benefits, and to recognize the importance of helping urban areas move towards greater reliance on public transit for mobility. The nascent transportation movements, now mainly found in selected urban centres, will therefore need to add the federal level to their political activities. Political effectiveness at the federal level appears to require the same coalition building necessary at the urban level, and also the same access to resources to be politically effective as well.

The former Progressive Conservative government, relished rejecting public opinion. In 1990, VIA Rail (Canada's national passenger train corporation, the equivalent to AMTRAK) was gutted in the West and in the East. To meet the public outcry, the government announced a federal Royal Commission on Passenger Transportation. At its public hearings, the public clearly expressed its desire to have a national passenger train system, and that the public transportation system of Canada become truly intermodal. The commission's own public opinion research supported continuing a federal subsidy to trains.[25] In its final report, all of that was ignored; there was no consideration of public transportation as a network, and it was recommended that VIA Rail be phased out over a 10-year period if it could not become wholly self-financing.

Because of a recent change in government, there is again an opportunity for public input into transportation decision making. The situation is very fluid. The new government is under pressure to provide jobs and appears ready to do this through the laying of more concrete for highways and airports. Just before the cancellation of a major military helicopter contract, which would have provided many jobs in the Montreal area and elsewhere, some politicians in Quebec were calling for the construction of a high-speed train line between Windsor, Toronto, Ottawa, Montreal and Quebec City as a replacement project. That would impact on the need for airport expansion. The Liberals, in the election, promised to review the previous government's privatization of Terminals 1 and 2 at Pearson Airport in Toronto; a key element of this project was major runway expansion. The Liberals also promised support for commuter rail lines during the election. The missing promise was support to upgrade the existing national passenger train system. Huge chunks of Canada's national freight railway system, significant parts of which are used for passenger services, are threatened by closure because economic policy favours trucking. Canada is signatory to an international convention to reduce greenhouse gases. Dealing with these transportation issues will set the direction of public transportation in Canada for a long time to come. Will, then, the new federal government include public input into the decision making regarding these issues, or shut it out, as did the last government? Listening and acting on the advice of transportation movements offers an opportunity to the new government to make genuinely popular decisions which are significant departures from both the old-style decision making (élitist) and the standard programmes favouring cars and planes.

Commentary

In closing, I would like to reflect on several aspects of my paper, firstly on the notion of a transportation movement. It is indeed novel to add transportation to the list of movements with which most of us are familiar. We could, of course, debate the extent to which the transportation movement is separate, or is simply an aspect of environmentalism generally. I would also note that there are as significant differences between Canada and the U.S. with regard to transportation movements as with respect to environmental ones. The population and wealth differences between the U.S. and Canada are important here. For instance, the Sierra Club in the U.S. has some 400,000 members, in comparison to perhaps 4,000 in Canada. There is a U.S. passenger-train advocacy group, the National Association of Railroad Passengers, with some 12,000 members; its counterpart in Canada, Transport 2000 Canada, has under 2,000 members. These size differences do mean a great deal in terms of the ability to mobilize politically, with the Canadian groups having very limited resources by comparison. Coalition formation is a feature of the U.S. environmental scene, as it is in Canada. In the U.S. there are coalition transportation groups as well, but similar coalitions in Canada are rare and are not strong.[26]

I do not want to attribute the weaker environmental and transportation movements in Canada to some notion of late or retarded development. Rather, I believe that there is at work here an important difference in political structure that has an impact on citizen mobilization. I am suggesting that, in comparison to the U.S., in Canada, there is much less of an opportunity structure, at both the provincial and federal levels of government, to facilitate the formation of new social movements.

I have made reference several times to the élite nature of decision making in Canadian politics. In the parliamentary system, the fusion of the executive and the legislative produces decision making that centres on the party leader and his or her cabinet. The senior government bureaucrats also become part of this decision-making nexus. Given this focus of power, party discipline excludes the incorporation of grass-roots policy input through individual legislators and privileges special-interest groups that get access at the top. I explicitly exclude social movement organizations as special interest groups, though such movements, if they are able to muster the resources, given this kind of structure, would also lobby at the top. In short, the U.S. political system provides multiple entry points for grass-roots input, whereas in Canada the entry points are much fewer in number.

My argument, thus, is that an important element in the underdevelopment of the transportation movement in Canada, especially at the national level, is the restricted system of access to decision making. The corollary to restricted access is that very few battles are ever won by the grass-roots when they attempt to influence decisions. With few battles ever being won, citizens see no use in joining advocacy organizations. The contrast between Canada and the U.S. in this respect is striking. In the U.S., the National Association of Railroad Passengers has played a key role in keeping up support for AMTRAK in Congress, when the White House was hostile to it (under the Reagan and Bush regimes). Recently a very large coalition of groups, the Surface Transportation Policy Project, was instrumental in shaping and gaining support for the path-breaking federal ISTEA legislation.[27] There has never been anything like this level of citizen lobbying in Canada on transportation, to my knowledge.[28]

If élite politics discourages the formation of new social movements on the Canadian scene in a general sense, the groups that do exist are still a resource to be tapped. Government leaders and top bureaucrats occasionally recognize that they are trapped in their own structures that cannot accommodate a new reality. They could seek the support of social movements to break a log-jam, but still may not be able to go too far in this regard for fear of reprisal by the élite interests that surround them. Specifically, these movements, both the transportation and the environmental ones, are a resource to help break out of the fly-drive, mega-project syndromes of conventional transportation politics.[29]

Finally, in this paper I have not tried to bring to bear the academic apparatus of social-movement analysis with respect to the transportation movement. I have not appealed to the old collective-behaviour tradition, or to the more recent resource-mobilization theories that trace how shared grievances become expressed in organizational forms. Nor have I tried my hand at some version of "new social-movement" theory that tries to understand social movements in a postmodern world characterized by a social diversity and multiple grievances beyond the economic. This is a direction in which I ought to move in my analysis. The transportation groups that I know have started from different points — an orientation towards the rail mode, as bicycle users, concern for air quality, an interest in urban design, environmentalism in general etc. — but from whatever starting point, and thinking in system terms, these diverse individuals have been driven to see the links between transportation and land use, come to question auto-dependency, and to desire and envision something quite different as to what the city might be. Knowledge of, or travel to, Europe helps form this awareness of the alternatives. This kind of thinking surely borders on the counter-hegemonic, to use a Gramscian term, or is a form of counter-power, an "insurrection of subjugated knowledges" as Foucault might put it. There is here a potential for a more thorough transformative politics.[30] Among the members of the transportation movement, there is a very strong personal commitment to changing the local, city and regional quality of life. It is difficult not to be over dramatic, but, given the huge contribution of transport to the production of greenhouse gases, how can these concerned citizens not think that saving of the planet may also be what is really at stake when they act locally?

DISCUSSION QUESTIONS

1. Describe the characteristics of a social movement. Consider what a transportation movement might be and the similarities and differences of such a movement compared to other kinds of environmental movements.
2. Given the goal of achieving less auto-dependent and less auto-dominated urban centres, how and why have professional bureaucracies, such as planning departments and transportation ministries, presented barriers to reaching such a goal?
3. Compare and contrast élite decision making with decision making based on public participation. Draw parallels between decision making in transportation and other institutional areas of community (e.g., health, energy etc.).
4. Why is it that round tables are not likely to be effective forces for social change? Examine the underlying assumptions of round tables as social processes. Consider the differences between vested and economic interests, and community interests, and how these interests are represented. Are round tables, for example, likely to affect legislation and regulation?
5. Conceptualize the difference between a mega-transportation/transit project and a locally oriented, smaller-scale transportation/transit project. What are the strengths and weaknesses of each as forms of capital investments? How do such projects come into being, and whose interests are served? How do competing notions of vision and supporting rhetorics relate to these very different ways of reshaping the city?
6. The case is being made that an inclusionary decision-making process leads to more innovative, public decision making, that will ultimately serve the goal of reducing auto-dependency. What are the implications of such a proposition for citizens and citizen participation in decision

making? What does this require of citizens in terms of knowledge, skills, values and commitment? In other words, why is leaving decision making to politicians and experts not enough?

Endnotes

1. The acceptance of élite domination in Canadian society, compared to the more democratic United States, has been argued by the U.S. sociologist Seymour Martin Lipset (1963), a position not without its critics (Romalis 1972; Truman 1974).
2. The passage of the National Transportation Act on August 20, 1987 has lead to a considerable degree of transport deregulation at the federal level.
3. The term "new social movements" has been used to describe the movements of recent decades, movements that are, it is argued, much different from the class-based political and labour movements of the past. The literature on this subject is large and international. For a recent overview, see Carroll (1992); for a critical review of new social-movement theory specifically, see Weir (1993).
4. In 1991, the U.S. Congress scrapped the old rigid highway-funding programme and replaced it with new legislation providing for flexibility in the use of highway funds for a broad range of transportation projects (Campaign for New Transportation Priorities 1993). For more on the transportation coalition that helped shape, and fought for, this legislation (ISTEA), see endnote [27]. In France, one public transportation advocacy group is the Fédération des Usagers des Transports, 15, rue de l'Echiquier, 75010 Paris.
5. For a quantitative assessment of differences between cities in the U.S. and Canada, see Goldberg and Mercer (1986). For a comprehensive overview of environmental issues in the U.S., see Gore (1992).
6. Some policy and legislative efforts in the U.S. are attempting to come to grips with the consequences of cars and deurbanization. Strong clean-air legislation exists. Transit investments are seen in many places as both helping to improve the air-management planning and arresting the erosion of the city form. Some jurisdictions even link land use and transit planning together. Portland, Oregon appears to be exemplary in these respects (Corbett 1992; see also the "Making the Land Use, Transportation, Air Quality Connection" or LUTRAQ studies by the 1000 Friends of Oregon, 534 S.W. Third Avenue, Suite 300, Portland Oregon 97204). There is even some recognition now in the U.S. that European and Japanese intercity train solutions can substitute for short-haul air trips and airport expansion (Vranich 1991).
7. Quoted at page 4 in the Ontario Round Table on Environment and Economy, *Challenge Paper* (1991).
8. The concept of a round table on environment and economy stems from the United Nations' Brundtland Commission report entitled *Our Common Future* published in 1987. In response to this report, the Canadian Council of Resource and Environment Ministers (now the Canadian Council of Ministers of the Environment) established a National Task Force on Environment and Economy which, in a report, in September of 1987, urged the creation of national and provincial round tables to bring together senior decision makers from all sectors of society to work towards a multisectoral consensus on sustainable development. Ontario was quick off the mark with its round table which went public with its 1991 *Challenge Paper*.

   The Ontario Round Table may have been part of the former Ontario Liberal government's preparations for an election that would have highlighted their environmental record. The New Democratic Party replaced the Liberals after the Ontario election in September, 1990. The Round Table continued under the new government and the sectoral reports were issued in early 1992. The sectoral task forces were charged with reporting to the

Round Table based on six principles set out in the Round Table's earlier *Challenge Paper*.

9. See the Ontario Round Table on Environment and Economy, *Sectoral Task Force Report: Transportation* (1992), hereafter referred to as the Sectoral Report.
10. The trucking industry made representations to the sectoral task force claiming that permitting longer trucks on Ontario's roads would improve the environment by increasing truck productivity. At the time the trucking industry was engaged in a major lobbying effort to obtain legislation to allow 53-foot trailers, an effort which will likely be successful, though opposed by the railway industry, with some opposition from an automobile association and from one transportation advocacy group.
11. Sectoral Report, p. 22. A joint federal, Quebec and Ontario study of high-speed trains in the Windsor–Quebec corridor was underway at the time of the sectoral task force.
12. Quoted at page 18 in the Sectoral Report. The Sectoral Report cites Newman and Kenworthy's landmark 1989 study to make the point that "at a residential density of 30–40 people per hectare a less auto-based, more transit-oriented urban transport tends to occur" (p. 18), but neglects Newman and Kenworthy's major finding that high-road congestion is associated with greater use of transit.
13. The Round Table was almost cancelled after the NDP became the Ontario government in September of 1990. Lobbying on the part of the former and present executive directors of Pollution Probe convinced the NDP to continue the Round Table.
14. One tract issued in the battle against the Spadina Expressway was Nowlan (1970). On the reform period in Toronto politics, see Magnusson and Sancton (1983), Sewell et al. (1971), and Sewell (1993). The effort to keep streetcars was fought by a small citizen's group called the Streetcars for Toronto Committee which, with the support of the City of Toronto, was able to change the position of the general manager of the TTC resulting in the retention, and later modernization, of the streetcar system. The modernization plan included converting some low-ridership streetcar lines to trolley-coach operation. The fight to save the streetcars is briefly mentioned in Bromley and May (1973: 126–28). For greater detail, see Chapter 15 in Fischler (1979).
15. The first new post-World War II light-rail line was built in Edmonton, opening in 1978, followed by LRT lines in San Diego (1981), Calgary (1981), Buffalo (1984), Pittsburg (1985), Portland (1986), Vancouver (1986), Sacramento (1987), San Jose (1987), Los Angeles (1990), Baltimore (1992) and St. Louis (1993). Trolley buses can be found in San Francisco, Seattle, Vancouver, Edmonton, Hamilton, Ont. (recently mothballed), Toronto (recently cut), Dayton, Ohio, Boston, and Philadelphia. New trolley bus systems are being considered in California for Los Angeles, Long Beach, Oakland and Sacramento.
16. Bicycle-user groups in the City of Toronto number at least 200 and they range in size from 2 to 150 members. The BUG at the Canadian Imperial Bank of Commerce in Toronto has 125 members, all linked through the firm's electronic mail messaging system.
17. The oldest public transportation advocacy group in Toronto is Transport 2000 Ontario. The two recent environmental transportation groups are Environmentalists Plan Toronto, and the bicycle-oriented Transportation Options. The mainline environmental organizations with strong transportation interests are Greenpeace and the Sierra Club. Two organizations interested in air quality, which, of course, strongly relates to transportation, are the Lung Association and Pollution Probe.
18. Keynote speakers at the first general meeting included an official from the Ministry of Transportation and the noted architect Jane Jacobs.
19. The issues and goals of the Transit Integration Task Force are outlined in a discussion paper entitled *Beyond the Periphery* produced by the Ontario Ministry of Transportation (1993).
20. The latter is a short extension of the Bloor-Danforth subway on the east end which is built on a guideway and uses linear induction motor technology. It is called the RT (for

Rapid Transit) and is the same technology used in Vancouver on its Skytrain, a largely elevated, rapid-transit line.

21. The problems indicated in this paragraph are developed in a series of small papers authored by Richard Gilbert of the Canadian Urban Institute (2nd floor, West Tower, City Hall, Toronto, Ontario, Canada M5H 2N2), and in a report with the subject "The Metro Rapid Transit Expansion Programme (RTEP) and Related Issues" prepared by the City of Toronto Planning and Development Department and the Department of Public Works and the Environment, dated August 26, 1993.
22. Formally, it is the TTC and Metro government that are the proponents of Let's Move, but the province will be the source of 75 per cent of the needed construction funds.
23. Vanderwagen (1992) discusses the dismal consequences of badly planned mega-project rapid transit in Edmonton and Montreal.
24. In Europe, Werner Broeg assembled research that shows that people he calls "opinion shapers" overestimate the contribution of the private car to everyday mobility and, furthermore, that most citizens believe policy makers do not correctly assess their views regarding transport planning which are more favourable towards substituting public transit for car use. Broeg is associated with Socialdata, Institut fuer Verkehrs- und Infrastruktur-forschung Gmbh, Hans-Graessel-Weg 1, 8000 Muenchen 70, Germany.
25. This research, and how the full results were selectively presented to fit the biases of the royal commission, is documented in the paper presented at this panel by John N. Dunbar entitled *What You Don't Know Hopefully Won't Hurt: Developing Public Support For Passenger Trains in Canada.*
26. Currently, there is a coalition of groups in the Maritime Provinces fighting the abandonment of the Canadian Pacific Railway's mainline from Sherbrooke, Quebec, through the state of Maine, to Saint John, New Brunswick which also carries VIA Rail passenger train service.
27. ISTEA stands for the Intermodal Surface Transportation Efficiency Act, passed by Congress in December of 1991. A key transportation coalition that shaped and supported this act was the Surface Transportation Policy Project. In 1990 its steering committee included the following organizations: America's Coalition for Transit Now!; American Institute of Architects; American Planning Association; Bicycle Federation of America; Campaign for New Transportation Priorities; Center for Neighborhood Technology; Energy Conservation Coalition; Environmental and Energy Study Institute; Environmental Consortium for Minority Outreach; Environmental Defense Fund; Friends of the Earth, National Association of Regional Councils; National Growth Management Leadership Project; National Trust for Historic Preservation; National Wildlife Federation; Rails-to-Trails Conservancy; Scenic America; and the Surdna Foundation Inc. The Campaign for New Transportation Priorities is itself a coalition which includes the Institute for Transportation and Development Policy, the Bicycle Federation, the Amalgamated Transit Union, Scenic America, the Sierra Club, the National Association of Railroad Passengers, Clean Air Council, Friends of the Earth, and Environmental Action.
28. On influencing politicians and bureaucrats, consult Thompson and Stanbury (1979), Campbell and Szablowski (1979), Malvern (1985), Sawatsky (1987), and Pross (1992).
29. At the municipal level the analogue to the fly-drive syndrome is the mentality that defines transit as a supplement to the automobile or, worse still, as a social service to those unfortunates unable to own a car.
30. For a thorough overview of contemporary social-movement theory, see the introductory chapter by William K. Carroll in Carroll (1992). The Foucault quotation is found on page 81 of Foucault (1980).

Bibliography

Bromley, John F. and Jack May. 1973. *Fifty Years of Progressive Transit: A History of the Toronto Transit Commission.* New York: Electric Railroaders' Association.

Campaign for New Transportation Priorities. 1993. *Citizen's Guide to ISTEA: Transit and Rail Funding and Public Participation Requirements.* Washington, D.C.: Campaign for New Transportation Priorities.

Campbell, Colin and George J. Szablowski. 1979. *The Superbureaucrats: Structure and Behaviour in Central Agencies.* Toronto: Gage Publishing Limited.

Carroll, William K., ed. 1992. *Organizing Dissent: Contemporary Social Movements in Theory and Practice.* Toronto: Garamond Press.

Corbett, Judith. 1992. *Portland's Livable Downtown: From Auto-Dependence to Pedestrian Independence.* Washington, D.C.: The Surface Transportation Policy Project.

Fischler, Stanley I. 1979. *Moving Millions: An Inside Look at Mass Transit.* New York: Harper & Row.

Foucault, Michel. 1980. *Power/Knowledge: Selected Interviews & Other Writings 1972–1977.* Edited by Colin Gordon. New York: Pantheon Books.

Goldberg, Michael A. and John Mercer. 1986. *The Myth of the North American City: Continentalism Challenged.* Vancouver: University of British Columbia Press.

Gore, Albert. 1992. *Earth in the Balance.* Boston: Houghton Mifflin.

Lipset, Seymour Martin. 1963. *The First New Nation: The United States in Historical and Comparative Perspective.* New York: Basic Books. (A 1979 edition has been published by Norton.)

Magnusson, Warren and Andrew Sancton, eds. 1983. *City Politics in Canada.* Toronto: University of Toronto Press.

Malvern, Paul. 1985. *Persuaders: Influence Peddling, Lobbying and Political Corruption in Canada.* Toronto: Methuen

Newman, Peter W.G. and Jeffrey R. Kenworthy. 1989. *Cities and Automobile Dependence: A Sourcebook.* Sydney: Gower.

Nowlan, David and Nadine Nowlan. 1970. *The Bad Trip: The Untold Story of the Spadina Expressway.* Toronto: New Press and House of Anansi.

Ontario Ministry of Transportation. 1993. *Beyond the Periphery: Co-ordinating Public Transit in the Greater Toronto Area. A Discussion Paper.* Toronto: Queen's Printer for Ontario.

Ontario Round Table on Environment and Economy. 1991. *Challenge Paper.* Toronto: Queen's Printer for Ontario.

———. 1992. *Sectoral Task Force Report: Transportation.* Toronto: Queen's Printer for Ontario.

———. 1992. *Restructuring for Sustainability.* Toronto: Queen's Printer for Ontario.

Pross, A. Paul. 1992. *Group Politics and Public Policy.* 2d ed. Toronto: Oxford University Press.

Romalis, Coleman. 1972. A Man of His Time and Place: A Selective Appraisal of Lipset's Comparative Sociology. In *Perspectives in Political Sociology*, ed. Andrew Effrat, 211–31. Indianapolis: Bobbs-Merrill. (With a reply by Lipset and a rejoinder by Romalis.)

Sawatsky, John. 1987. *The Insiders: Government, Business, and the Lobbyists.* Toronto: McClelland and Stewart.

Sewell, John. 1993. *The Shape of the City: Toronto Struggles with Modern Planning.* Toronto: University of Toronto Press.

Sewell, John et al. 1971. *Inside City Hall: The Year of the Opposition.* Toronto: A.M. Hakkert Ltd.

Thompson, Fred and W.T. Stanbury. 1979. *The Political Economy of Interest Groups in the Legislative Process in Canada.* Montreal: Institute for Research on Public Policy, Occasional Paper No. 9.

Truman, Tom. 1974. A Critique of Lipset's Article Value Differences: The English-Speaking Democracies. In *Sociology Canada: Readings*, ed. C. Beattie and S. Crysdale, 40–61. Toronto: Butterworth and Co. Ltd.

Vanderwagen, Joell. 1992. *Transit in Canada*. 2d ed. Toronto: Greenpeace.

Vranich, Joseph. 1991. *Supertrains: Solutions to America's Transportation Gridlock*. New York: St. Martin's Press.

Weir, Lorna. 1993. Limitations of New Social Movement Analysis. *Studies in Political Economy* 40(Spring): 73–102.

World Commission on Environment and Development (Brundtland Commission). 1987. *Our Common Future*. New York: Oxford University Press.

PRESS ANALYSES AND ENVIRONMENTAL MANAGEMENT

Robert Tessier

Université du Québec à Montréal

This paper offers some reflections on the significance of the results of press analyses and the usefulness of these results in environmental management.[1] To this end, the conclusions reached by 30 years' of research into relationships between media content, the environment and society will be discussed. First, the type of analysis results will be examined. In order to facilitate understanding, the controversy surrounding the Great Whale River hydroelectric project in Quebec will be used as an example.

The Hydro-Québec project has generated complex debate involving a large number of players. The project is situated on land belonging to the Cree Indians, who oppose its realization for various reasons. One of the questions that may be asked about media content is: Do Quebec newspapers allow for the Cree point of view? To what extent are other points of view favoured? For exploratory purposes, a detailed analysis was done of how events surrounding the Great Whale project were viewed, based on the headlines of 54 articles that appeared in the daily newspapers *La Presse* and *Le Devoir* in 1990. The main hypothesis was that press headlines were generally not neutral and that, the majority of the time, they suggest a particular way of viewing the facts reported, even when they headline descriptive articles. Consider the following headlines: "Great Whale project to harm the Cree" and "Cree go all out to stop Great Whale." In the first headline, the accent is on the fate of the Cree as a result of the project. In the second, the fate of the project, because of Cree actions, is emphasized. The object of concern described by each headline is different, and is not necessarily dependent on the situation or event being described. The same article could very well be found under either headline. What does the angle of the story depend on?

I believe it depends on the values held by the newspapers regarding the issue. If, for one reason or another, the paper sees the Great Whale project as important, it will generally emphasize the positive or negative consequences on the project by various groups. On the other hand, if the paper sees the Cree as important, headlines will accent the consequences suffered by this group. If this hypothesis is true, Quebec newspapers should favour a relatively stable point of view, guided by relatively coherent values, throughout their headlines concerning Great Whale, rather than diversified and changing values responding to alternating points of view on current events. Analysis of the headlines demonstrate this.

Of 54 headlines analysed, none presented the event from the angle of promotion or protection of the Cree themselves. At the same time, the Great Whale project appears 10 times, as something one should be concerned about because of its importance. Foreign sales of Quebec electricity, which in part justify the project, appear prominently five times, and Hydro-Québec, the state corporation responsible for the project, appears three times in the same important position. Other players, whether attacked or promoted, such as governments or "Green" groups active in the matter, are presented only once each. Only the role of the federal government is given any weight, and that indirectly as regards its responsibilities in environmental matters (twice in *La Presse*).

Opponents of the project are nearly always presented as negative agents: they attack things that appear elsewhere as strong values when the entire group of headlines is taken into consideration. Even where the most current object of concern — the procedure chosen to evaluate the project, which appears in an important position 14 times out of 54 — it is the Quebec government that plays the most leading role, with five mentions in favour of one form of evaluation, and insisting that the project be accepted quickly. The accent on the "evaluation" of the project can be understood insofar as the project itself is never clearly presented as having a positive function relative to anything else. On several occasions, the project even appears negatively against other values, principally the notion that it might place Québécois society in debt for generations, or that it might weaken Quebec's competitive advantage by providing energy across the border. The project also appears negatively in relation to the only environmental value mentioned in these headlines (which appears only twice): the balance of wildlife on aboriginal land — a balance that is more likely to be affected by the road that will bring thousands of white hunters after the project is constructed, than by the project itself. Somewhat surprisingly, Hydro-Québec, in this case, is presented as coming to the defence of this value, while the Cree are shown as contradicting themselves on this point because they are actually in favour of a road that would have comparable effects on another part of their land.

These results are intriguing in that they demonstrate a particular way of viewing the players and the important elements of the issue in press headlines at any given moment. The viewpoint is neither that of the Cree nor of ecological groups. It is, of course, more sympathetic to promoters of the project, but it does not exactly reflect their point of view, either. The positive value of the project is not always apparent, and at times doubts are even highlighted. In spite of everything, the viewpoint in question is coherent in its way of posing the problem of the Great Whale project. It includes a judgement as to what merits attention in this matter. This point of view belongs to no one in particular — it could be said to apply to the whole of Quebec society, which observes what happens by way of media interpretations and through its own values. A cohesive force seems to be at work in the press discourse revealed by these headlines.

This exploratory study has focused on headlines from two francophone daily newspapers in Quebec. The picture should, of course, be completed with other studies focusing on headlines concerning Great Whale projet from anglophone newspapers in Quebec, other Canadian provinces or the United States. The choice of newspaper for this kind of study depends upon the society one wishes to understand. The problem of selection results from the fact that no society is perfectly homogeneous. Even within francophone Québécois society, represented here by the dailies *La Presse* and *Le Devoir*, differences in content have been uncovered. *La Presse*, for example, attaches importance to the federal government. Ultimately, each newspaper is aimed at a particular "society" whose members have certain common values. These common values are what readers recognize as interesting in the content of a daily newspaper. We will see later at what point the question of common values is fundamental to the concerns that are conveyed by the press in a given society. The challenge of analysis thus consists of choosing daily newspapers in which the content is fairly similar in order to uncover the values particular to a given society. A good way to ensure that the society being studied corresponds as closely as possible to global society, rather than a particular subgroup, is to select newspapers representative of the political majority in that society.

In Quebec, this majority is made up of francophones whose political aspirations to become a "distinct society" are acknowledged and involved in the debate surrounding the Great Whale project. In fact, the question of powers is at the heart of the debate. Some claim that Quebec wants to use its hydroelectric projects to finance independence; the Cree demand that Ottawa, the only government they consider themselves historically linked to, intervene, and also blame Ottawa for forgetting its obligations to try to keep Quebec in Canada. Federal organizations are involved in evaluations of Hydro-Québec projects (an economic lever many Québécois identify with) because of the resulting exports of electricity to the United States; and, after a long debate with the Quebec government, Ottawa participated in the evaluation process of the Great Whale project. All these elements demonstrate that we are dealing with a debate between many societies having different political agendas. A whole group of information media corresponds to each of these societies. This exploratory study has attempted to uncover one "member" of this group particular to one of the societies active in the Great Whale issue. The remainder of this article treats the link between "political agenda" and the content of information media in society.

Given this, the following question may be asked: which reality can these results be considered to express, and how useful does knowledge of this reality have for environmental management? In order to reflect on this question, it is helpful to consider research done on the significance of media content and the environment.

The Political Agenda-Setting Function of Press Content

One approach to media analysis that has had remarkable success over the last few years is centred on the placing of issues dealt with by the media on the political agenda.[2] As with many other issues (Lazar 1991: 150–52), it has been demonstrated (Salwen 1986) that the general public's perception of the importance of an environmental problem is directly linked to the amount of coverage given to it by the media. The length and frequency of this coverage affects people's judgements about what priority the problem should have on the government's agenda. To this end, the media does not tell people what they should think, but rather what they should *think about*, in general, or regarding a particular issue (McCombs and Shaw 1972). This function is fully compatible with what I have observed concerning press headlines: by bringing attention to certain aspects of a question, headlines participate in the influence of media content on political agenda setting. These results are thus precursors of developing public opinion. In this context, these results could be useful for environmental management, if it is the case that environmental management is carried out primarily through public choices, and that a substantial electioneering component comes into play in the behaviour of political decision makers.[3] Media content on an environmental question thus deserves attention from interested parties because it plays a key role in the social regulation of the problem in question. Pressure groups already seem to be aware of this: they often use the media to publicize their events and objectives. For "Green" groups, it is[4] an eminently political action. Many analysts (Ullmann 1985; Pasquero 1989) believe that this is how such groups give rise to new societal awareness which leads to new regulations. It is even thought that only the possibility of constraining intervention by the state as a result of fears provoked by the media, has motivated business to take the lead in the fight against global problems like acid rain, the diminishing ozone layer and the greenhouse effect (Salles 1993; Roqueplo 1988). But this is a double-edged sword: even if the role of the media is undeniable in the promotion of environmental awareness, campaigns can also be orchestrated[5] *against* environmentalist demands — during a recession, for example. It also appears that public communication strategies were used by the industrial sector, which took the debate initiative in order to orient political agendas (Salles 1993: 109) towards energy or technological choices that were in its own interest (Salles 1993; Roqueplo 1988; Faucheux and Noël 1990; Hourcade, Mégie, and Theys 1991).

Media content can thus be considered a political, even an economic, tool for the social players involved in conflicts over regulation of environmental problems, and therefore as a tool for management of these conflicts. Press analyses such as this could be useful for strategic purposes to measure the presence of specific points of view latent in public opinion. This involves bringing certain types of concerns to public attention, so that they will eventually find their way to the political level. Such an approach could benefit research results, and may be considered complimentary to those obtained within the framework of the "agenda-setting" hypothesis. A model for the diffusion of an ideology has already been validated (Strodthoff, Hawkins, and Schoenfeld 1985). According to this model, in the treatment of information concerning social causes, the media evolves through phases of clarification, legitimation, and routinization, which favour the emergence of a social movement involved with a particular issue. The conditions for efficient articulation between the local media and the national and international media in such a process have also been observed (Faupel, Bailey and Griffin 1991). Such insights can give rise to rather machiavellian, practical applications. One thus finds methods for public-image management (Dagenais 1991) and marketing being used, for instance, to sell a business project both at home and abroad through global-communication strategies (Beauchamp 1991). From this perspective, the environment can, whether we like it or not, be an object of management, like any other object and businesses, which already effectively use press review analysis to influence matters that promote their interests.

However, for those who believe that environmental issues deserve a more democratic hearing, studies show that public preoccupations are not as controllable as research done within the agenda-setting hypothesis framework tries to show. Because of this and other factors yet to be mentioned, the political, institutional and practical results that some would like to obtain by manipulation of opinion through the media, prove to be, in large part, uncertain.[6]

The Limits of Press Analysis on Environmental Themes

It has been established (Hansen 1991) that object of research — which is to describe relationships between the environment, media and society — has had some difficulty explaining the variations between public interest and media content.[7] This is especially true of the agenda-setting approach. Issues that receive strong media attention do not always become important issues for the public. For an example of media coverage that had no great impact on public opinion and political decision making, see Protess et al. (1985) on the effects of a series of articles on rape in the Chicago *Sun Times*. To explain this,[8] supporters of the hypothesis introduce a "presetting" of public awareness in relation to certain contents as a necessary factor in agenda setting. In his classic study of the preoccupations of the 1960s, Funkouser (1972) explained the differences between public opinion and the content of widely circulated magazines by the media's tendency to fabricate artificial news — pseudo-events that do not attract attention precisely because they are too far from what is, in fact, a double "reality" — people are sensitive to what they consider, *a priori*, the "real" and "essential" events in their society's evolution. The media, in turn, are experts in identifying issues that correspond to judgements widely shared in a society. Up to two years in advance, the press reported more on real and essential events, than during the most important facts of these events themselves. From this point of view, the bet often won by the press concerning the actual or latent events that already potentially concern the public determines the political agenda-setting function of press content.[9] If not, there would be no way to simultaneously explain the limitations encountered and the convincing results of the majority of research conducted according to the agenda-setting hypothesis.

The results of a press analysis, like the ones presented here, take on meaning in the framework of this logic. The headlines analysed specifically suggest an angle according to which an event or situation is important. As we will see, in the majority of cases, press content responds to what is already important to the public. I believe headlines are particularly significant in this respect because they are often formulated in such a way as to dramatize a situation. That is to say, they present a situation by indicating that something important is in danger, or has overcome great difficulties against all odds. It can thus be said that in the majority of cases, the objects of concern expressed in press headlines correspond to important public values. These values change — this is evident if one compares press themes over a 10-year period — but as demonstrated by the limits encountered by agenda-setting-directed research, these values cannot be manipulated by the media. These values influence media content as well as the political choices that follow.

This latter hypothesis[10] is clearly important. It confronts two major difficulties that should be overcome, given the implications of the agenda-setting hypothesis for environmental management.[11] One is due to the fact that it assumes the existence of a collective, homogeneous judgement in a world generally perceived as individualistic and pluralist. The other is due to the fact that the resulting approach relegates the effects of the media on awareness to a secondary level. However, much other research places the accent on the often-perverse effects on behaviour, although these effects are uncertain (Cumberbatch and Howitt 1989; Katz 1990).[12]

The first argument is refuted by the positive results of agenda-setting directed research. In order to bring together results that confirm the hypothesis, its supporters differentiate, in a quasi-ontological manner, the individual order from the collective order, by clearly demonstrating that respondents' answers to questions concerning each of these orders are completely different. Answers that correspond to social interests are those that correspond to newspaper content (Lazar 1991: 151–52). What must therefore be accepted, and the postulate without which the convincing results of the agenda-setting hypothesis cannot be explained, is the existence, for the majority of individuals and parallel to their individual interests, of collective interests that are largely shared with their fellow citizens.[13]

This can be explained, in part by the fact that everything is relative to a citizen within a society in perpetual need of political choices, and not simply to an isolated individual; people are influenced by the media in their appreciation of problems needing solutions for the greater good of society, and not simply for themselves. It is as if people have a double-sided personality, one private and the other public,[14] and the latter acquiesces to media content on the issue much of the time. The content thus seems to participate in a common personality, because the media are not only expressions of our collective life, it is also through the media that others can become part of ourselves by presenting themselves within the public forum, and thus to each of us.[15] In the case of environmental issues, it has been well established (Galle 1990) that pollution, first perceived as a local problem, now finds itself a "social problem" on the media stage (Galle 1993: 122). I believe that headline formulation is part of the concrete mechanisms through which the press succeeds in expressing contents which are not exclusively of individual or even local interest, but of societal interest. Because these headlines function to attract public attention, the values they express thus seem to be more fruitful, and more useful, for understanding "the social dimension" of environmental problems that environmental managers are, in practice, obligated to take into account because of the tenure and political scope of this dimension.

The Concrete Context of Social Regulation of Environmental Problems

Taking into account the interests at work in social reality of the battles for the regulation of environmental problems, we begin to see that the will to attain certain objectives by imposing them institutionally, is subject to failure because of the multiple negotiations necessary in practice in the field, on the local, as well as the larger, scale (Galle 1993). Each time, it seems that those who are animated by strong convictions concerning what should be, and who want to see their ideals made concrete, are excluded from affecting this social regulation, as are those who seek only to protect or advance their interests without compromise. In modern social organization — and this, of course, applies to environmental management — the interdependence of the realization of ideals, like that of protection of interests, in practice obliges the sharing of power and the legitimacy of interventions. The media and the general public are thus in the position of a more-or-less informed and timid observer of the results of the quasi-perpetual negotiation that is done in the field between norms to be imposed, the collectives in place, and the social roles to protect. Because of this distance, they work (Galle 1993) within the global process of social regulation of an environmental problem, a planning role more efficient than all regulation of exchanges. They participate in a fundamental way in defining the minimal framework within which negotiation must occur so that the results are socially acceptable. That is, they help define the common values to be respected.

As media content reflects in numerous cases, public indignation and scandal, and not simply geophysical criteria, are at the root of many behaviour changes in the environment (Galle and Pagès 1991). At the present time, as Galle notes (1993: 124), "something more" than the risk of ecological imbalance "regulates" a society. This "something," which is otherwise independent of individual wills, is the social-environmental demand. The determining factors for all public intervention touching on the environment are regulated by the interaction of social values with, on the one hand, the events they call attention to, particularly in the natural environment, and on the other hand, with the coherence experts conferring on these facts through their discourse, which becomes the reference of rationality in the decision process of a society. This will continue until a new evolution of values and environmental demand is set in motion by one or more factors, including notions of thinking and management finally chosen (Galle 1993: 121).[16] Finally, it is important to note how the press is implicated in this process, a process that would be more efficient than any other environmental-management framework, and what this function implies for useful press analyses in this area.

If what may be deduced from the results of research directed by the agenda-setting hypothesis is reliable, media content on environmental themes would, in large part, be a reflection, if not the realization, of the process that involves a collective judgement of value and reality leading to political choices. Most global communication concentrates on some theme through the media in a society, and it is by the way of the media that society thinks about its action through the values most common to its members. According to the hypothesis often verified by agenda setting, these values are based in society's future choices which, in the majority of cases, determine newspaper content. Information found in the press concerning the regulation process of environmental problems — whether about the object (that is, conflicts and agreements between the participants), or about regulation factors (that is, pertinent events, expert opinion on their subject), or the decision process, or the policies chosen — will be relative to the "process" in question if they are related to the "values" in question. However, headlines, as we have seen, are signs of the presence of values that call attention to certain things in a society. They are a way of uncovering these values, as well as the way in which they guide and bias society's attention to other regulatory elements in the case of an environmental problem. Press data analysis could be useful information for individuals or groups interested in a particular environmental issue. These analyses underline the minimal moral framework within which any action on this subject should be placed to produce results, and it can only be changed from within. This framework is changing, and its modifications should be followed for each separate case. It may also be confirmed, as some have begun to show (Salles 1993; Tessier 1992; Tessier 1994), that social awareness and the realizations that follow, proceed by certain steps made up of characteristics that can guide efficient environmental management in a society.

DISCUSSION QUESTIONS

1. How can the values promoted by newspapers be seen through their headlines?
2. Give some reasons why it can be argued that media content may be used as a political and economic tool by various social players involved in conflicts over the regulation of environmental problems.

3. How do supporters of the "agenda-setting" hypothesis explain that issues receiving strong media attention do not always become important issues for the public?
4. With reference to research on the social reality of battles over regulation of environmental problems, what kind of citizens are excluded from influencing this social regulation, and why?
5. What topics covered in the press define the minimal moral framework able to guide efficient environmental management in a given society?

Endnotes

1. For a complete view of the problem of environmental management, see Mermet (1992).
2. For a comprehensive overview of research on the effects of the media and the place of the "agenda-setting" hypothesis, see Katz (1990). The results of this research may be understood in relation to what is known in press analysis as the "agenda setting" hypothesis. My examination will start by considering what confirms this hypothesis, and the practice that results from the hypothesis. Next, I will demonstrate how this course of analysis encounters limits at the empirical level, where theoretical meaning acts retroactively on application possibilities—even on the significance of research results that, in practice, are counter to the "agenda-setting" hypothesis. Finally, it will be shown that these opposed practices also encounter limitations that are otherwise explicable by the social-regulation dynamic environmental problems. The nature of environmental issues permits an explanation of both the convincing results of the agenda-setting hypothesis and its limitations, while at the same time offering a more precise meaning, and one which may be of use in environmental management.
3. This has been demonstrated elsewhere (Sasseville, Boucher, and Nobert 1987) in environmental matters and in correlation with newspaper content (Boucher and Cosme 1987).
4. As demonstrated (Lowe and Morrison 1984).
5. As has been shown (Lowe and Morrison 1984).
6. However, press analyses do not lose their pertinence and possible usefulness for environmental management. Further explanation of certain limitations mentioned will clarify this.
7. Real control of one by the other proves to be impossible because the mechanisms are not well understood.
8. These differences can be of a variety of orders. The results obtained by Hill (1985), for example, suggest that the influence of televised news on political agenda setting increases according to certain viewer characteristics, whether they are already aware of a subject by way of the written media, and whether they are college educated. This kind of result indicates that "agenda setting" by the media is dependent on "presetting" in a targeted population.
9. The fact that the press itself contributes to these preoccupations is not important in this context because this effect is now exterior to the press. Being exterior, it is also beyond control of the totality of events that will occur in the society. The coverage is directed to where such events actually happen.
10. This hypothesis tries to subordinate that of agenda setting by giving meaning to its negative results without completely contradicting it.
11. In effect, the presence of particularly strong values in a society at a given time, relative to a given environmental issue establishes, in practice, a framework within which environmental management must operate in order to be socially and politically acceptable.

12. However, that a school of thought gaining strength (Proulx 1990), insists that the media primarily reflect general opinion that they cannot influence, and according to which, the recipients appropriate their content much more than is generally believed. See the collected texts by Balle and Padioleau (1973).

13. However, it is worthy to mention that there is a difficulty in my approach. This approach consists of taking note of the social significance of media content in environmental management. As it is the case here with agenda-setting-directed research, this postulate can subordinate that of research (done along entirely different lines) that criticizes and seeks to modify the practices by which the media influence public opinion. These studies generally emphasize the fact that the way the media treat environmental issues, by situating them in a vast and unreachable universe, while at the same time, because of various organizational restraints, relating them to the world of power (De La Haye 1985: 70, 218), disenfranchises citizens (Shoenfeld, Meir, and Griffin 1979; Beauchamp 1987: 157–58; Dumas 1992) and brings them only the obligation to be "futilely well-informed" (Wiebe 1973), without personal involvement in action. The more people are informed by the media about environmental matters, the less they act privately in this area, and the more they favour societal solutions (Novic and Sandman 1974). This situation was interpreted in terms of a possible social dysfunction of the media, but there is another way to understand it. In effect, it confirms the fruitfulness of the postulate of the agenda-setting hypothesis, whereby media content concerns not private life, but public life, which has different priorities. Thus, deciding to ride a bicycle, for example, when the majority of people do not, is a derisionary act, considering all that is at stake in society, unless the act is done in public as a political statement. In practice, this type of behaviour is referred to as the "problem of assurance" (Salles 1993: 109): to know that a change of options, given its individual "costs," only has meaning in a context of interdependence for research purposes; if the individual is assured, others will be also. Definitively, one comes back to the necessity for a kind of collective sentiment regarding a problem or its solution, within which an isolated act rarely has significance. The difference between the two approaches is that one assumes the split between individual and collective priorities, which is the position I am attempting to support, while the other seeks to go beyond this split, by assuming that modifications to journalism would produce changes in citizens' behaviour, particularly in their private lives. Content analyses of this material touch on journalistic ethics, in general (Bernier 1992), and on environment (Lamay et al. 1991), as a function of environmental education of the general public (Eveslage 1991), and a necessary reorganization of mass media's current production in this respect (Althoff, Graig, and Stuckey 1973; Rubin 1973).

    At the environmental management level, one may wonder about the capacity of the media to implement the necessary changes, and about the value of such changes, when the media, like the behaviour of people, are also subject to the weightier structures of our societies, such as the market economy, for example. One may also question, not only the ethical value, but the practical value of imposing such changes of form and foundation in the name of very specific convictions, concerning the way people should think and act. Moreover, the successes of the past few years in major environmental issues, demonstrate that this type of problem is not managed with any degree of certitude (Salles 1993).

14. Although the supporters of the agenda-setting hypothesis do not refer to this, it is a postulate of classical sociology found especially in Durkheim, from his doctoral thesis ([1893] 1960: 74) until towards the end of his career (1914). For a discussion of the basic theoretical context of this postulate, see Prades (1989).

15. Mumper (1987) demonstrated how taking other circumstances, such as support from a specialized community, into account, problems long considered personal, such as child

abuse, have become public questions, susceptible to government intervention over time, thanks to the media.

16. As Mayer (1991) has shown, with the evolution of consumer concerns between 1960 and 1987, such interactions led to a complexification of the agenda-setting hypothesis, by introducing an effect of the questions raised by political decision makers regarding media content in the process through which media influence public opinion that then, in turn, has an effect on governmental policies. Practical solutions adapted to this complex dynamic have been proposed to effectively counter social problems such as terrorism (Crelinsten 1989).

References

Althoff, P., W.H. Graig and F. Stuckey. 1973. Environmental Pollution Control Attitudes of Media Managers in Kansas. *Journalism Quarterly* 50(4): 666–72.

Balle, F. and J.-G. Padioleau, eds. 1973. *La sociologie de l'information. Textes fondamentaux.* Paris: Larousse.

Beauchamp, C. 1987. *Le silence des médias. Les femmes, les hommes et l'information.* Montréal: Éditions du remue-ménage.

———. 1991. Pouvoir, organisation, communication et projet d'entreprise. In *Communication publique et société. Repéres pour la réfexion et l'action*, ed. M. Beauchamp, 372–98. Boucherville, Que.: Morin.

Bernier, M.-F. 1992. Les «incontournables» en éthique du journalisme. *Communication* 13(1): 201–20.

Boucher, V. and D. Cosme. 1987. Opinion publique et comportement électoraliste: analyse du traitement par la presse écrite du problème des pluies acides. In *Utilité de l'opinion publique dans les choix publics : le cas du contrôle des émissions acidogènes*, ed. Sasseville, Boucher et Nobert, 1–45 (Appendix). Rapport scientifique no. 193. Ste-Foy, Que.: INRS-eau.

Crelinsten, R.D. 1989. Terrorism and the Media: Problems, Solutions, and Counterproblems. *Political Communication and Persuasion* 6(4): 311–39.

Cumberbatch, G. and D. Howitt. 1989. *A Measure of Uncertainty. The Effects of the Mass Media.* London/Paris: John Libbey.

Dagenais, B. 1991. Les institutions et la gestion de leur image publique. In *Communication publique et société. Repéres pour la réfexion et l'action*, ed. M. Beauchamp, 337–71. Boucherville, Que.: Morin.

De La Haye, Y. 1985. *Journalisme, mode d'emploi. Des maniéres d'écrire l'actualité.* ELLUG-Pensée Sauvage.

Dumas, B. 1992. La presse et la question environnementale: le cas des pluies acides. In *Gestion de l'environnement, éthique et société*, ed. J.A. Prades, R. Tessier and J.-G. Vaillancourt, 93–126. Montréal: Fides.

Durkheim, É. [1893] 1960. *De la division du travail social.* Paris: P.U.F.

———. 1914. La dualisme de la nature humaine et ses conditions sociales. *Scientia* XV: 206–21.

Eveslage, T. 1991. Shared Goals of Social Studies and Scholastic Journalism: Enhancing the Environment and Effectiveness of Citizenship Education. In *Proceedings of 74th the Annual Meeting of the Association for Education in Journalism and Mass Communication* Part. XVII: Miscellaneous Studies, Section B.

Faucheux, S. and J.-F. Noël. 1990. *Les menaces globales sur l'environnement.* Paris: La Découverte.

Faupel, C.E., C. Bailey and G. Griffin. 1991. Local Media Roles in Defining Hazardous Waste as a Social Problem: The Case of Sumter County, Alabama. *Sociological Spectrum* 11(4): 293–319.

Funkouser, G.R. 1972. Trends in Media Coverage of the Issues of the 60's. *Journalism Quarterly* 50: 533–38.

Galle, M. 1990. *Le processus de décision en matière de pollution. Une étude du jeu conflictuel comme mode de régulation.* Ph.D. diss., Université Paris I Panthéon-Sorbonne.

———. 1993. La régulation conflictuelle des pollutions. *Natures, sciences, sociétés* 1(2): 118–27.

Galle, M. and J.-P. Pagès. 1991. La presse locale et le nucléaire: une étude de cas. *Séminaire du groupe de recherche SORISTEC, Médias et risques technologiques*, 7–8 March, Chantilly.

Hansen, A. 1991. The Media and the Social Construction of the Environment. *Media, Culture and Society* 13(4): 443–58.

Hill, D.B. 1985. Viewer Characteristics and Agenda Setting by Television News. *Public Opinion Quarterly* 49(3): 340–50.

Hourcade, J.-C., G. Mégie and J. Theys. 1991. Modifications climatiques et évaluation des politiques énergétiques: comment gérer l'incertitude? *Actes du 14e Congrès de la Conférence mondiale sur l'énergie*, Montréal.

Katz, E. 1990. À propos des médias et de leurs effets. In *Technologies et symboliques de la communication: colloque de Cerisy*, ed. L. Sfez and G. Coutlée, 273–83. Grenoble: Presses Universitaires de Grenoble.

Lamay, C.L. et al. 1991. *Media and the Environment.* Washington, D. C./Covelo, Calif.: Island Press.

Lazar, J. 1991. *Sociologie de la communication de masse.* Paris: Armand Colin.

Lowe, P. and D. Morrison. 1984. Bad News or Good News: Environmental Politics and the Mass Media. *Sociological Review* 32(February): 75–90.

Mayer, R.N. 1991. Gone Yesterday, Here Today: Consumer Issues in the *Agenda Setting* Process. *Journal of Social Issues* 47(1): 21–39.

McCombs, M. and D. Shaw. 1972. The Agenda-Setting Function of Mass Media. *Public Opinion Quarterly* 36: 176–84.

Mermet, L. 1992. Stratégies *pour la gestion de l'environnement. La nature comme jeu de société?* Paris: Éditions L'Harmattan.

Mumper, M. 1987. Review Essay: Understanding Policy Agendas. *Social Science Journal* 24(1): 83–86.

Novic, K. and P.M. Sandman. 1974. How Use of Mass Media Affects Views on Solutions to Environmental Problems. *Journalism Quaterly* 51(3): 448–52.

Pasquero, J. 1989. Gérer stratégiquement dans une économie politisée. *Gestion* 14(3): 116–28.

Prades, J.A. 1989. Autonomie individuelle et solidarité collective. Problématique socio-éthico-religieuse et modernité chez Durkheim. *Ethica* 1(1): 41–61.

Protess, D.L., D.R. Leffe, S.C. Brooks and M.T. Gordon. 1985. Uncovering Rape: The Watchdog Press and the Limits of Agenda Setting. *Public Opinion Quarterly* 49(1): 19–37.

Proulx, S. 1990. De la métaphore télégraphique à celle de la conversation: représentations du pouvoir des médias et modèles de communication. In *Technologies et symboliques de la communication: colloque de Cerisy*, ed. L. Sfez and G. Coutlée, 283–97. Grenoble: Presses Universitaires de Grenoble.

Roqueplo, P. 1988. *Pluies acides: menaces pour l'Europe.* Paris: Economica.

Rubin, D.M. 1973. Access to Environmental Information: Orchestrated Confusion in Atomic Energy. In *Mass Media and the Environment — Water Resources, Land Use and Atomic Energy in California*, ed. D.M. Rubin and D.P. Sachs, chap. 6. New York: Praeger.

Salles, J.-M. 1993. Les enjeux économiques des risques globaux d'environnement. *Natures, sciences, sociétés* 1(2): 109–17.

Salwen, M. B. 1986. Time in Agenda Setting: The Accumulation of Media Coverage on Audience Issue Salience. *36th Annual Meeting of the International Communication Association.*

Sasseville, J.-L., V. Boucher and M. Nobert. 1987. *Utilité de l'opinion publique dans les choix publics: le cas du contrôle des émissions acidogènes*. Rapport scientifique no. 193. Ste-Foy, Que.: INRS-eau.

Shoenfeld, C., R.F. Meir and R. J. Griffin. 1979. Constructing a Social Problem: The Press and the Environment. *Social Problems* 27(1): 38–61.

Strodthoff, G.G., R.P. Hawkins and A.C. Shoenfeld 1985. Media Roles in a Social Movement: A Model of Ideology Diffusion. *Journal of Communication* 35(2): 134–53.

Tessier, R. 1994. L'approche sociologique en éthique appliquée: comprendre et poursuivre la lutte aux pluies acides au Québec. In *Instituer le développement durable, Éthique de l'écodécision et sociologie de l'environnement*, ed. J.A. Prades, R. Tessier and J.-G. Vaillancourt, 77–96. Montréal: Fides.

———. 1992, *L'émergence d'une éthique sociétale de l'environnement. Une analyse de presse sur une thématique sacrée: les pluies acides au Québec (1981–1990)*, Ph.D. diss., Université du Québec à Montréal.

Ullmann, A.A. 1985. The Impact of the Regulatory Life Cycle on Corporate Political Strategy. *Academy of Management Review* 28(1): 140–54.

Wiebe, G.-D. 1973. Mass Media and Man's Relationship to His Environment. *Journalism Quarterly* 50(3): 426–32, 446.

INDEX

LIST OF CONTRIBUTORS

David Cameron David Cameron is working on his Ph.D. in Sociology at the University of Waterloo. He has conducted research on the social and political consequences of environmental contamination. Other areas of research include research methodology, technology and the commercialization of sexuality. He received his M.A. in Political Science in 1992.

Michael Clow Michael Clow is an Associate Professor of Sociology at St. Thomas University in Fredericton, New Brunswick. He has an Honours B.Sc. in Physics from St. Francis Xavier University (1975) a Masters in Arts in Political Science from Dalhousie University (1979), and a Ph.D. in Political Science from York University (1989). He has been a member of the Sociology Faculty at St. Thomas University since 1986, and is Adjunct Professor in the International Development Studies Programme at St. Mary's University, Halifax, Nova Scotia.

Éric Darier Éric Darier has taught environmental studies at several Canadian universities. He obtained a Master's degree in European Studies at the University of Reading (England) and a Ph.D. in Political Science at McGill University. His doctoral thesis was a theoretical investigation of how Michel Foucault's concepts can be applied to Canadian environmental policies in the period 1966 to 1990.

Charles Gordon Charles Gordon is Associate Professor of Sociology and Architecture and Associate Director of the Institute of Interdisciplinary Studies at Carleton University. He received his B.A. at Amherst College and his Ph.D. from the University of North Carolina. His research interests include studies of the design professions, the relation between social organization and the built environment, standardization and regulation of the built environment, building use and building safety, and the built environment within material culture.

Allan Greenbaum Allan Greenbaum has studied biology, philosophy, and law. He is currently a Doctoral candidate in Sociology at York University where he co-designed and co-teaches a course on social conflict and environmental law in the Division of Social Sciences. His research interests include environmental thought, the sociology of environmental aesthetics, and the sociology of cosmological beliefs. He is a co-founder of the Environmental Studies Association of Canada (ESAC).

Louis Guay Louis Guay studied sociology at the Université de Montréal where he received a B.Sc. and a M.Sc. degree. He did his Ph.D. at the London School of Economics in urban and regional planning. Since 1977, he has been Associate Professor in the Department of Sociology at the Université Laval. He has taught, worked, and published in urban sociology and planning, and in environmental sociology. His current research is on the production, diffusion and utilization of scientific messages with special reference to global climate change. He is now Vice-Dean of Research at the Faculté des sciences sociales, Université Laval.

Paul Harpley Paul Harpley (B.Sc., M.A), President of the South Lake Simcoe Naturalists of Georgina, Ontario, has been directing field research through the naturalists in Landscape Ecology, Environmental Perception, and Breeding Birds project. He has undertaken research on environmental perception of the Lake Simcoe watershed since 1993. His interests include examining the relationship between humans and nature in Western culture and how this is linked to planning and land-use issues. He is currently Project Manager for the Metropolitan Toronto Zoo and is involved in planning exhibits consistent with his research. Harpley is a well-known Ontario naturalist and nature artist.

Melody Hessing Melody Hessing is a sociologist who lives in Vancouver where she teaches sociology of women and sociology of the environment courses at Douglas College. She has studied, taught and published in the area of sociology and environment for many years on themes including ecofeminism, public participation and environmental decision making, the regulatory state and women's "dual labour." She finds respite from the sociological in the observation of "natural" ecosystems, and continues to explore the intersection of the social/ecological in her work.

Donna E. Hughes Donna Hughes is a Toronto-based researcher specializing in interdisciplinary issues related to the spatial and temporal decentralization of information work. She hold an undergraduate degree in sociology from Glendon College, York University, and a Masters degree in Environmental Studies — specializing in telework — also from York University. She has been an intermittent consulting researcher with the Centre for Information Technology Innovation of Industry Canada since 1990, conducting empirical and theoretical research on telework.

Michael D. Mehta Michael Mehta is an environmental sociologist who specializes in risk, environmental attitudes, social movements and quantitative methodology. His academic background includes a B.A. in Psychology, a Masters in Environmental Studies, and Ph.D. in Sociology — all from York University. He is a co-founder of the Environmental Studies Association of Canada (ESAC), and co-editor of this book. His present work deals with the democratic control of technology, specifically how risk and public policy relate to hazardous facilities like nuclear power plants and how communications systems such as the Internet can bring the public back into the decision-making arena. Dr. Mehta is currently a Post-Doctoral Fellow at the Environmental Policy Unit at Queen's University.

Robert Milne Robert Milne (B.Sc., M.Sc.) has lectured extensively in geography and environmental studies at the University of Guelph where his field studies in geomorphology and landscape ecology are particular areas of interest. Milne is a research partner in the South Lake Simcoe Naturalists of Georgina, Ontario, studying breeding bird populations, spatial diversity, and human perception issues in the Lake Simcoe watershed. He currently resides in Eastern Ontario near Kingston.

Joseph Moore Joseph Moore is a Doctoral candidate in Sociology at McMaster University with interests in labour and environmental movements. He is currently working on a thesis which examines the intersection of labour and environmental movements during land-use disputes on Vancouver Island from 1960 to the present.

Éric Ouellet Éric Ouellet is completing a Ph.D. in Sociology at York University. He received a B.A. and M.A. degree in Political Science at Université Laval, Quebec City. His research interests include environmental organization, organizational theory, and sociology of scientific knowledge. His present work deals with the social construction of consultants' expertise in organizational settings. He is a co-founder of the Environmental Studies Association of Canada (ESAC), and co-editor of this book.

Ted Schrecker Ted Schrecker is Associate Director (Environmental Ethics) of the Westminster Institute for Ethics and Human Values in London, Ontario, and also teaches in the Department of Political Science at the University of Western Ontario. He has more than 16-years' experience in the environmental policy field as a legislative researcher, academic, and consultant.

Ralf Starkloff Ralf Starkloff has a B.A. in Sociology from the University of Massachusetts at Boston and a M.A. and Ph.D. degree in Sociology from Brandeis University. He has taught courses in community organization, introductory sociology, and the sociology of power relations. His interests include environmental sociology, political sociology and social movements and the sociology of work and organizations.

Robert Tessier Member of the GREIGE (Groupe de recherche interdisciplinaire en gestion de l'environnement), Robert Tessier teaches social sciences at UQAM (Université du Québec à Montréal). He specializes in the sociology of values and press analysis. He is co-editor of many books dealing with issues related to environmental sociology including *Environnement et développement* (1991), *Gestion de l'environnement, éthique et société* (1992), *Instituer le développement durable* (1994) and *Aspects sociaux des précipitations acides au Québec* (1994).

David Tindall David Tindall is an Assistant Professor at the University of British Columbia in the Department of Forest Resources Management and the Department of Anthropology and Sociology. He has expertise in environmental sociology, social movements, and sociological theory, and has an interest in a number of theoretical and substantive areas including environmental thought/theory, social impact assessment, the problem of common property, social networks, the sociology of resource communities and collective action.

Anton H. Turrittin Anton Turrittin is Associate Professor of Sociology, Faculty of Arts, at York University. His research interests have centred on issues of social inequality including topics such as work and industry, social mobility and the link between education and occupations. A long-standing concern with public transportation has recently led him to consider the relationships of city form, automobile-dependency, public transit, public policy and larger environmental questions. His academic work has been combined with advocacy in the community.

Jean-Guy Vaillancourt Jean-Guy Vaillancourt (Ph.D., University of California, Berkeley, 1975) was born and raised in the Sudbury region of Ontario. He is now Professor of Sociology at the Université de Montréal. He is the author of many books, articles and book chapters in the areas of ecosociology, religion, war and peace and politics. He was city councillor in Dunham, Quebec, from 1976 to 1980, member of the board of directors of Oxfam Quebec, the Canadian Institute for International Peace and Security and the European University Centre for Peace Studies in Austria. His current research focuses on the human dimensions of environmental change and on the environmental movement in Quebec. He has recently co-edited three books on sustainable development and one on the social aspects of acid rain. He was guest editor for the special issue of *Ecodecision* magazine on implementing sustainable development in early 1995.